Applications of High-Field and Short Wavelength Sources

Applications of High-Field and Short Wavelength Sources

Edited by

Louis DiMauro
Brookhaven National Laboratory
Upton, New York

Margaret Murnane
University of Michigan
Ann Arbor, Michigan

and

Anne L'Huillier
Lund Institute of Technology
Lund, Sweden

Plenum Press • New York and London

Library of Congress Cataloging in Publication Data

Applications of high-field and short wavelength sources / edited by Louis DiMauro, Margaret
  Murnane, and Anne L'Huillier.
        p.   cm.
    "Proceedings of the Optical Society of America Conference on Generation and Applications
of High-Field and Short Wavelength Sources, held March 20–22, 1997, in Santa Fe, New Mexico"—Verso t.p. .
    Includes bibliographical references and index.
    ISBN 0-306-45909-4
    1. Lasers—Congresses. 2. Radiation sources—Industrial applications—Congresses. 3. Laser
pulses, Ultrashort—Congresses. 4. X-ray lasers—Congresses. I. DiMauro, Louis. II. Murnane,
Margaret. III. L'Huillier, Anne. IV. Optical Society of America. V. Optical Society of America
Conference on Generation and Applications of High-Field and Short Wavelength Sources (7th:
1997: Santa Fe, N.M.)
    TA1673.A58 1998   95-35190
    621.36′6—dc21   CIP

Proceedings of the Optical Society of America Conference on Generation and Applications of
High-Field and Short Wavelength Sources, held March 20 – 22, 1997, in Santa Fe, New Mexico

ISBN 0-306-45909-4

© 1998 Plenum Press, New York
A Division of Plenum Publishing Corporation
233 Spring Street, New York, N.Y. 10013

http://www.plenum.com

10 9 8 7 6 5 4 3 2 1

Printed in the United States of America

PREFACE

The Optical Society of America Conference on *Applications of High Fields and Short Wavelength Sources*, held in Santa Fe, New Mexico, USA, from March 20–22, 1997, was an exceptionally exciting conference. This conference was the seventh in a series of topical conferences, held every two years, which are devoted to the generation and application of high field and short wavelength sources. The meeting was truly international in scope, with equal participation from both within and outside of the US.

In the past two years, there has been dramatic progress in both laser and x-ray coherent sources, both fundamental and applied. The 1997 meeting highlighted these advances, which are summarized in sections 1 and 2 of this volume. Terawatt-class lasers are now available in the UV or at high repetition rates. Michael Perry (LLNL) presented a keynote talk on petawatt class lasers and their applications in inertial confinement fusion, while Jorge Rocca (Colorado State University) presented a keynote talk on tabletop soft-x-ray lasers. Generation and measurement techniques are becoming very sophisticated throughout the UV and x-ray region of the spectrum, and coherent sources have been extended to wavelengths below 30Å. Phase control in the x-ray region is also now possible, and new phase-matching schemes in the UV have been experimentally demonstrated. It is clear that a new field of x-ray nonlinear optics will develop rapidly over the next few years. Other forefront work was reported in the keynote addresses given by Erik Johnson (BNL) and Irene Nenner (France) on the capabilities and applications of the next-generation synchrotron and accelerator-based light sources.

Exciting advances in plasma-based x-ray and ion sources are also reported in section 3, while section 4 covers the physics of laser-plasma interactions. New debris-less plasma-based x-ray sources are now available, while cluster sources have proven useful for both x-ray and high-energy ion generation. Detailed and accurate diagnostics have also increased our understanding of laser-solid-density plasma interactions. Section five illustrates the tremendous progress in our understanding and use of high field interactions, and the physics of atoms and molecules in intense laser fields. Several groups reported new understanding of strong-field multi-photon processes, and new methods to trap, accelerate and guide electrons in strong laser fields were presented.

Interdisciplinary work in particular was highlighted at the meeting, because of the increasing applications of high field science in chemistry, materials science, laser-based particle accelerators, and extreme ultraviolet lithography. Keynote speakers in these areas were Richard Haight (IBM), who presented work on applications of soft-x-ray light in materials and chemical dynamics, and Jeff Bokor (UC Berkeley), who reviewed progress on extreme ultraviolet lithography. Other exciting work covered novel applications of hard x-rays in microscopy, angiography, radiology, time-resolved diffraction, and x-ray imaging.

We would like to take this opportunity to thank all the conference participants, and contributing authors to these proceedings, for making the Seventh OSA Conference on *Appli-*

cations of High Field and Short Wavelength Sources a great success. In particular, we would like to thank Lisa Myers from the OSA, for coordinating the meeting. Finally, we gratefully acknowledge support for the meeting from the Air Force Office of Scientific Research, Brookhaven National Laboratory/Associated Universities, Department of Energy Basic Energy Sciences, Lawrence Livermore National Laboratory, Positive Light, Spectra-Physics and Kapteyn-Murnane Laboratories LLC. We would also like to take this opportunity to give the next conference co-chairs, Eckhart Foerster, Wolfgang Sandner and David Meyerhofer, our best wishes for the eighth conference of this series, which will be held in Germany in 1999.

Louis DiMauro
Anne l'Huillier
Margaret Murnane

CONTENTS

1. High-Power Laser Sources

2. Ultrafast Coherent UV and X-Ray Sources

3. Novel Short Wavelength Sources

4. Ultrashort-Pulse Laser Plasma Interactions

6. Applications of Short Wavelength Sources

THE PRODUCTION OF PETAWATT LASER PULSES

M.D. Perry, B.C. Stuart, D. Pennington, G. Tietbohl,
J. Britten, C. Brown, S. Herman, J. Miller, H.T. Powell,
B.W. Shore and V. Yanovsky

Lawrence Livermore National Laboratory
PO Box 808, L-439, Livermore, CA 94550
e-mail: perry10@llnl.gov

INTRODUCTION

Chirped-pulse amplification applied to broad-bandwidth solid-state lasers has created a revolution in the production and use of terawatt and now petawatt class lasers.[1,2] The concepts and technology contributing to this revolution have evolved continuously since the early 1970's. Following the grating compressor work of Treacy[3], Bischell[4] and others described the application of chirped-pulse amplification to Nd:Glass lasers. This was followed by a large amount of work on fiber-grating pulse compression for communication research.[5] In 1985, Strickland and Mourou combined many of these ideas into the first practical demonstration of chirped-pulse amplification with a solid-state laser.[6] Following this initial demonstration, rapid developments in technology such as the stretcher design of Martinez[7] led to small scale systems capable of terawatt[8] and multiterawatt pulses.[9-11] Occurring in parallel with the development of chirped-pulse amplification technology using Nd:Glass lasers, was the development of the new laser material, titanium-doped sapphire.[12] The commercial availability of this unique laser material dramatically propelled the revolution in CPA based solid-state lasers. An overwhelming majority of CPA lasers now employ Ti:sapphire either throughout the entire laser system or at least as the oscillator material.[13] These early developments and the large amount of effort that has gone into the laser technology in recent years have culminated in high pulse energy systems producing pulses with a peak power of 125 TW[14] and very short-pulse systems producing multiterawatt pulses which only contain a few optical cycles.[15-18] Here, we describe the limits of CPA technology in the context of a large scale system producing pulses with a peak power exceeding 1.25 petawatts (1250 TW).

All laser systems are limited in peak power by the phenomena of self-focusing and self-phase modulation arising from the intensity dependent refractive index, $n = n_0 + \gamma I$, where $\gamma = 4\pi n_2 / n_0 c = 4.19 \times 10^{-3} \, n_2/n_0$ (cm^2/W) with n_0 and n_2 the linear and nonlinear refractive index (esu), respectively. The nonlinear index results in a phase retardation, $B(r,t)$,

$$B(r,t) = \frac{2\pi}{\lambda} \int \gamma \, I(r,t) \, dz$$

where the integration is performed over the optical path length through the laser system. As an example, a laser pulse propagating through 10 cm of fused silica at a power density of 10 GW/cm^2 accumulates 2 radians of nonlinear phase. This nonlinear phase results in both degradation of the spatial quality of the beam by self-focusing and of the temporal and

Applications of High Field and Short Wavelength Sources
Edited by DiMauro *et al.*, Plenum Press, New York, 1998

spectral quality by self-phase modulation. The technique of chirped-pulse amplification was developed to overcome these limitations by stretching the pulse prior to amplification thereby reducing the peak power in the amplifiers. Even with CPA, the temporal quality of petawatt class lasers will still be limited by the effects of nonlinear phase.[19]

Petawatt lasers can be produced by either moderate energy systems operating at extremely short pulse lengths (e.g, 20 J in 20 fsec) or by high energy systems operating at subpicosecond pulse widths. The choice will be dictated by the applications of the system. In 1992, we embarked on a project to develop a laser capable of producing petawatt pulses in order to examine the fast ignitor concept for inertial confinement fusion.[20] This application requires high pulse energy in addition to the short pulse duration. The essential idea is to pre-implode a deuterium-tritium capsule to an isochoric (uniform density) condition. At the point of maximum compression, irradiate the side of the imploded core with a laser pulse much shorter than the hydrodynamic disassembly time of the irradiated spot ($\tau \approx R_{spot}/v_s \leq 10$ psec). Hot electrons (200 keV<E<1 MeV) generated by the interaction of the intense (10^{19} - 10^{21} W/cm^2) light with plasma rapidly equilibrate in the dense fuel. The energy equilibration of the electrons raises the overall ion temperature to 5-20 keV initiating fusion burn. High laser pulse energy is required in order to produce enough hot electrons to heat a sufficient number of ions to initiate a self-sustaining fusion burn. In addition, nearly perfect beam quality is required in the laser in order to achieve a small spark region. To meet the conditions necessary to address these issues and serve as a general facility for experiments in intense laser matter-interaction, the Petawatt laser was designed to produce ≈ 1 kJ pulses with a pulse duration adjustable between 0.4 and 20 ps and sufficient beam quality to produce an irradiance of 10^{21} W/cm^2 when focused at f/3.

The laser system begins with a Kerr-lens mode-locked Ti:sapphire oscillator producing 105 fsec pulses at 1054 nm. A single pulse is selected from the mode-locked train and stretched to ≈ 3 nsec prior to amplification. The stretcher is formed from a single 1480 l/mm grating, a 2.1 m focal length aspheric lens, a 15.2 cm diameter high reflector placed at the focal plane of the lens and a gold coated retroreflector to provide vertical displacement of the beam for a total of eight passes through the stretcher (figure 1). A total of 16.5 nm of bandwidth is passed by the stretcher assembly.

In order to maintain sufficient bandwidth to achieve ≈ 400 fsec pulses upon recompression, this pulse is amplified in a series of two Ti:sapphire regenerative amplifiers. The first regenerative amplifier is a linear cavity design formed with two 5 m radius of curvature mirrors separated by 1.2 m. Similar to a Kerr-lens mode-locked oscillator, the cavity is designed for dynamic TEM$_{00}$ stability when accounting for the combination of pump-induced thermal lensing and gain guiding along with self-focusing in the laser rod and Pockels cells.[21] The 2.5 cm long Ti:sapphire laser rod is 0.15 wt % doped with a figure of merit of $\alpha_{532}/\alpha_{1053}$>200. The nonlinear coefficients used were γ=3.3x10^{-16} cm^2/W for sapphire and γ=3.0x10^{-16} cm^2/W for the KD*P Pockels cells. The Pockels cells were originally placed 2 cm from the cavity end mirrors. Since the stretched pulse was ≈ 2 nsec in duration (reduced by spectral narrowing in the regenerative amplifier), the pulse overlaps with its reflection in the Pockels cells. This overlap results in an increase in the effective B integral in the Pockels cells by a factor of three. By moving the Pockels cells to 25 cm from the end mirrors, this overlap was eliminated and the accumulated B integral in the regenerative amplifier was reduced from 2 to 0.6 radians at an output energy of 7 mJ.

The S-polarized seed pulse is introduced into the regenerative amplifier cavity by reflection off the Brewster cut rod. The rod is pumped from both ends by ≈ 150 mJ at 532 nm with a 3 mm diameter Gaussian beam (F_{pump}=4.7 J/cm^2). Immediately following the pump, the single pass gain of the rod is 1.5 (σ_e = 2.6x10^{-20} cm^2 @ 1054 nm) while near 800 nm, the gain is over 100. Suppression of cavity oscillation near 800 nm is achieved by utilizing end mirrors with a reflectivity less than 10% for wavelengths below 970 nm and a unique polarizer design. A net gain of nearly 10^8 to 6.7 mJ is obtained after 114 passes through the crystal. A short cavity was required in order to overcome the depletion of stored energy due to spontaneous emission during the 440 nsec build-up time in the cavity. The spectrum is reshaped by multiple passes through the polarizer to a nearly perfect Gaussian distribution with an 8.0 nm fwhm centered at 1054 nm.

The pulse is further amplified to $\approx$ 50 mJ in a ring-regenerative amplifier.[22] The TEM_{00} ring cavity is formed by two thin-film polarizers, two 35° mirrors and a 10 m focal length lens. The P-polarized pulse enters the ring cavity through a thin-film polarizer and then passes through a $\lambda/2$ plate. The now S-polarized pulse passes through the lens and Pockels cell and reflects off the output polarizer. The Ti:sapphire rod is pumped on both sides by 400 mJ at 532 nm in a 4 mm diameter spot. The Gaussian mode size of the amplified beam in the crystal is 2.4 mm ($\text{fw} 1/e^2$). After initial passage through the Pockels cell, a $\lambda/2$ voltage is applied which cancels the action of the $\lambda/2$ plate thereby trapping the pulse inside the cavity. The pulse makes 12 passes through the crystal and is then switched out of the cavity by dropping the Pockels Cell voltage back to ground. The pulse then passes through the output polarizer. The spatial distribution of the pulse exiting the ring amplifier is over 98% Gaussian. Spectral narrowing in the linear and ring regenerative amplifiers reshapes both the spectrum and temporal distributions to nearly perfect Gaussians with a fwhm of 7.0 nm and 1.4 nsec, respectively. The B-integral in the ring amplifier is $\approx$0.5, yielding a total value for the 10 Hz section of 1.1 radians. These pulses are compressed to a duration of $\approx$280 fsec in good agreement with the calculated value accounting for self-phase modulation on the stretched pulse.[19]

Further amplification occurs in mixed phosphate glass rod amplifiers. The pulse passes through a serrated aperture (6 mm ID, 8 mm OD) which serves as the object plane for the relay imaged system. Approximately 50% of the pulse energy is passed by the serrated aperture. Passage of this beam through a pinhole at the focus of the M=2 telescope produces a beam with a central Gaussian profile out to 12 mm and an approximately eighth-order super-Gaussian profile beyond this. The central Gaussian profile compensates nearly exactly for the radial gain profile of the two 19x240 and one 45x240 mm Nd:Phosphate glass rod amplifiers producing an eighth-order super-Gaussian beam at the output. The 19 mm diameter rod amplifiers contain APG-1 (3% doping) and LG-750 (2% doping) Nd:phosphate glass and are pumped by six 200 mm long flashlamps. These amplifiers produce a small signal gain of 16 for 3000 J pump. Up to 2 J can be extracted from these amplifiers in the stretched, 12 mm diameter pulse with typical operation of 1.5 J. The final 45 mm diameter rod amplifier contains LHG-8 (0.8 % Nd) Phosphate glass and is pumped by twelve Xenon (300 Torr) flashlamps, each 200 mm in length providing a small signal gain of 7 for 15 kJ electrical energy delivered to the flashlamps. Although up to 15 J has been extracted typical operation is 12 J in a diffraction-limited 38 mm diameter beam. Gain narrowing in the Nd-glass amplifiers reduces the spectral width to 5.0 nm and the stretched pulse to 900 psec (fwhm).[9]

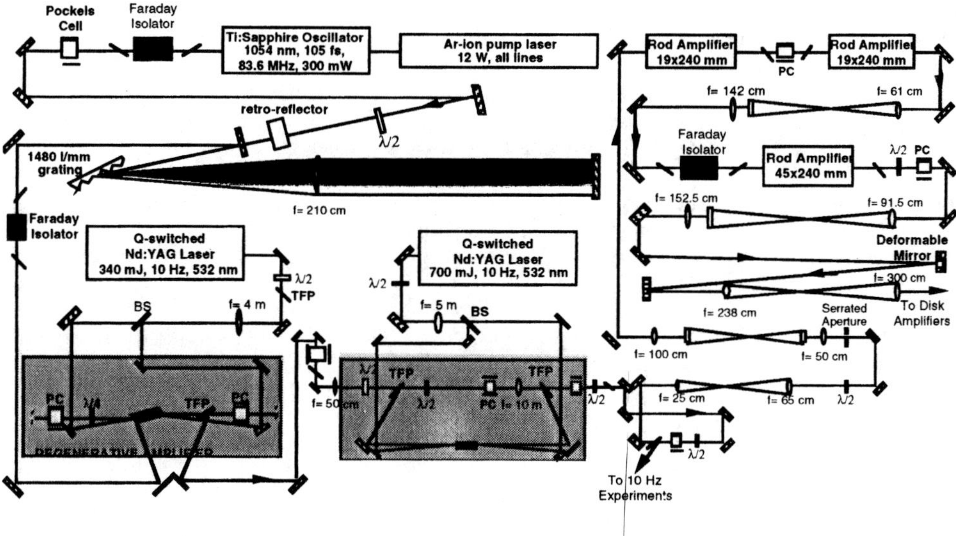

Figure 1: Schematic of Petawatt front-end producing $\approx$15 J in a 2 nsec stretched pulse.

The pulse is then transported by three relay imaging telescopes to the disk amplifier section of the Nova laser. Here, the pulse is amplified by a series of disk amplifiers to a maximum of 700 J (for 46 cm beam) and 1120 J for the full aperture (58.3 cm) beam (figure 2). Near diffraction-limited beam quality is achieved by limiting the beam to the central 80% of these amplifiers. Further gain narrowing in these phosphate glass amplifiers reduces the spectral width to 4.0 nm and reduces the stretched pulse duration to 800 psec. The nonlinear phase accumulated in the disk amplifier section is ≈ 2 radians. Since the pulse is strongly-chirped in this section, the central part of the pulse (near the carrier frequency) experiences a greater phase retardation than the early or later part of the pulse where the intensity is less. This self-phase modulation effect limits the ability to recompress the pulse and results in temporal wings appearing on either side of the compressed pulse at the percent level a few hundred femtoseconds before the arrival of main pulse.[19]

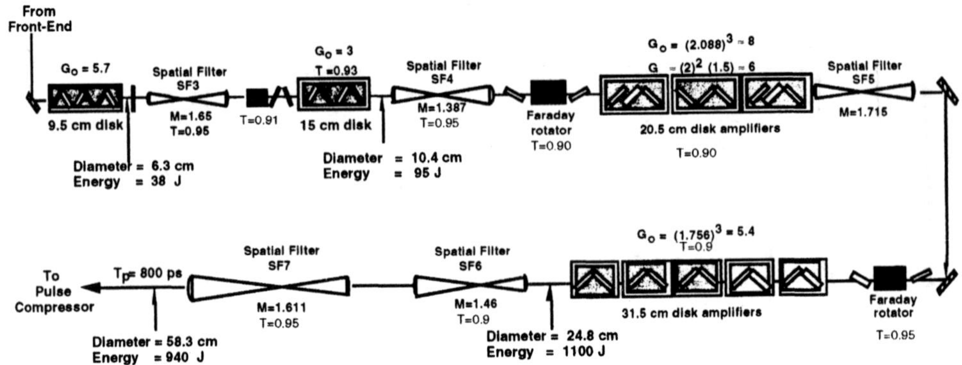

Figure 2: Disk amplifier section (Only three of the 31.5 cm amplifiers are energized).

Following amplification, the chirped pulse is compressed to a pulse duration which can easily be adjusted from 0.43 to 20 ps. Pulse compression occurs in vacuum with a pair of large aperture diffraction gratings arranged in a single pass geometry and a compressor throughput of 84% (**figure 3**). For the data reported here (from May 1996), 75 cm diameter gratings were used which limited the maximum pulse energy to 620 J in a 46 cm beam. These gratings are currently being replaced with 94 cm diameter gratings which will enable operation at 920 J with a 58.3 cm diameter beam. In most CPA systems the beam ellipticity and spatial chirp associated with a single-pass compressor geometry would be intolerable. However, in the limit that the diameter of the beam is much larger than the dispersed length, these effects are negligible. With the 46 cm diameter beam, we observe $\approx 5\%$ ellipticity in the compressed beam due to the use of a single pass compressor geometry.

One of the key technologies required to achieve petawatt class laser pulses was the development of large area, high damage threshold diffraction gratings. We have developed numerous types of compression gratings. Holographically produced metallic gratings are by far the simplest to produce. Commercially available metallic gratings are available up to the 40 cm size but with a damage threshold substantially below the theoretical damage threshold. We have investigated the fundamental mechanism of damage of metallic gratings and have been able to achieve damage thresholds of >0.42 J/cm^2 for 100 fsec pulses [figure 4], a value close to the theoretical limit.[23] We have also examined multilayer dielectric reflection gratings[24] and fused silica transmission gratings.[25]

Multilayer dielectric (MLD) reflection gratings can achieve up to 97% diffraction efficiency, have negligible absorption and exhibit a higher laser damage threshold than metallic gratings for pulses with a duration longer than ≈ 100 fsec.[24] While these gratings are superior in performance to gold gratings, they are difficult to fabricate in sizes larger than ≈ 20 cm. We have achieved a diffraction efficiency of 95% with fused silica

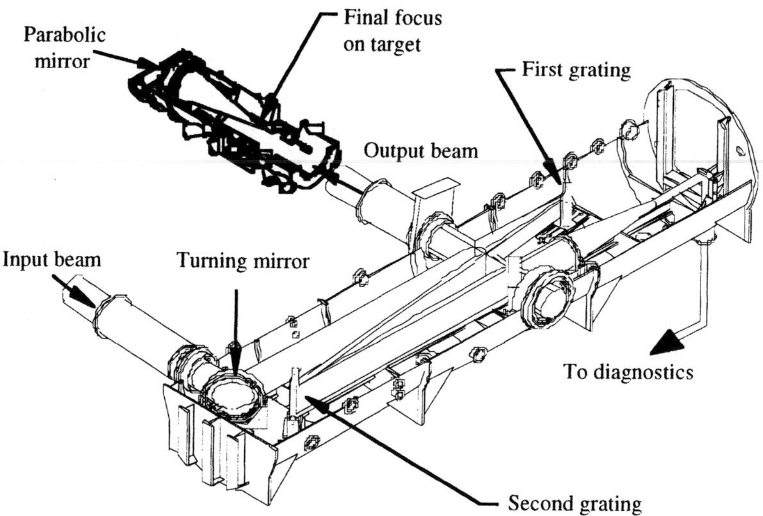

Figure 3: Pulse Compressor and single-beam target chamber.

transmission gratings. These gratings offer an even higher damage threshold than multilayer dielectric gratings. However, the effects of self-focusing and self-phase modulation associated with transmission of the compressed pulse through the grating substrate place a limit of ≈ 100 GW/cm^2 on the use of transmission gratings. Operation of these gratings at ≈ 1 J/cm^2 would therefore limit the achievable pulse duration to ≈ 10 psec.

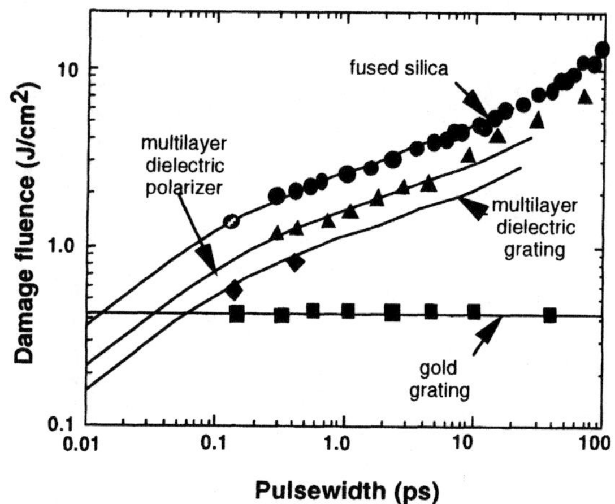

Figure 4: Damage threshold of gratings and other optical components

Because of the cost and difficulty associated with scaling MLD gratings, we constructed the pulse compressor with gold coated gratings. These gratings are 1480 l/mm and exhibit a diffraction efficiency of over 94% at 1054 nm.[26] The 75 cm grating pattern is produced initially in photoresist using interference (holographic) lithography. The size of these gratings currently limits the energy output of the compressor to a maximum of 650 Joules in a 46.3 cm diameter beam ($F_{normal}=0.39$ J/cm^2) with normal operation limited to ≈ 550 J, 75% of the damage threshold (0.33 J/cm^2). The diffracted wavefront exhibits a peak to valley modulation of 0.1 μm and 0.09 μm off of the first and second gratings, respectively with a gradient error of < 0.03 μm per cm. These gratings were replaced in Dec. 1997 with 94 cm diameter gratings which will enable operation at >900 J.

Most CPA laser systems are designed to produce the minimum pulse duration upon recompression. This is accomplished by setting the compressor to cancel the dispersion of the stretcher and material. Since the compressor cannot cancel the dispersion of the stretcher and material in the laser system exactly, the system is designed to minimize the residual phase, $\delta = \phi_{com}(\omega) + [\phi_{str}(\omega) + \phi_{mat}(z,\omega)] = 0$. These phase functions are often written in a Taylor series expansion,

$$\phi_{mat}(z,\omega) = \beta_1(\omega-\omega_0)\,z + \frac{\beta_2(\omega-\omega_0)^2\,z}{2!} + \frac{\beta_3\,(\omega-\omega_0)^3\,z}{3!} + \frac{\beta_4\,(\omega-\omega_0)^4\,z}{4!} + \dots$$

where $\beta_n = [\partial^n k/\partial\omega^n]_{\omega=\omega_0}$. The stretcher/compressor combination in the Petawatt laser is designed to correct for chromatic aberration and material dispersion in the system up to third-order (i.e., only the fourth-order terms in δ remain). This is sufficient for pulses of duration greater than 100 fsec. Output pulse characteristics of the petawatt beam under current operational conditions are shown in figure 5. The fine scale structure in this figure is the result of individual elements of the CCD array used and serves as a calibration. Third-order autocorrelation measurements show the pulse sits atop a background which is 10^5 below the main pulse and extends for several nanoseconds.

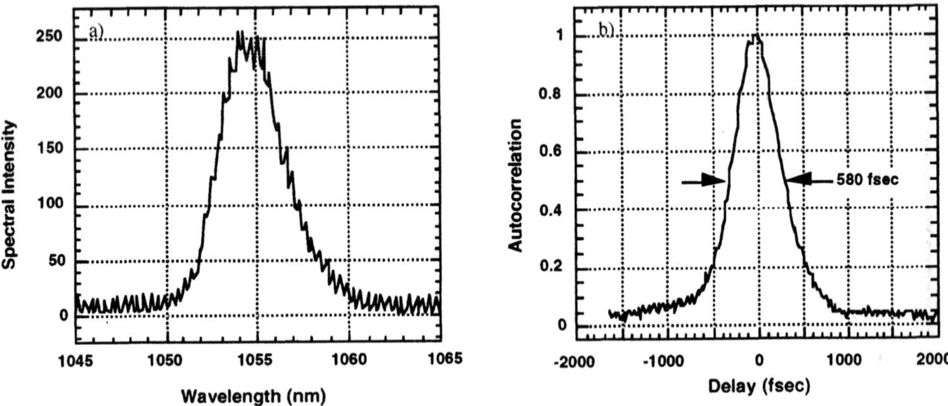

Figure 5: Spectrum and autocorrelation of 620 J Petawatt shot (May 1996). The deconvolved pulse duration (fwhm) is 430±20 fsec with a time-bandwidth product of $\Delta\nu\Delta\tau=.43$

The system was designed to perform target experiments either in an independent target chamber or with the Nova laser system. Focusing is accomplished using an on-axis parabolic mirror alone (f=180 cm), or in conjunction with a secondary "plasma" mirror (fig. 6).

The far-field beam distribution obtained using an equivalent plane image and conventional focusing corresponds to a 15x25 µm Gaussian distribution in target chamber containing ≈60% of the beam energy (figure 7). On this shot, the peak power was 1.1 PW (550 J in 480 fsec), corresponding to $4\pm1\times10^{20}$ W/cm^2. This beam quality is degraded by thermal distortion in the amplifiers and is extremely sensitive to grating alignment, groove rotation and stress in the substrate. Distortion in the disk amplifiers can be corrected by using a deformable mirror (adaptive optics) in the front-end of the laser system. This system will be activated in late 1997 and should enable $>10^{21}$ W/cm^2.

The Cassegrainian/plasma mirror geometry (figure 6b) is required for target shots when the laser is operated at the shortest pulse durations. In this case, transmissive shields cannot be used to protect the parabola from target debris since the power density on the debris shield is ≈0.7 TW/cm^2. With a 1 cm thick, fused silica debris shield, the B-integral accumulated upon two passes through the shield would be 28 radians! This destroys the beam quality and results in damage to the debris shield. Experimentally, we have found that

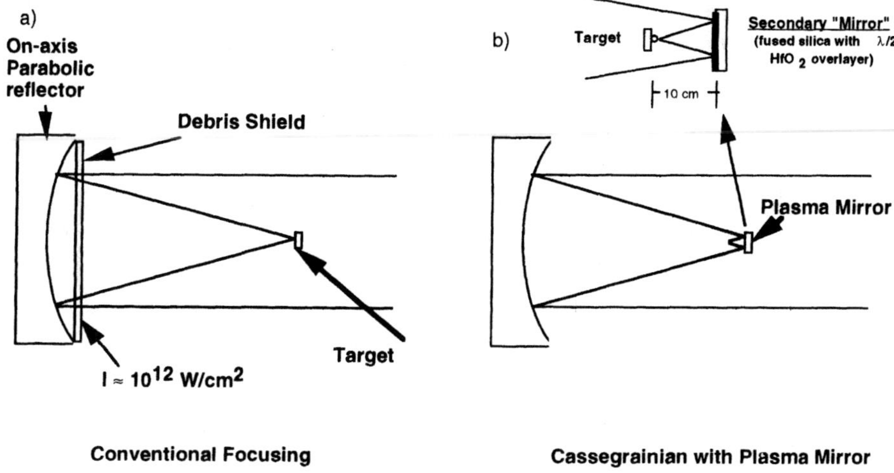

Figure 6: Focusing systems employed on the Petawatt laser for a) long (>5 psec) pulses, and b) for short (0.4-2 psec) pulses

our standard 1 cm debris shields cannot be used with the Petawatt for pulse durations less than $\approx$ 5 psec due to degradation of beam quality resulting in severe distortion in the far-field. The Cassegrainian geometry overcomes this problem by eliminating a line of sight for target debris to the primary parabola. However, in this geometry, the intensity on the surface of the secondary mirror is such that it is converted to a plasma during the laser pulse. Specifically, for irradiances $> 10^{14}$ W/cm^2, short pulse radiation creates a critical density plasma on the surface of a dielectric substrate, with a reflectivity approaching 95% (figure 8).[26] The data in figures 8-9 were acquired in small scale (2-3 mm beam size on mirror) using the Petawatt front-end. For incident pulses on the order of 500 fs, the plasma has insufficient time to undergo hydrodynamic expansion, producing a reflected wavefront comparable to the original optical surface (figure 9). Approximately 93% of the reflected energy is contained within the central spot indicating negligible distortion from the plasma surface. This novel targeting system enables the production of high contrast pulses, with an easily varied effective focal length as well as off-axis experiments.

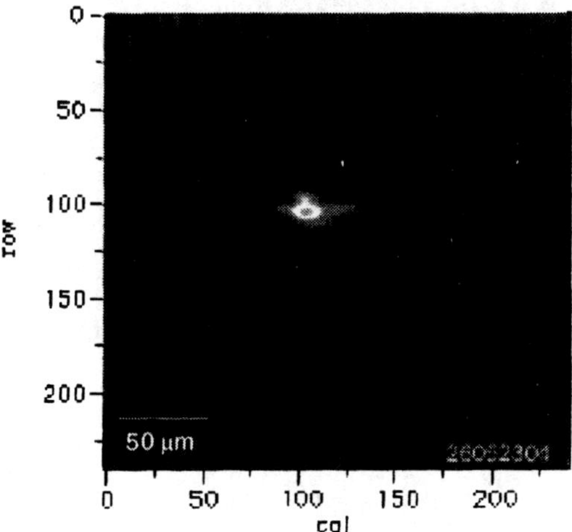

Figure 7: Equivalent plane image of the far-field beam distribution at 550 J (May 1996).

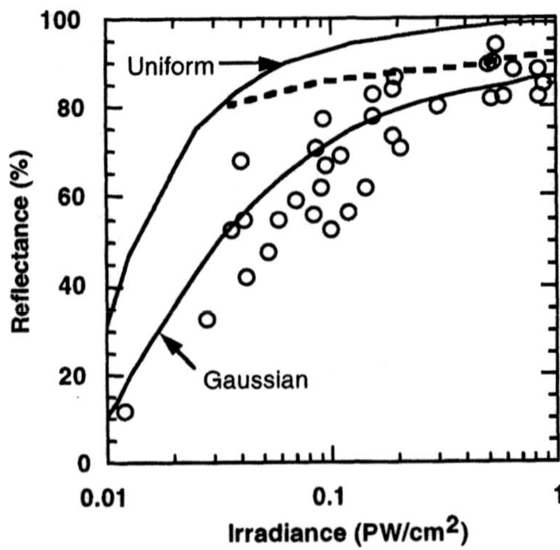

Figure 8: Reflectivity of the plasma formed on a fused silica surface with 450 fsec pulses. The solid curves are the calculated reflectivity for a uniform and Gaussian spatial distribution, respectively.

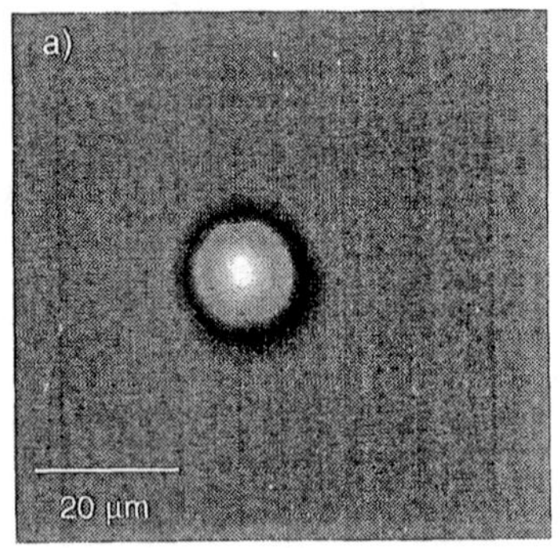

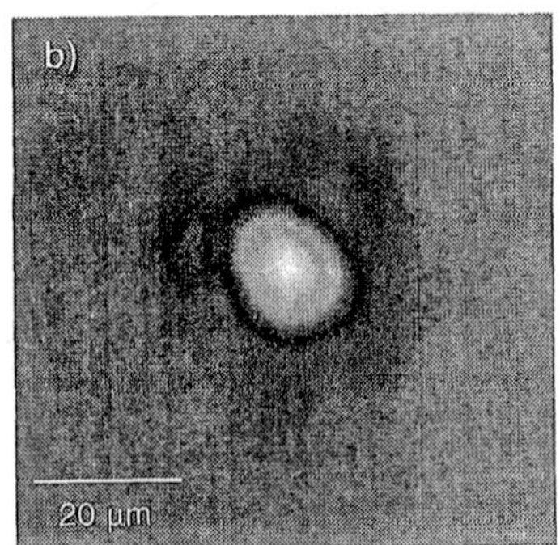

Figure 9: Far-field distributions from the secondary mirror with a) polished fused silica flat, and b) plasma mirror from SiO_2 surface at 90% reflectivity.

In conclusion, we have developed a system which currently produces a maximum of 600 J in a 450±20 fsec pulse (1.3 PW) in a 46 cm beam. Nominal operation is limited to 550 J. Pulse energy is limited by the 75 cm diameter diffraction gratings which exhibit a damage threshold of 0.42 J/cm^2. These gratings are currently being replaced by 94 cm gratings which should enable operation over 900 J.

Many people contributed to the successful development of this system. Particular recognition is due to the Nova engineering staff and W. Olsen for their help in construction and activation. This research was performed by LLNL for the US Department of Energy under contract W-7405-Eng-48.

References:

1. M.D. Perry, and Gerard Mourou, "Terawatt to Petawatt Subpicosecond Lasers," *Science*, **264**, 917 (1994).
2. C. Joshi and P.B. Corkum, *Physics Today*, January 1996.

3.   E.B. Treacy, IEEE J. Quan. Elec., **5**, 454 (1969).
4.   R.A. Fisher and W.K. Bischell, IEEE J. Quan. Elec., **11**, 46 (1975).
5.   D. Grischkowsky and A.C. Balant, App. Phys. Lett., **41**, 1, (1982).
6.   D. Strickland and G. Mourou, Opt. Comm., **56**, 219 (1985).
7.   O.E. Martinez, IEEE J. Quan. Elec., **23**, 1385 (1987).
8.   P. Maine, D. Strickland, P. Bado, M. Pessot, and G. Mourou, IEEE J. Quan. Elec., **24**, 398 (1988).
9.   M.D. Perry, F.G. Patterson, R. Ettlebrick and J. Weston, Opt. Lett, **15**, 381 (1990); F.G. Patterson, M.D. Perry and J.T. Hunt, J. Opt. Soc. Amer. B, **8**, 2384 (1991).
10.   C. Sauteret, et al, Optics Lett., **16**, 238 (1991); C. Rouyer, et al, Optics Lett., **18**, 214 (1993).
11.   K. Yamakawa, H. Shiraga, and Y. Kato, Optics Lett., **16**, 1593 (1991).
12.   P.F. Moulton, J. Opt. Soc. Amer. B, **3**, 125 (1986).
13.   D.E. Spence, P.N. Kean, and W. Sibbett, Optics Letters, **16**, 42 (1991).
14.   B.C. Stuart, M.D. Perry, J. Miller, G. Tietbohl, S. Herman, J.A. Britten, C. Brown, D. Pennington, V. Yanovsky and K. Wharton, "125-TW Ti:Sapphire/Nd:Glass Laser System," Optics Letters, **22**, 242 (1997).
15.   J.P. Chamberet, et al, Optics Letters, **21**, 1921 (1996)
16.   J.P. Zhou, C.P. Huang, M.M. Murnane and H.C. Kapteyn, Optics Letters, **20**, **64** (1995).
17.   C.P. Barty, J. Squier, K. Wilson,
18.   C. Spielmann, Optics Letters, (1996)
19.   M.D. Perry, T. Ditmire, and B.C. Stuart, "Self-Phase Modulation in Chirped Pulse Amplification," Optics Letters, **19**, 2149 (1994).
20.   M. Tabak, M.D. Perry, J. Hammer, W. L. Kruer, S. C. Wilks, J. Woodworth, E. M.Campbell, and R. J. Mason, *Phys. Plasmas*, **1**, 1626 (1994).
21.   B. Stuart, S. Herman and M.D. Perry, IEEE J. Quan. Elec., **31**, 528 (1995).
22.   M.D. Perry, T. Ditmire, and D. Strickland, Optics Letters, **17**, 601 (1992).
23.   R. Boyd, J.A. Britten, B.W. Shore, B. Stuart, and M.D. Perry, "High Efficiency Metallic Gratings for Laser Applications," Applied Optics, **34** 1697 (1995).
24.   M.D. Perry, J.A. Britten, R.D. Boyd, B. Shore, C. Shannon, and E. Shults, Optics Letters, **20**, 940 (1995).
25.   H.T.Nguyen, B.W. Shore, S.J. Bryan, J.A. Britten, and M.D. Perry, "High-efficiency fused silica transmission gratings," Optics Letters, **22**, 142 (1997).
26.   J.A. Britten, M.D. Perry, B.W. Shore, and R.D. Boyd, "Universal grating design for pulse stretching and compression in the 800-1200 nm range," Opt. Lett., **21**, 540 (1996).
27.   M. D. Perry, V. Yanovsky, M. Feit, and A. Rubenchik, "Plasma mirrors," *Phys. Plasmas*, submitted, 1997.

MULTITERAWATT ULTRAVIOLET LASERS

Fiorenzo G. Omenetto,[1,2] Keith Boyer,[1] Tom Nelson,[1]
James W. Longworth,[1] W. A. Schroeder[1] and C. K. Rhodes[1,3]

[1]Laboratory for Atomic, Molecular and Radiation Physics, rm.2136,
   University of Illinois at Chicago, 845 W. Taylor Street, Chicago,
   Illinois 60607-7059.
[2]Università di Pavia, Dipartimento di Elettronica, Via Ferrata 1, 27100
   PAVIA, Italy and INFM, sezione di Pavia.
[3]TARA, University of Tsukuba, Tennodai 1-1-1, Tsukuba, Ibaraki Japan

INTRODUCTION

The advances in ultrashort pulse generation and chirped pulse amplification (CPA) techniques have been the driving force in the development of laser systems with peak powers at and beyond the Terawatt (10^{12} W) level[1-7]. Most high power systems developed to date, work in the infrafed, and Terawatt and Petawatt (10^{15} W) class lasers have been demonstrated in Ti:Sapphire, Nd:glass and Cr:LISAF based systems. These systems are moving towards shorter pulse durations, now commonly of the order of tens of femtoseconds, and design goals are heading towards high repetition rate (kHz)[8] multiterawatt systems. The progress in this field has certainly been inspired by the improvements in solid-state laser materials for ultrashort pulse generation. An analogous statement does not generally hold, however, for the ultraviolet wavelengths.

Efforts directed towards the development of ultrahigh peak power systems in this spectral region can be justified by the expectation of observing a distinct set of physical phenomena caused by the different properties of the stimulating radiation, among which we mention, *inter alia* (a) the greater energy per photon (b) the tighter focus obtainable and (c) the shorter period of the stimulating electric field. A means to obtain TW-class lasers in the ultraviolet is offered by the combination of excimer lasers and ultrafast oscillators. Efforts in this direction have been undertaken by a number of groups [9,10]. In the present paper, the laser system operating at the University of Illinois at Chicago will be outlined, followed by an overview of promising new solid-state laser materials directly emitting in the UV, and the first tests performed on them.

THE LASER SYSTEM

The laser currently operational is a hybrid TiAl$_2$O$_3$/KrF* system. The initial pulses are obtained with a Ar-ion pumped KLM Ti:Sapphire oscillator optimized for stable operation at λ=745 nm with a single-plate birefringent filter. The average mode-locked power is 350 mW (P$_{pump}$=6.7 W) for a pulse duration of t = 85 fs. Amplification in the red is performed through a conventional CPA technique. Gold-replica gratings (groove density of 1200 lines/mm) are used in a single grating stretcher/single grating compressor

Applications of High Field and Short Wavelength Sources
Edited by DiMauro *et al.*, Plenum Press, New York, 1998

combination (stretch factor ~1000). The previously employed dye preamplifier/triple pass Ti:Sapphire combination[11] has been substituted by a home built regenerative amplifier[12]. Pumping is achieved by means of a Continuum Q-switched Nd:YAG, furnishing 25 mJ of energy to the amplifier crystal. Typical energy outputs are of 1.3 mJ (with fluctuations of ±5% caused primarily by instabilities in the pump energy) at a central wavelength of 745 nm. This gives us, as expected, considerable improvement over the dye-cell arrangement in terms of energy stability, beam profile and gain. The repetition rate of the amplified pulses is kept to 2 Hz for convenience, given the limitations on the repetition rate of the final excimer amplifier stage. After recompression of the pulses from the regenerative amplifier, frequency conversion takes place in two KDP crystals of thickness 3 mm and 1 mm, respectively, for doubling and mixing of the 372 nm radiation with the residual 745 nm fundamental. Seeding energies of at least 10 µJ from the tripler are desirable, to guarantee a good contrast ratio between the signal and the amplified spontaneous emission (ASE) in the KrF* amplifier. Average output energies from the tripler are of 18 µJ at 248 nm (for a 0.7 mJ 745 nm input), which are sent to seed the excimer preamplifier. To ensure good beam quality, the beam is spatially filtered in vacuum after the tripler by focusing it into a 35 mm pinhole which reduces the seed energy to 12 mJ. The first UV amplification stage is performed in a KrF* excimer (Lambda Physik 201 MSC EMG) arranged in a double-pass off-axis geometry[12]. The device is operated with a 22 kV discharge voltage and a pressure of 2000 mbar (100 mbar F_2, 120 mbar Kr, 1800 mbar Ne). The off-axis angles in the two paths are designed to achieve optimum amplification over the whole beam dimension. The output energies are measured to be an average of 1.5 mJ. The duration of the 248 nm pulse, after the double-pass excimer, obtained through a two-photon fluorescence (TPF) measurement gives a value of 160 fs[13]. The beam is recollimated after amplification and the contrast ratio (signal to ASE) at this point is measured to be in excess of 10^3. The beam is then directed through a telescope beam expander before entering the final amplification stage which consists in a large aperture (10 cm) excimer amplifier[14]. This device is operated at relatively low pressure and low gain in order to reduce wave-front distorsions and ASE. The pulses, produced at a repetition rate of 0.4 Hz exhibit a 10-shot average energy above 0.25 J, with peak recorded shots of 0.75 J. Characterization and optimization of the spectral and temporal features of the final pulse are presently underway. From the previous performance of the system , there is reason to believe that there will not be considerable broadening of the pulse through the final amplification stage furnishing evidence of a laser system that can reach the 2-3 Terawatt power level in the UV region of the spectrum (248 nm). Preliminary observations have shown improved experimental conditions, with a more homogeneous beam profile and near diffraction limited focal spot (measured to be < 3 µm in diameter in the target chamber) indicating improved efficiency in the delivery of the stimulating radiation for high-intensity physics experiments.

SOLID-STATE ULTRAVIOLET LASERS

The possibility of extending high-peak power laser technology to the ultraviolet involves rare-earth doped colquiriites such as Cerium doped LiCAF[15,16] and Cerium doped LiSAF. For the sake of this discussion, we will examine the latter, but the considerations hold true, in general, for the other crystal host as well.

Ce:LiSAF has an emission bandwidth located between 285 and 295nm[17], hence theoretically capable of supporting pulses as short as 10 fs. A number of technical issues, inherent to operation at shorter wavelengths, and to the performance of the host and other materials (especially in the amplification stages) under high ultraviolet fluences, remain to be verified experimentally. The advantages, however, would be considerable and would entail, among other things, reduced size of the laser system, phase control of the generated pulses, and the possibility of supporting shorter pulse duration in comparison to the hybrid system described above.

The design of the system is based upon the application of Kerr Lens Modelocking (KLM) to a Ce:LiSAF crystal. This well established technique requires *(i)* a gain bandwidth sufficient to support a short optical pulse, *(ii)* no significant non-linear absorption (i.e. two-photon absorption) at the oscillating wavelength and *(iii)* a suitable non-linear refractive index n_2. By examining the properties of the various components in the melt, an estimate of the two-photon absorption edge for LiSAF can be inferred and is conservatively estimated to be at about 125 nm, or 9.9 eV, indicating that there will be no significant two-photon absorption. Furthermore, from this value for the band gap, we can estimate the non-linear index of refraction (n_2) of LiSAF at 290 nm using the theoretical method developed by M. Sheik-Bahae *et al.*[18]. A value $n_2 \sim 1.4 \times 10^{-20}$ m^2/W (positive) is obtained (comparable to n_2 for Ti:Sapphire at 800 nm, *i.e.* $\sim 3 \times 10^{-20}$ m^2/W[19]).

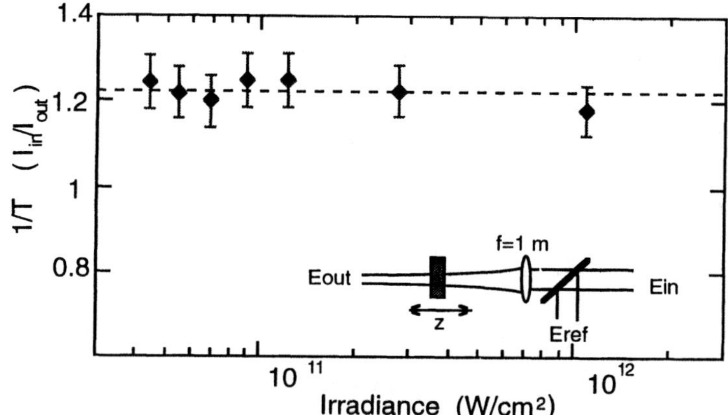

FIGURE 1. Inverse transmission data for a 1 mm Ce:LiSAF sample. Each data point is the average of 100 shots to account for energy fluctuations in the beam. The dashed line represents the best fit to the data (slope = 0)

In order to confirm these assumptions, a series of tests was performed on a 1 mm-thick sample of Ce:LiSAF (1% CeF$_3$, Lightning Optical Co.) to verify the non-linear absorption properties of the material. A conventional transmission measurement was performed by using 248 nm, 160 fs pulses, as illustrated in fig. 1. The best fit to the inverse transmission data yields β=0, where β is the two photon absorption coefficient, as described by Taylor et al.[20], in support of the predicted value for the material's band-gap. Furthermore, inspecting the sample after irradiation, no damage is detectable.

The design of the cavity can be based on the common geometries employed in the infra-red (i.e. Ti:sapphire oscillators), with one of the main points being the appropriate choice of the suitable pump laser. A first order solution could be the use of the fourth harmonic of a CW mode locked Nd:YLF (263nm), which is a close match to the absorption profile of the active Ce^{3+} ion and provides, with proper cavity length matching, gain modulation that can assist the KLM mechanism (synchronous pumping). On the other hand, the progress in diode pumped solid-state IR sources, such as the ones based on Nd-doped YAG, and YVO$_4$, offer a number of alternatives which can deliver, after frequency conversion, more pump power in aid of the non-linear modelocking process as well. By comparison of the parameters estimated for Ce:LiSAF with those of a typical hard-aperture KLM Ti:Sapphire laser, we note that the critical power for self-focusing, P_c, is significantly lower (P_c prop. λ^2/n_2) thus indicating the possibility of operation at lower pump power levels. The projected energy output, assuming a 100 fs pulse duration (a conservative estimate) yields an average power output of ~30mW for KLM operation with ~ 0.5nJ/pulse[21,22]. Modelling of the cavity, using an ABCD matrix code similar to the one first employed by Magni *et al.*[23] has been performed[24] and shows favorable regions for mode-locking.

Once these pulses have been generated, amplification would be possible by the application of standard Chirped Pulse Amplification (CPA) techniques. Recently, Q-switched operation of Ce:LiCAF has been demonstrated to yield 14 mJ pulses at 290 nm, indicating the possibility of pushing performances of these materials towards TW-power levels, in conjunction with appropriate pulse compression techniques.

It should be pointed out that the output of every stage of a UV-solid-state system of this kind, is readily applicable to numerous areas of scientific activity, aside from the specific use of the system as a whole for high-intensity experiments, among which the generation of Xrays from atomic rare gas clusters[25]. The pulses generated by the oscillator alone or with the first amplifier stage, would allow <u>coherent control</u> experiments with tuneable UV pulses. Fluorescence spectroscopy of proteins and nucleic acids, applications to photocatalysis related research and manufacture of pharmaceuticals are well suited to ultrafast optical pulses in the ultraviolet. The system as a whole would also provide a compact, high powered UV source for use in many other applications such as specialized forms of lithography or microfabrication.

It is auspicable to expect the same trend that revolutionized IR sources, to extend to the ultraviolet region of the spectrum. If supported by consistent progress on the materials[26], the path to the availability of new versatile sources at shorther wavelengths and to new powerful tabletop systems will be open.

ACKNOWLEDGEMENTS

Support for this research was provided under contracts with SDI/NRL (N00014-93-K-2004), ARO (DAAH04-94-G-0089) and the University of California/Lawrence Livermore National Laboratory (B328353).

REFERENCES

1. Ditmire, T. and Perry M., Optics Letters, **18** (6), 426 (1993)
2. Barty, C. P. J., Guo, T., LeBlanc, C., Raksi, F., Rose-Petruck, C., Squier, J., Wilson, K. R. , Yakolev, V. V. and Yamakawa, K., Optics Letters, 21 (9), 668-670 (1996)
3. Zhou, J., Huang, C. P., Shi, C., Murnane, M. M. and Kapteyn, H. C., Optics Letters, **19** (2), 126-128(1994)
4. Sullivan, A., Bonlie J., . Price D. F and White, W. E., Optics Letters, **21** (8), 603 (1996)
5. Beaud, P., Richardson, M., Miesak, E. J. andChai, B. H. T., Optics Letters, **18** (18), 1550-1552 (1993)
6. Antonetti, A., Chambaret, J. P., Cheriaux, G, Curley, P. F., Darpentigny, G., LeBlanc, C. and Salin, F. : *Ultrafast Phenomena 8*, , (OSA, Washington D.C. 1996) p.160
7. Y. Nabekawa, K. Kondo, N. Sarukura, K. Sajiki and S. Watanabe, Optics Letters, **18** (22), 1922-1924 (1993)
8. Backus, S., Durfee, C., Mourou, G., Kapteyn, H., and Murnane, M., Optics Letters, **22** (16), 1256-1258 (1997)
9. Ross, I. N., Damerell, A. R., Divall, E. J., Evans, J., Hirst, G. J., Hooker,C. J., Houliston, J. R., Key, M. H.,Lister, J. M. D., Osvay, K., Shaw., M. J., Optics Comm., **109**, 288-297 (1994)
10. Szatmári, S., Schäfer, F. P., Müller-Horsche, E. and Mückenheim, W, Optics Comm., 63, 305 (1987)
11. Bouma, B., Luk,T. S., Boyer, K.and Rhodes, C. K., JOSA B, **10** (7), 1180-1186 (1993)
12. Nelson,T., Omenetto, F. G., Longworth, J. W., Schroeder, W. A., and Rhodes, C. K., Applied Optics, in press
13. Omenetto, F. G., Boyer, K. , Longworth, J. W. , McPherson, A. , Schroeder, W. A., Rhodes, C. K.," Applied Optics,**36** (15), 3421-3424, 1997
14. Luk, T. S., McPherson, A., Gibson, G., Boyer, K. and Rhodes, C. K., Optics Letters, **14**, 1113-1115 (1989)
15. Dubinskii, M. A. et al., *J. Mod. Opt*, **40**, 1, (1993)
16. Dubinskii, M. A. et al., Optics Letters, **22** (18), 994-996, (1997)
17. Marshall, C. D., Speth, J. A., Payne, S. A., Krupke, W. F., Quarles, G. J., Castillo, V., and Chai, B. H. T., J. Opt. Soc. Am. B **11**, 2054 - 2065, (1994)
18. Sheik-Bahae, M., Hutchings, D. C., Hagan, David J., and Van Stryland, E., IEEE J. Quant.Elect. **QE-27**, 1296 - 1309 (1991)
19. Salin, F., Squier, J., and Piché, Opt. Lett. **16**, 1674 1676 (1991)
20. Taylor, A. J., Gibson, R. B., and Roberts, J. P. , Optics Letters **13** (10), 814-816, 1988

21. Omenetto, F. G., Nelson,T., Longworth, J. W., Schroeder, W. A., and Rhodes, C. K., *"Design and Analysis of a New Solid-state Deep UV Laser based on Ce:LiSAF,"* in Laser Spectroscopy, Ed. M. Inguscio, M. Allegrini, A.. Sasso, World Scientific, Singapore, p. 360-361, 1995

22. Nelson,T., Omenetto, F. G., Longworth, J. W., Schroeder, W. A., and Rhodes, C. K., , *"Design and Analysis of an all Solid-State UV Laser based on $Ce^{3+}LiSrAlF_6$,"* in "High Power Lasers", NATO-ASI series, Ed. R. Kossowsky, M. Jelinek, R. F. Walter, Kluwer Ac., London, p. 177-,184 1995.

23. Magni,V., Cerullo, V. and De Silvestri, S., Optics Comm., **101**, 195 (1993)

24. Omenetto, F. G., Boyer, K. , Longworth, J. W. , McPherson, A. , Schroeder, W. A., Rhodes, C. K. *"High Intensity UV Sources: Multiterawatt Ultraviolet Laser System and Next Generation Sources"* in proceedings Intl. Conference on Superstrong fields in Plasmas, Varenna, Italy, August 1997, in press

25. McPherson, A., Thompson, B. D., Borisov, A. B., Boyer, K., and Rhodes, C. K. (1994) Multiphoton-induced X-ray emission at 4-5 keV fromXe atoms with multiple core vacancies., Nature **370**, 631 - 634.

26. Dubinskii, M. A., "Recent Developments in Ce-activated Tunable Solid-State UV Lasers," in *Conference on Lasers and Electro-Optics, Vol.10 of Technical Digest Series* (Optical Society of America, Washington D.C., 1997), CThN1 p. 404

0.27 Terawatt laser system at 1 kHz

Sterling Backus, Charles G. Durfee III,
Henry C. Kapteyn, and Margaret M. Murnane

Center for Ultrafast Optical Science
University of Michigan
Ann Arbor, MI 48109-2099

INTRODUCTION

Recent advances in high-average-power, high-repetition-rate, ultrashort laser designs has allowed these systems to approach the 5W average power level at a fraction of a terawatt peak power[1]. Amplified spectral bandwidths of greater than 80nm have been generated, which can support pulse durations below 17fs[2, 3]. In this paper, we describe such a laser system, capable of delivering the highest average power femtosecond pulses to date. Thermal lensing in the amplifier system was eliminated by cryogenically cooling the Ti:sapphire crystal. The output pulses from our system were fully characterized using the technique of transient grating and second harmonic frequency resolved optical gating. We demonstrate experimentally that the pulse duration at the output is limited only by fifth-order dispersion.

TERAWATT CLASS AMPLIFIER DESIGN

The low energy seed pulses are derived from a kerr-lens modelocked Ti:sapphire oscillator, which is capable of producing pulses with spectral bandwidths greater than 140nm[4]. The oscillator spectrum is tailored to compensate as much as possible for the spectral shaping that the pulse experiences as a result of gain shaping and gain narrowing during the amplification process. The pulses are then stretched to 100ps using a pair of 1200g/mm gratings, in an all-reflective pulse stretcher. The pulse repetition rate is reduced to 1kHz before injection into the amplifier chain.

The laser system requires two stages of amplification to reach output energies of several millijoules per pulse. The first stage uses a ring design, where the multiple passes through the ring cross in the crystal at the focus[5, 6]. It produces 1.4mJ of amplified output energy using 8.5W of pump power from a Quantronix 527 Nd:YLF laser. The second stage uses a similar ring mirror configuration to that of the first stage, as shown in Fig. 1. However for the second stage, the crossing point of the various beam passes through the ring is moved out of the focal region, to a point where the beam diameter is 1.2mm. The Ti:sapphire crystal is then placed at this enlarged crossing point to prevent damage. The

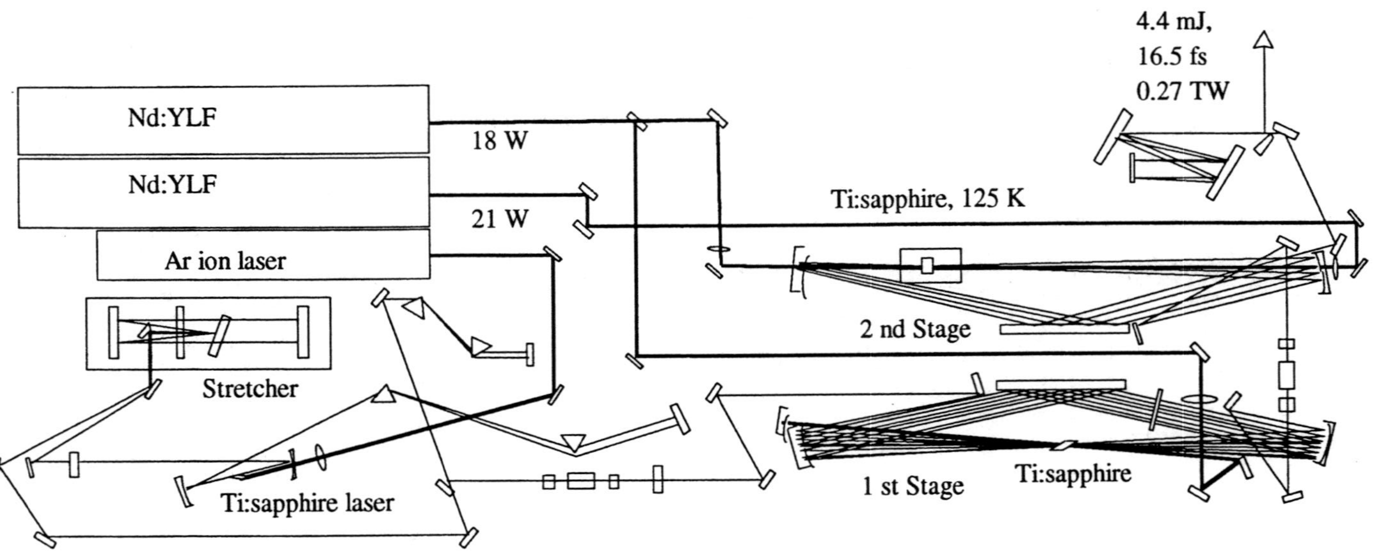

Figure 1. Schematic of laser system layout on a 4' x 10' optical table.

ring design allows for this flexibility of moving the crossing point of the multiple passes to anywhere within the cavity. It has the added advantage that the pump-mode size can be adjusted without the need for changing any optics, since the pump beam is often incident through one of the end focusing mirrors of the ring.

Thermal Lensing

One of the major problems with high average power ultrafast systems is thermal lensing induced by the pump light within the amplifier crystal. This can lead to severe miss-match between the pump mode and laser mode, leading to a degradation in the amplifier efficiency[7]. For regenerative amplifiers, this thermal lens can be compensated by using corrective optics. However, this approach works over a limited pump power range, and has only been demonstrated to date for pulse durations of approx. 40fs. Moreover, high order aberrations induced by the thermal lens are much more difficult to correct.

Our approach has been to use the desirable material properties of Ti:sapphire at low temperatures to eliminate the thermal lens[1]. Sapphire is a very hard material with high thermal conductivity, which is very temperature dependent. At room temperature, the thermal conductivity is 46W/m*K; however at 125K, the conductivity jumps to over 200W/m*K. The thermal lens is also a strong function of the change of refractive index as a function of temperature (dn/dT). By cooling the Ti:sapphire, (dn/dT) reduces from 13×10^{-6} /K at 300K, to 3×10^{-6} /K at 125K[8].

By cooling the Ti:sapphire to 125K, the thermal lens induced by 23W of pump light is reduced from 45cm at 300K, to 5.6m at 125K, essentially removing the major effects of the thermal lens from the second amplifier stage. This preserves good pump mode overlap. The second stage is pumped by the residual 6.5W of pump light from the first Nd:YLF laser, and 20W from a second Nd:YLF laser. With an injection energy of 700µJ, and 22W of pump light absorbed in the amplifier crystal, we extract up to 6.5W of amplified 800nm light, giving an extraction efficiency of $\approx$30%. Upon compression, the useful output energy is 4.4mJ. Our grating compressor consists of two 1200 g/mm gratings, with a total throughput efficiency of 70%. The pulses were fully characterized using second harmonic frequency optical gating (SHG-FROG)[9, 10]. Figure 2 (c) shows a retrieved pulse of 16.5fs in duration, together with the spectral transform limit, which are in excellent agreement.

Laser System Model

In order to predict the output characteristics of the amplifier chain, we developed a complete model of the system, which includes gain narrowing, spectral shaping, material dispersion, and gain dispersion[11]. Our model simulates the propagation and amplification of a stretched chirped pulse, and we take care to calculate the total spectral phase of the entire optical system so that we can model it accurately. Thus, we must include not only the material spectral phase, but the phase imparted to the pulse by the reflective optics in our amplifier. Figure 2(a) shows the measured and calculated group delay for the amplifier, and Fig. 2(b) shows the measured and calculated spectrum after a gain of 7×10^6, while Fig. 2(c) shows the measured and calculated pulse intensity profile. Our model thus predicts the output power, spectral shape, and group delay of the system with high accuracy.

The beam profile in the far field is excellent, and focuses to the diffraction limit. Our gratings undergo thermal distortion when the average incident intensity exceeds 5W/cm^2. Therefore, the beam is enlarged to 6mm ($1/e^2$) radius after the second amplifier stage, resulting in an average intensity of 1.6 W/cm^2 on the grating, which is below the

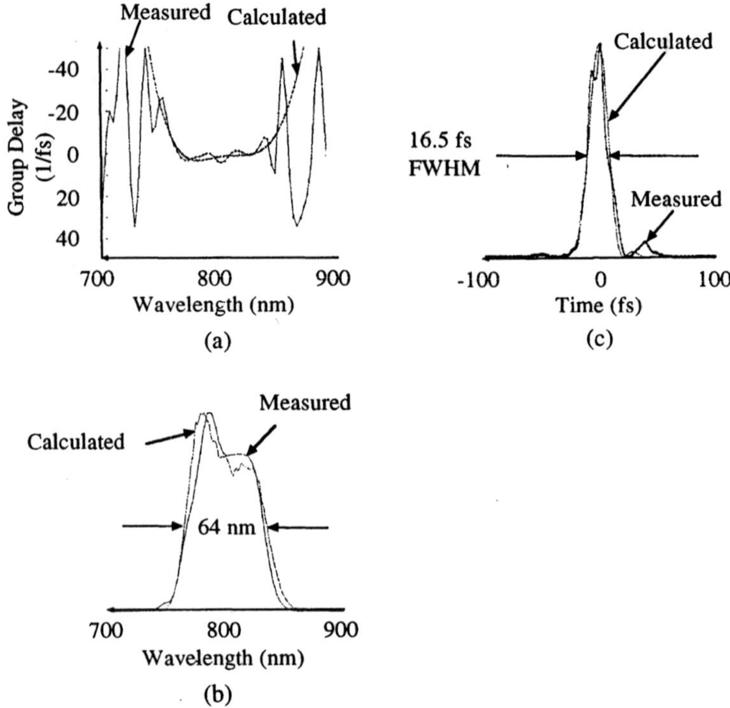

Figure 2. Output of laser; (a) Group delay, (b) spectrum, (c) pulse intensity profile.

distortion threshold. The non-linear phase shift, or B-integral, is calculated to be < .2 for the entire system due to the low amount of material in the beam path.

CONCLUSION

In conclusion, we have generated high peak and average powers simultaneously from a kilohertz repetition rate Ti:sapphire system. The output pulses have been fully characterized using SHG-FROG, and have a duration of 16.5fs, with an energy of 4.4mJ. This corresponds to a peak power of 0.27TW, and an average power of 4.4W, for the system. Thermal lensing was controlled by cooling the Ti:sapphire crystal to 125K. Finally, we developed a model that predicts the group delay, spectral shaping, and output characteristics of our system, which are in excellent agreement with our experimental measurements. We believe this work represents the highest peak power kHz system demonstrated to date.

We are currently constructing the next generation high peak power kHz laser system. It will utilize all-solid-state diode-pumped laser technology for the pump lasers. The system will occupy a 5' x 5' section of an optical table, and produce 10fs, 50mJ pulses at 500 Hz (25W average power), and will deliver 5TW of peak power. Such high average power lasers will enable many applications such as the generation of compact vacuum ultraviolet and soft-x-ray sources[12-14], laser-induced processing of materials, x-ray microscopy, and ultrafast x-ray spectroscopy of molecules and solids.

We gratefully acknowledge support for this work from the Department of Energy and the National Science Foundation. H. Kapteyn acknowledges support from an Alfred P. Sloan Foundation Fellowship.

REFERENCES

1. S. Backus, C. Durfee, M. M. Murnane, *et al.*, Optics Letters **22**, 1256 (1997).
2. S. Backus, C. Durfee, M. M. Murnane, *et al.*, Review of Scientific Instruments **to be published** (1997).
3. C. Barty, G. Korn, F. Raksi, *et al.*, Optics Letters **21**, 219 (1996).
4. C. Durfee, S. Backus, K. Read, *et al.*, JOSA B **submitted** (1997).
5. S. Backus, J. Peatross, C. P. Huang, *et al.*, Optics Letters **20**, 2000 (1995).
6. C. Hillimann, O. Seddiki, J. F. Morhange, *et al.*, Opt. Comm. **59**, 52 (1986).
7. W. Koechner, *Solid-State Laser Engineering* (Springer-Verlag, Heidelberg, 1996).
8. A. DeFranzo and B. Pazol, Applied Optics **32**, 2224 (1993).
9. K. W. DeLong, R. Trebino, J. Hunter, *et al.*, JOSA B **11**, 2206 (1994).
10. G. Taft, A. Rundquist, M. M. Murnane, *et al.*, JQE Special Topics **2**, 575 (1996).
11. C. G. Durfee, S. Backus, M. M. Murnane, *et al.*, JQE **submitted** (1997).
12. S. Backus, J. Peatross, E. Zeek, *et al.*, Optics Letters **21**, 665 (1996).
13. C. G. Durfee, S. Backus, M. M. Murnane, *et al.*, Optics Letters **22**, 1565 (1997).
14. Z. Chang, A. Rundquist, H. Wang, *et al.*, Physical Review Letters **79**, 2967 (1997).

DETERMINATION OF THE DURATION OF UV FEMTOSECOND PULSES.

J.-F. Ripoche, B. S. Prade, M. A. Franco, G. Grillon, R. Lange and A. Mysyrowicz

Laboratoire d'Optique Appliquée
ENSTA-École Polytechnique-CNRS, Batterie de l'Yvette,
91761 Palaiseau, France
tel.: (33) 1.69.31.02.20; fax: same
e-mail: mysy@enstay.ensta.fr

The determination of the duration of visible or near-infrared optical pulses with subpicosecond duration has become a routine operation in ultrafast time-resolved spectroscopy. Several experimental methods have been demonstrated, which all rely on the same principle, transforming a time measurement requiring an ultrafast detector into a length measurement using a slow detector[1]. By far the most commonly used technique is the measurement of the sum frequency radiation which yields the autocorrelation of the pulse from which the pulse duration is extracted, assuming the pulse timeshape known. Recently, more sophisticated methods have been developed to obtain more precise knowledge of the pulse characteristics, such as its phase and amplitude[2,3,4]. However, to extend such measurements to the ultraviolet (UV) domain is a nontrivial task[5,6], for instance because of phase-matching or group velocity dispersion issues.

In this paper, we present an approach based on the same general principles, which has the advantage that it can be easily implemented in the UV (and in principle in the far UV as well). It is based on spectral analysis following cross-phase modulation (XPM). Since the non-linear process is described by a symmetric term of the third-order non-linear susceptibility tensor, it does not require frequency up-conversion nor any phase-matching conditions but it is still limited by group velocity dispersion.

Applications of High Field and Short Wavelength Sources
Edited by DiMauro *et al.*, Plenum Press, New York, 1998

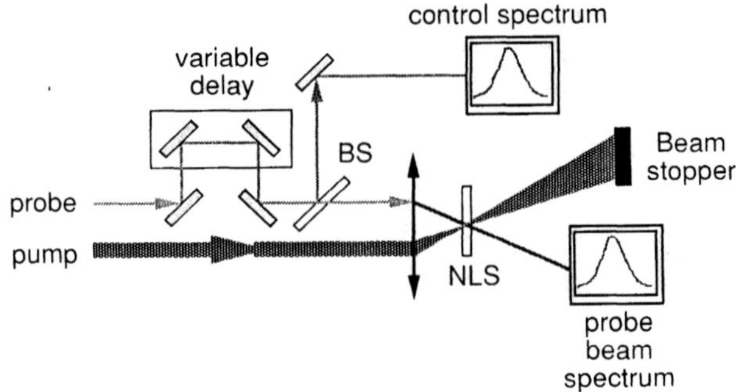

Figure 1. Experimental set-up. NLS: non-linear sample. BS: beam splitter.

The principle of the method is shown in figure 1. The pulse to be analysed is divided in two parts: the pump beam containing most of the energy and a weak probe beam. After passing one of the beams through an optical delay, both beams are made to overlap in a common region of a non-linear sample. The sample consists of any transparent, centrosymmetric medium with instantaneous Kerr response, here a thin plate of fused silica. To get sufficient intensity, the beams are focused into the sample.

The spectrum of the probe beam is then recorded in the presence of the pump beam as a function of time delay between pump and probe. The pump pulse induces a change of the probe spectrum, which is a sensitive function of the time delay during the pulse overlap. As we show now, by simply recording the change in the centre of gravity of the probe beam spectrum, the derivative of the autocorrelation function is directly obtained[7]. The centre of gravity of the probe spectrum is given by:

$$g(\tau) = \frac{\int \nu |F(\nu,\tau)|^2 \, d\nu}{\int |F(\nu,\tau)|^2 \, d\nu} \qquad (1)$$

where τ is the delay between the pulses, ν is the frequency and $F(\nu,\tau)$ the Fourier transform of the probe beam complex electric field in the time domain.

When propagating through the non-linear sample illuminated by the intense pump pulse, the probe beam acquires a non-linear phase and one can express its electric field as:

$$E(t,\tau) = E(t,-\infty)\exp\left(i\frac{2\pi}{\lambda} n_2 z I_p(t-\tau)\right) \qquad (2)$$

where λ is the laser central wavelength, I_p the pump intensity, n_2 the sample non-linear index and z its thickness. Noting that the denominator in (1) is the fluence of the pulse W, and integrating by parts, one finds that $g(\tau)$-g_0 is proportional to the derivative of the autocorrelation function $A(\tau)$:

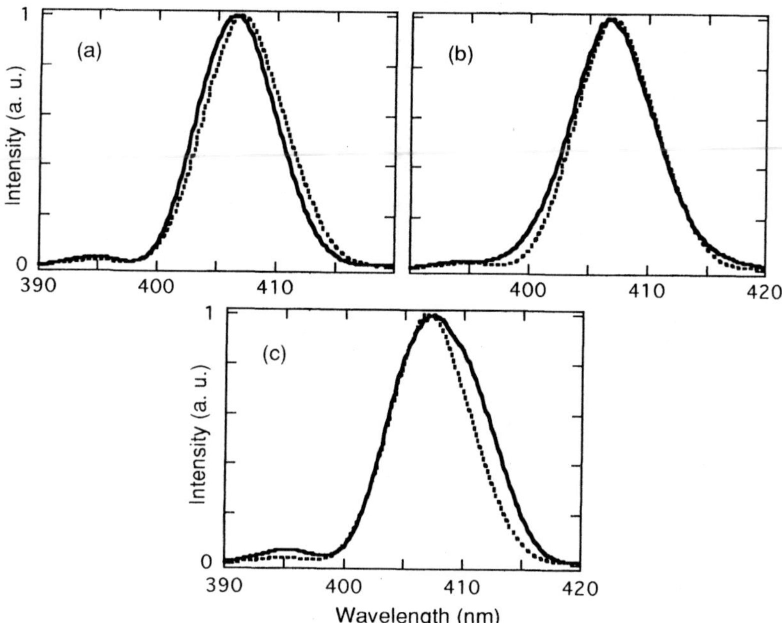

Figure 2. Experimental cross-phase modulation spectra of the probe beam from the second harmonic of a Ti:sapphire chain recorded at different time delays, without the pump beam (dashed line) and in presence of the pump beam (solid line). (a): t=-40 fs. (b): t=4 fs. (c): t=+40 fs.

$$g(\tau) - g_0 = \frac{n_2 z}{\lambda E} \frac{dA(\tau)}{d\tau} \qquad (3)$$

with $g_0 = g(\tau = -\infty)$.

To test the method we have performed two experiments. The first with near IR and visible optical pulses already well characterised by other standard techniques. We obtained a good agreement between the different methods. The second with UV pulses obtained by harmonic generation in a doubling crystal, for which the results could be compared to predictions since the pulse duration at the fundamental frequency was known.

We first used a Ti:sapphire oscillator-amplifier providing up to 10 mJ in about 30 femtoseconds at 800 nm and the second harmonic was generated using a 200 μm thick BBO crystal. Typical cross-phase modulation experimental spectra are shown in figure 2. The first (last) picture shows very clearly that blue (red) components are added to the spectrum when the pump intensity decreases (increases) with time. When the two pulses arrive at the same time on the sample, one only observes spectral broadening and the centroid of the probe beam spectrum remains unchanged.

Figure 3 displays the shift of the centre of gravity of the probe spectrum as a function of time delay at 400 nm. For each point of the curve a selection criterion is applied: the control spectrum energy and centroid must fit into a predetermined window (±5% for the

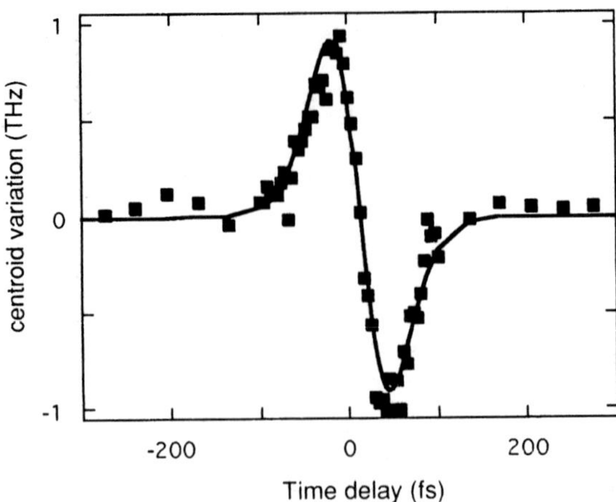

Figure 3. Change of the centre of gravity of the probe spectrum at 400 nm as a function of time delay between pump and probe in a 0.5 mm plate of fused silica (Suprasil). The full line is a fit assuming an $(\text{sech})^2$ pulse shape, the retrieved pulse duration is 50 femtoseconds.

energies and ±0.2 THz for the centroids). The data were fitted to equation (3) using a Levenberg-Marquardt method. Assuming a pulse shape of the $(\text{sech})^2$ form, we obtain a pulse duration t_{FWHM} (full width at half maximum) of 50 femtoseconds. The obtained pulse duration is in good agreement with simulations taking into account group velocity and phase mismatch in the doubling crystal.

We also studied the second harmonic of an amplified CPM dye laser working at 615 nm with an energy of 2 mJ per pulse and pulse durations as short as 80 femtoseconds. It was obtained with a KDP crystal (100 μm thickness) yielding 10 μJ per pulse in the UV. The obtained pulse duration (126 femtoseconds assuming a sech^2 temporal profile) is in good agreement with what is derived from the fundamental pulsewidth (196 femtoseconds).

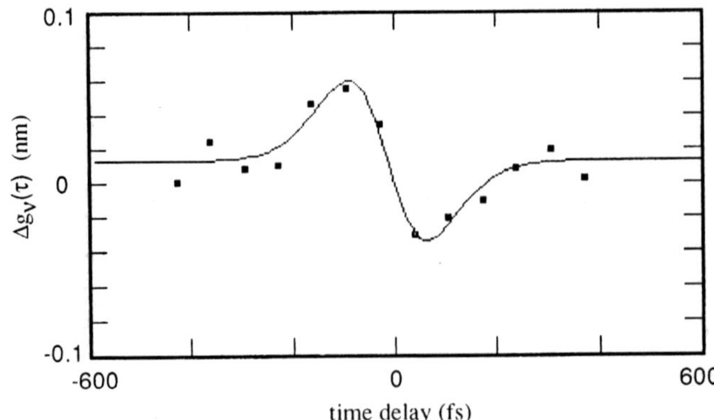

Figure 4. Change of the centre of gravity of the probe spectrum at 307 nm as a function of time delay between pump and probe in a 3 mm plate of fused silica (Suprasil). The full line is a fit assuming an $(\text{sech})^2$ pulse shape, the retrieved pulse duration is 126 femtoseconds.

One could extend this method further in the UV to vacuum-ultraviolet (VUV) domain by using, for example, a cell filled with gas (Xenon) as the non-linear sample. The lower non-linear index n_2 in gases, due to the lower density of the medium, can be compensated for, to some extent, by a long interaction length since the dispersion is small.

In conclusion, we have demonstrated a general method for measuring the derivative of the autocorrelation function of a femtosecond pulse which is valid from the near-infrared to the ultraviolet.

REFERENCES.

1. E. P. Ippen and C. V. Shank, Picosecond techniques and applications, in: *Ultrashort Light Pulses*, S. L. Shapiro ed. Springer - Verlag, Berlin (1977).

2. D. J. Kane and R. Trebino, Characterization of arbitrary femtosecond pulses using FROG, *IEEE. J. Quantum Electronics*. 29:571 (1993).

3. M. Franco, R. Lange, J.-F. Ripoche, B. Prade and A. Mysyrowicz, Characterization of ultra-short pulses by cross-phase modulation, *Opt. Commun.*, 140:331 (1997).

4. K. Chu, J. Heritage, R. Grant and W. White, Temporal interferometric measurement of femtosecond spectral phase, *Opt. Lett.*, 21:1842 (1996).

5. D. J. Kane, A. J. Taylor, R. Trebino and K. W. DeLong, Single-shot measurement of the intensity and phase of a femtosecond UV laser pulse, *Opt. Lett.*, 19:1061 (1994).

6. J. M. Schins, P. Breger, P. Agostini, R. C. Constantinescu, H. G. Muller, A. Bouhal, G. Grillon, A. Antonetti and A. Mysyrowicz, Cross-correlation measurements of femtosecond extreme-ultraviolet high-order harmonics, J. Opt. Soc. Am. B 13:197 (1996).

7. J.-F. Ripoche, B. Prade, M. Franco, G. Grillon, R. Lange and A. Mysyrowicz, Determination of the duration of UV femtosecond pulses, *Opt. Commun.*, 134:165 (1997).

SATURATED TABLE-TOP SOFT X-RAY LASERS BY DISCHARGE EXCITATION

J.J. Rocca, F.G. Tomasel, J.L.A. Chilla, M.C. Marconi[1], V.N. Shlyaptsev, C.H. Moreno[1], B.R. Benware, and J.J. Gonzalez.

Department of Electrical Engineering
Colorado State University
Fort Collins, CO 80523

[1] Permanent Address: Departamento de Física, FCEN, Universidad de Buenos Aires.

INTRODUCTION

We review the rapid progress in the development and study of compact and practical ultra-short wavelength lasers based on discharge excitation. The field has advanced from the first observation of large soft x-ray amplification in a capillary discharge created plasma[1], to the demonstration of an extremely compact saturated laser at 46.9 nm [2]. Other papers in these Proceedings and references therein address the recent progress in the development and application of laser pumped soft x-ray lasers[3-5].

In the capillary discharge soft X-ray laser scheme an elongated, needle-like plasma with remarkable axial uniformity is created by a fast current pulse that excites pre-ionized material contained in a capillary channel defined by insulating walls. In the fast capillary discharges we have successfully used for excitation of collisional soft x-ray lasers, the electromagnetic forces of the rapidly rising current pulse that compress the plasma creating a shock wave. The current distribution is influenced by the presence of a wall plasma created from wall material ablated by plasma radiation and heat conduction[2,6]. Following the initial phase of the discharge, the soft x-ray emitting region of the plasma, that initially fills the majority of the capillary volume, is rapidly compressed to form a plasma column 200-300 μm in diameter[2,7,8]. The optimum plasma temperature and density for lasing by collisional excitation are achieved several nanoseconds before the stagnation of the plasma column, when the first compression shock wave reaches the axis. Subsequently, the plasma density continues to increase as the plasma stagnates, and laser action ceases due to increased refraction and collisional thermalization. The capillary discharge structure has the advantages of providing a very good initial plasma symmetry, and a short compression time which leads to stable plasma columns. This has allowed us to generate hot and dense plasma columns that are axially uniform, with a very large length-to-diameter ratio (1/d=500-1000) and length up to 20 cm [9].

In the next Section we summarize the results that lead to the development of an extremely compact 46.9 nm Ne-like Ar laser capable of emitting pulses of up to 25 μJ in energy. Section III discusses measurements of the near-field and far-field spatial distributions of this laser as a function of discharge parameters, and Section IV summarizes results of the study of its spatial coherence. Section V discusses the first demonstration of lasing in a capillary discharge plasma by collisional excitation in ions ablated from a solid target, at 60.8 nm in Ne-like S

DEMONSTRATION OF A SATURATED DISCHARGE-PUMPED NE-LIKE AR TABLE-TOP SOFT X-RAY LASER

The first observation of large soft x-ray amplification in a discharge-pumped amplifier was realized in the J=0-1

Applications of High Field and Short Wavelength Sources
Edited by DiMauro *et al.*, Plenum Press, New York, 1998

line Ne-like Ar at 46.9 nm [1,7]. In that initial experiment a discharge current pulse, having a half-period of 60 ns and a peak current of approximately 40 kA, was used to excite Ar plasma columns in 4 mm in diameter channels drilled in polyacetal. The capillary channel was placed in the axis of a 3 nF liquid-dielectric circular-parallel plate capacitor. The capacitor was charged by a Marx generator and rapidly discharged through the capillary channel by closing a spark gap switch pressurized with SF_6. The gain at 46.9 nm was determined by measuring the integrated line intensity as a function of plasma column length. A gain-length product of $g*l = 7.2$ was obtained using plasma columns 12 cm in length. Subsequent experiments conducted utilizing longer plasma columns under optimized discharge conditions, and double-pass amplification using an Ir mirror yielded an effective gain-length product of 27, the largest reported to date for a table-top soft x-ray amplifier[2]. Spectra we obtained in the wavelength region corresponding to the J=2- 1 line of Ne-like Ar at 69.8 nm also showed a supra-linear increase with capillary length, indicative of gain[7]. However, the gain in this line was much smaller than for the J=0-1 line, and a significant increase in the amplification of this line is still required to make it of practical interest. Small amplification in this line in a capillary discharge was more recently also observed by Hildebrand et al.[10].

A major step in the development of compact ultra-short wavelength lasers consists in saturating the gain in the amplifier and in demonstrating substantial output energies. At this condition, which occurs when the laser intensity reaches the saturation intensity, an important fraction of the energy stored in the laser's upper level can be extracted. Depending on the specific characteristics of the amplifier, reaching the saturation intensity requires to overcome barriers that are imposed by the small gain volumes and short plasma lengths, by the short duration of the gain, or by plasma inhomogeneities and limiting refraction effects. To reach gain saturation in our discharge-pumped Ne like Ar laser we conducted experiments in longer plasma columns, and used an iridium mirror to implement double-pass amplification measurements. This resulted in the first clear observation of gain saturation in a table-top soft X-ray amplifier[2]. The experiments were conducted exciting 4 mm diameter capillaries filled with 700 mTorr of Ar with currents pulses of approximately 39 kA peak current having a half cycle duration of about 70 ns. A detection system consisting of a 2.2 m grazing incidence spectrograph with a microchannel plate (MCP) intensified CCD array detector was used to measure the relative variation of the laser energy as a function of plasma column length. Absolute measurements of the laser output pulse energy were performed using a fast vacuum photodiode having a calibrated Al photocathode[2].

The results of single pass amplification measurements for capillary plasma columns up to 15.8 cm in length are shown as open circles in Fig. 1. The energy of the laser pulse is observed to increase exponentially for lengths up to about 12 cm, where it begins to saturate. A fit of the data corresponding to plasma columns up to 11.5 cm with the Linford formula[11] yields a gain coefficient of 1.16 cm^{-1}. Saturation of the laser intensity is observed at gain-length products of about 14. Laser pulse energies of 6 µJ were measured to exit the 15.8 cm long plasma columns. Double-pass amplification experiments allowed us to substantially increase the laser pulse energy and to study the saturation behavior for significantly longer effective plasma column lengths. The double pass amplification measurements were performed using a flat iridium mirror for two different plasma column lengths: 9 and 14 cm. In the 9 cm long capillaries the laser intensity enhancement due to the mirror was measured to be 63x. In contrast, the enhancement observed in the 14 cm long capillaries was in the average only 8x. This behavior is indicative of saturation of the amplification in the second pass. The increase in the laser energy measured in the double pass experiments in the 14 cm long capillaries corresponds to laser

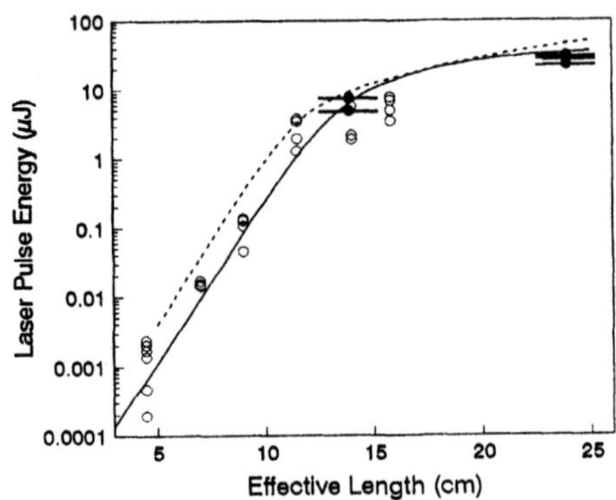

Figure 1. Measured and computed 46.9 nm laser output energy as a function of capillary discharge plasma column length. Single and double-pass measurements are indicated by open and full circles, respectively. The solid line is the result of simple radiation transport calculations assuming parabolic gain and density profiles The dashed line was computed with a hydrodynamic/atomic physics code

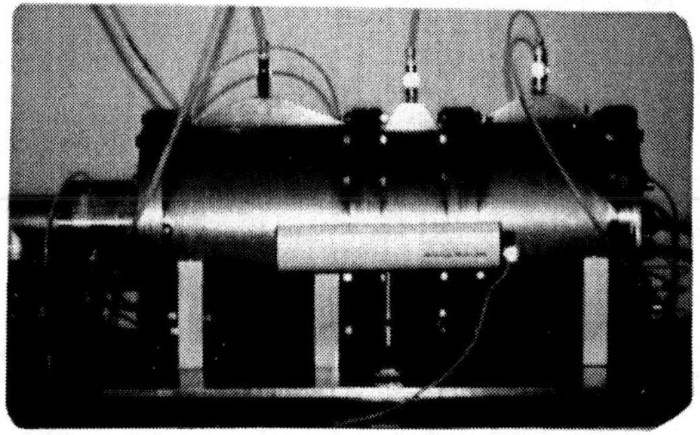

Figure 2. Photograph showing the size of the table-top capillary discharge pumped 46.9nm laser in comparison with a 5 mW He-Ne laser.

pulse energies of up to 30 μJ and to beam intensities larger than the computed saturation intensities of 56-78 MW/cm^2. The saturation intensity was calculated considering an ion temperature of 100 eV, and an effective-to-radiative lifetime ratio for the laser upper level between 20 and 30 for plasma densities of 5-8x 10^{18} cm^{-3}, as computed by our magnetohydrodynamic calculations for these discharge conditions[2].
The measured saturation behavior is in good agreement with the result of two independent radiation transport models for the capillary plasma columns. Computations of the variation of the laser energy as a function of plasma column length were conducted by solving the radiation transport equation for an inhomogeneously broadened transition, taking into account line narrowing, refraction effects, and gain saturation It should be noticed that consideration of refraction losses is essential to adequately describe the measured energy dependence on plasma column length. The solid line in Fig. 1 is the result of calculations performed assuming parabolic gain and density profiles, while the dashed line correspond to computations conducted for the time dependent electron density and gain profiles obtained from magnetohydrodynamic and atomic physics calculations conducted with the code RADEX[6,12]. We have also studied the influence of an externally applied axial magnetic field on the capillary discharge soft x-ray laser performance. An optimized magnetic field of about 0.15 T was observed to increase the uniformity and intensity of the soft x-ray laser beam by decreasing the plasma density gradients at the time of lasing[13].
Based on these results, we have recently succeeded in developing a very compact saturated 46.9 nm laser of size comparable with that of many widely utilized visible and ultraviolet gas lasers. This laser, illustrated in Fig. 2, generates sub-nanosecond pulses with energies up to 25 μJ and laser power up to 40 kW [14].

TWO-DIMENSIONAL NEAR-FIELD AND FAR-FIELD IMAGES.

Significant insight in the physics of a soft x-ray amplifier can be gained obtaining near-field images of its output. Recently, two-dimensional near-field imaging studies of the laser pumped soft x-ray amplifiers have been reported[15,16]. In this Section we discuss results of the first near-field imaging study of a capillary discharge pumped soft x-ray laser. To further characterize and understand this laser we also recorded the corresponding far-field patterns and compared the results with model calculations.
The experiments were conducted using the 46.9 nm Ne-like Ar capillary discharge laser shown in Fig. 2. The imaging set-up used to record the near-field patterns is schematically illustrated in Fig. 3. It uses a 150 cm radius of curvature mirror and a flat mirror to image the output of the laser into a two-dimensional soft x-ray sensitive detector. The mirrors are iridium coated. The setup imaged the output of the laser into the detector with a magnification of 8x. The detector consisted of a micro-channel plate followed by a phosphor screen, an image intensifier and a CCD array of 1024x1024 pixels.
A thin Al film was evaporated on the phosphor screen to avoid the detection of visible plasma radiation. The micro-channel plate was gated for about 5 ns to differentiate the laser radiation from the long lasting (>100 ns) soft x-ray radiation emitted by the discharge. The amplitude of the gate voltage pulse was maintained low (150 V) to avoid saturation of the detector and the linearity of the system was experimentally verified. The spatial resolution of the imaging system was determined to be approximately 16 μm by imaging an array of 25 μm diameter holes placed at the exit plane of the capillary channel. The same measurement allowed for an experimental verification of the magnification of the imaging system.
We recorded two dimensional near-field images of the output of 16.4 cm long Ar plasma columns generated in 4 mm diameter polyacetal capillaries as a function of Ar pressure. The laser was excited by 37 kA current pulses having a first half-cycle duration of about 72 ns. Figure 4a) shows the result of a series of measurements for Ar pressures between 500 and 750 mTorr. At the higher pressures (>650 mTorr) the near-field laser beam pattern has a single peak with maximum intensity at the center and monotonically decreasing intensity towards the

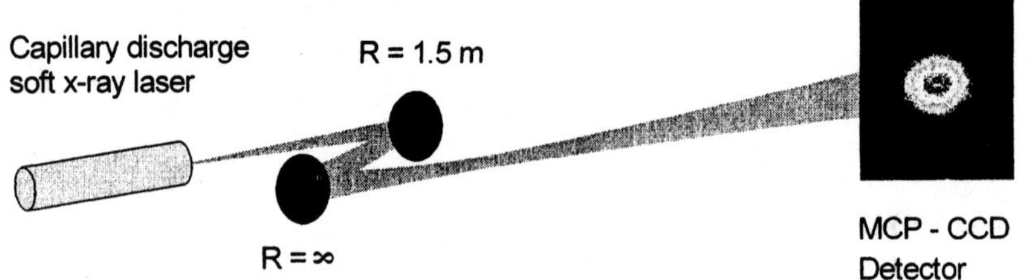

Figure 3. Schematic representation of the setup used to obtain two-dimensional near-field images of the output of the capillary discharge Ne-like Ar laser.

periphery. As the pressure is decreased, a ring structure is observed to develop towards the periphery, and simultaneously the beam size at the exit of the amplifier gets larger. The FWHM beam diameter is measured to increase from about 150 μm at 750 mTorr to about 300 μm at 500 mTorr. This, in combination with the far-field data discussed below, indicates the existence of larger plasma density gradients in the lower pressure discharges, that result in increased refraction of the amplified soft x-ray radiation. We have observed that the pressure at which the beam profile makes a transition from a single peak profile to a ring structure can vary from one capillary to another by as much as 100 mTorr. For a given discharge pressure the shape of the beam pattern was observed to be relatively independent of the current within the range investigated (32.5 to 41 kA). Figure 4b shows the far-field beam patterns corresponding to the same discharge conditions of the near-field patterns of Fig. 4a. As in the case of the near-field patterns, a transition is observed from a beam profile with a single peak to a ring profile as the pressure is decreased. In the far-field patterns a ring structure is again observed for the discharges at pressures <550 mTorr. This is accompanied by an increase in the beam divergence, which increases from about 2 mrad for discharges at 750 mTorr to 5 mrad for discharges at 500 mTorr. These far-field measurements confirm that refraction increases as the pressure is decreased.

These observations are in good agreement with hydrodynamic-atomic model computations for this Ne-like Ar capillary discharge-pumped amplifier. The calculations show that while the electron temperature and the gain coefficient increase as the pressure is decreased, refraction losses also increase drastically due to larger density gradients. The larger gradients associated with the lower pressure discharges are the result of both a higher electron density and a reduced plasma column diameter. At the lower pressures the maximum amplification occurs at higher electron densities, as a result of a higher electron temperature and reduced opacity. The increased electron temperature decreases collisional mixing between the laser levels, shifting the optimum electron density for amplification to higher values. The reduced opacity is the result of the smaller plasma column diameter, and of a higher plasma velocity that result in increased Doppler de-tuning. At pressures ≤ 550 mTorr the increase in refraction losses with decreasing pressure is a more important effect than the increase in the gain coefficient due to the increase in electron temperature, and the overall result is the degradation of the laser output at the lower pressures.

 Also at the lower pressures the duration of the gain is smaller and gradually approaches the propagation time of the photons along the amplifier, ~ 0.5 ns, further decreasing the amplification. The maximum output laser energy is therefore found at higher pressures (near 600 mTorr), where refraction is smaller and the gain duration reaches ~ 2.5 ns. At pressures >700 mTorr opposite effects than those at low pressure occur, and both the gain and refraction decrease. At these higher pressures the electron temperature and the optimum electron density decrease, while opacity increases, resulting in a smaller on axis gain. Smaller density gradients result in a reduced beam divergence, that starts to approach the value determined by the geometrical dimensions of the gain medium. However, this is accompanied by a large decrease of the laser output energy, that is the result of the decreased gain. It should be noticed that in general the near-field patterns of the discharge pumped Ne-like Ar amplifier are more homogeneous and have significantly less structure than those corresponding to the laser-pumped amplifiers[15,16]. In addition, it can also be concluded that, as in the case of laser-pumped soft x-ray amplifiers, refraction effects often overshadow the influence of the size of the gain region in determining the near-field pattern. A most dramatic illustration of this are our near-field measurements with decreasing pressure, in which the dimension of the near-field pattern is observed to increase remarkably in spite of the fact that the size of the gain region is decreasing.

STUDY OF THE SPATIAL COHERENCE OF THE NE-LIKE AR LASER

Good transverse spatial coherence will be essential in realizing the full potential of these sources in some important applications. Current soft x-ray lasers face the difficulty of achieving a good coherence without the aid of an optical cavity, mainly due to their short gain duration. Measurements of the spatial coherence of laser pumped soft x-ray lasers have been reported[17-19], but most of them correspond to a single amplifier length of the

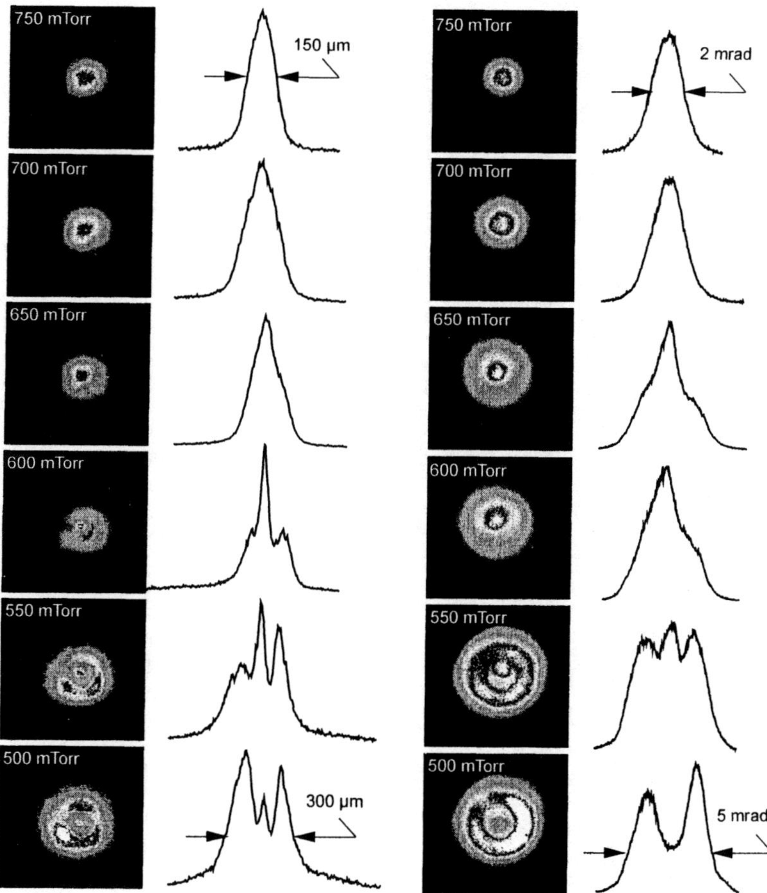

Figure 4. Near-field (left) and far-field (right) patterns of the Ne-like Ar laser beam output as a function of pressure. Diametral cuts with normalized intensities are shown at the left of each image. The measurements correspond to a 16.4 cm long, 4 mm diameter capillary excited by a 37 kA current pulse.

particular laser studied. Several theoretical studies have been conducted to understand the spatial coherence of soft x-ray amplifiers[20-24]. In general, they predict an improvement of the coherence with amplifier length. This is the result of a decreasing number of modes guided along the amplifier column due to gain guiding and refractive anti-guiding[20]. Such build-up of the coherence is essential in achieving very good spatial coherence in cavity-less soft x-ray amplifiers. However, this monotonic increase of the spatial coherence with amplifier lengths had not been previously experimentally observed in soft x-ray lasers.

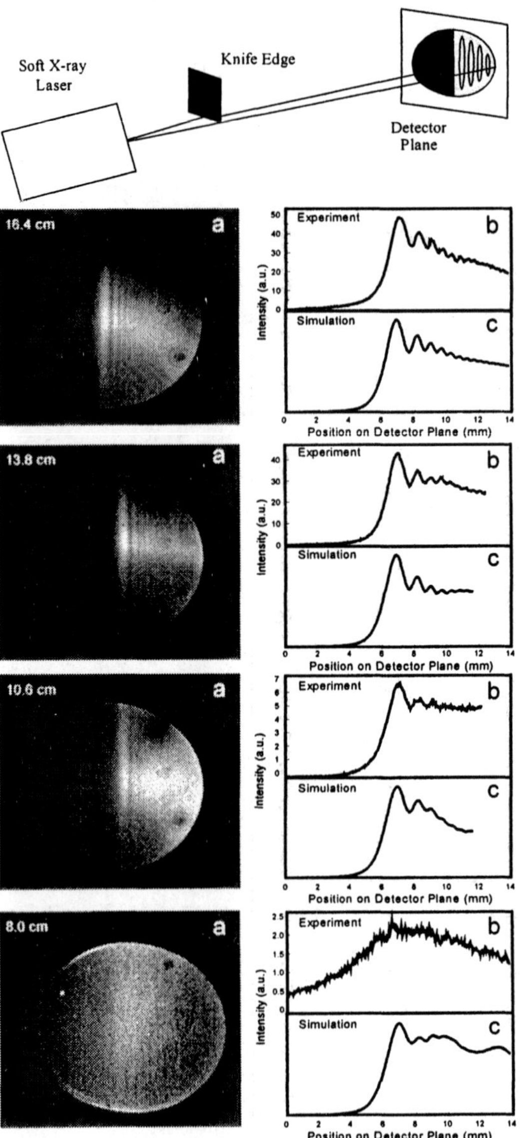

Figure 5. Top. Schematic representation of the set up used to measure the spatial coherence of the Ne-like Ar laser. The knife-edge and the detector were placed at 56 cm and 589 cm from the exit of the amplifier respectively. Bottom a) Measured diffraction patterns corresponding to capillaries with lengths between 8 and 16.4 cm. b) Cross sections of the diffraction patterns of a) obtained by vertically integrating 50 pixels of the CCD in the region of maximum fringe visibility c) Corresponding diffraction patterns computed using the result of the wave-optics model.

We have measured a monotonic increase of the spatial coherence as a function of plasma column length in a capillary discharge soft x-ray amplifier. The measurements were conducted in the 46.9 nm line of Ne-like Ar for capillary plasma column lengths up to 16.4 cm. The spatial coherence was measured recording the diffraction produced when the soft x-ray laser beam intersects a knife-edge (Fig. 5, top)[25]. This technique has been recently utilized to observe an improvement in the coherence of a laser pumped Ne-like Zn laser when a reflecting multilayer mirror was used[19]. It has the advantage of determining in a single shot the degree of coherence for any two points on the illuminated region of the detector plane that correspond to a line perpendicular to the knife edge. The diffraction fringes were recorded with a gated detector consisting of the combination of a gated MCP, phosphor screen, image intensifier, and a CCD array. To conduct the measurements of the spatial coherence as a function of plasma column length, the knife-edge was placed in a radial position respect to the beam. This gives a measurement of the coherence in the tangential direction. The measurements were conducted in 4 mm diameter polyacetal capillaries filled with Ar gas at pressures of 600 mTorr. The plasma was excited by current pulses of approximately 37 kA peak current, having first half cycle duration of about 70 ns.

Figures 5a and 5b show the measured diffraction patterns corresponding to capillaries with length between 8 and 16.4 cm. The improvement of the coherence with amplifier length is evident in the increased fringe visibility observed in the diffraction patterns corresponding to the longer plasma columns. The observed diffraction patterns compare well with those obtained numerically using the results of a wave optics model for the generation and propagation of the radiation in the capillary plasma column, shown in Fig. 5c). The model used to calculate the coherence of the beam produced by the capillary discharge soft x-ray laser is similar to that developed by Feit and Fleg[24], but takes into account the

temporal variation of the electron density and of the gain coefficient, and was extended to two dimensions.
The data of Fig. 5, and that corresponding to other similar series of measurements, were analyzed to quantify the coherence function and its dependence on amplifier length. The result is shown in Fig 6. The coherence is observed to increase monotonically with capillary length. Good agreement is observed with the result of wave-optics calculations, represented by a line in the same figure. This is, to our knowledge, the first observation of a monotonic increase of the transverse spatial coherence with amplifier length in a soft x-ray laser. For the longest capillaries studied, 16.4 cm, the coherence length in the tangential direction was determined to be about 4.5 mm in the detector plane situated at 5.89 m from the exit of the amplifier. This corresponds to an effective coherence angle of 0.8 mrad and to an effective source size of 26 µm FWHM. The beam is approximately 6 times diffraction limited in the tangential direction.
To study the uniformity of the coherence properties across the beam profile, we performed measurements placing the knife-edge in different positions relative to the laser beam. Measurements conducted placing the knife-edge along a diameter of the beam (on-axis) showed that the coherence is the same in the two orthogonal directions. Measurement off-axis showed that the coherence length in the radial direction is 30-50 percent shorter than in the tangential direction. The wave-model computations suggest that a likely cause of the observed anisotropy of the spatial coherence is the change of the electron density during the laser pulse. As curves of constant phase are circles concentric with the beam, the dephasing due to a change in the curvature of the wavefront is more significant off-axis and in the radial direction. This effect is clearly shown by our simulations that use parabolic profiles and a time varying electron density. It is nevertheless possible that the observed anisotropy could be caused by a non parabolic density profile in which the curvature is radialy dependent.

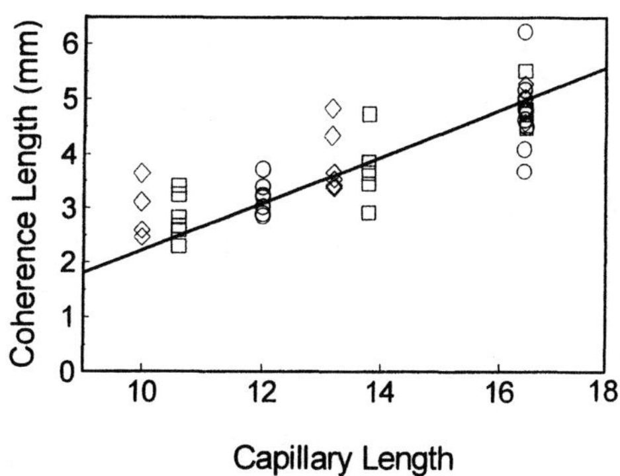

Figure 6. Variation of the coherence length in the tangential direction as a function of plasma column length. Different symbols correspond to series of measurements conducted in several different identical capillaries. The line is the result of wave-model calculations.

DEMONSTRATION OF LASING CREATED BY ABLATION OF A SOLID MATERIAL: AMPLIFICATION IN Ne-LIKE S.

The extension of the Ne-like Ar results to other wavelengths, in particular the search for gain in discharge-created plasmas at shorter wavelengths, will require in most cases the generation of plasma columns of elements that are solid at room temperature. In this Section we discuss the first demonstration of large amplification in a discharge-pumped ultrashort wavelength amplifier in an element that is solid at room temperature. A gain-length product of 7.5 was achieved in the J=0-1 line of Ne-like S. To obtain this result, we developed a method to generate vapor from solid materials by discharge ablation[8].
For this experiment our original discharge setup described in references 1 and 26 was modified as schematically illustrated in Fig 7 to allow for the injection of the sulfur vapor into the capillary channel through a hole in the ground electrode. The sulfur vapor was produced ablating the wall of a 5 mm diameter, 2 cm long secondary capillary channel, drilled in a sulfur rod, with a slow current pulse delivering 200 J in about 50 µs. The vapor generated by this capillary discharge was injected into the main capillary channel and was subsequently excited by a fast current pulse of 35-37 kA peak amplitude having a first half cycle duration of ~ 72 ns, to generate a narrow plasma column with the necessary conditions for amplification.

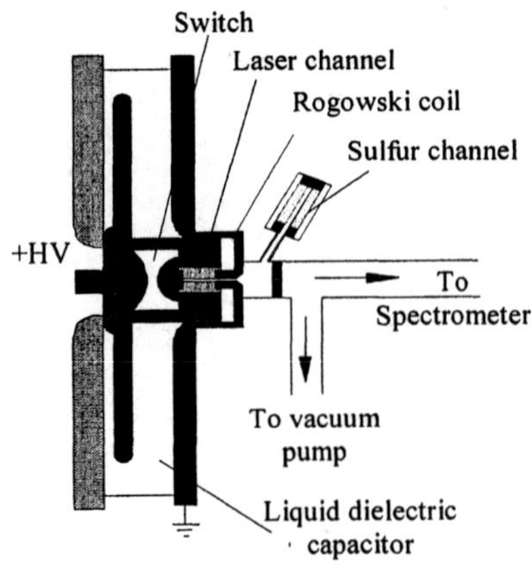

Figure 7. Schematic diagram of the discharge set up utilized to obtain lasing in Ne-like S.

Under optimized conditions, strong lasing can be expected to occur in the $3s^1P^0_1$ - $3p^1S_0$ line, as is the case in the discharge pumped Ne-like Ar laser. This J=0-1 line had been previously identified at 60.84 nm in spectra obtained from laser-created sulfur plasmas[27], and has recently been observed to lase in laser-created plasmas generated using the Asterix iodine laser facility at the Max Planck Institute for Quantum Optics[28]. Figure 8a shows a spectrum obtained under optimized laser conditions in the spectral region spanning from 58.8 to 61.2 nm using a tin filter. The spectrum corresponds to a 37 kA discharge through a 4 mm diameter, 16.8 cm long channel filled with 460 mTorr of sulfur vapor. The spectrum is completely dominated by the J= 0-1 line transition of Ne-like sulfur, which appears at 60.84 ± 0.015 nm. Another line of Ne-like sulfur (the $3p\ ^1P_1$ - $3d\ ^1P^0_1$, transition at 60.12 nm [27]), that in the absence of amplification should have similar intensity also falls in the spectral range of Fig. 8a. The fact that in our plasma column the intensity ratio of these two lines is observed to be at least 100 is clear evidence of amplification in the J = 0-1 line. The measured variation of the integrated line intensity of the J= 0-1 line of Ne-like sulfur as a function of plasma column length is shown in Fig. 8b. An increase of about 1.6 in the plasma column length is observed to increase the integrated intensity of the line by a factor 13. This corresponds to a gain coefficient of 0.45 cm^{-1}, and a gain-length product of 7.55 for 16.8 cm long capillaries. Strong amplification was observed for a broad range of pressures, from 300 to 700 mTorr, and for currents between 33 and 38 kA.

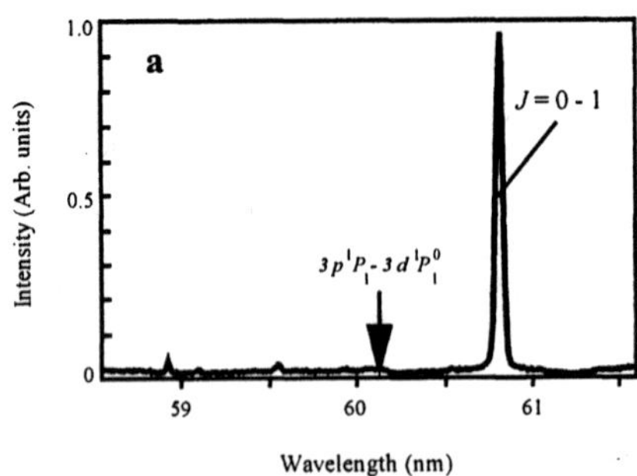

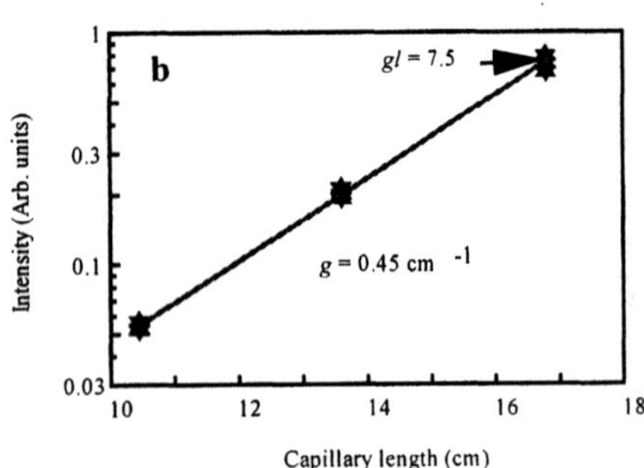

Figure 8 a. Spectrum of the axial emission of the sulfur plasma column in the region between 58.5 and 61.5 nm. The spectra corresponds to a 37 kA discharge trough a 4 mm diameter, 16.8 cm long capillary filled with 460 mTorr of ablated sulfur vapor. The $3s\ ^1P^0_1$ - $3p\ ^1S_0$ line of the same ion, that in absence of amplification should have similar intensity, is not observed. b) Variation of the integrated intensity of the J=0-1 line of Ne-like sulfur as a function of plasma column length. The line is a fit with the Linford formula, which results in a gain coefficient of 0.45 ± 0.01 cm^{-1}, and a gain-length product of 7.55± 0.15 for the 16.8 cm long column.

Lasing occurs shortly before stagnation of the plasma column, as in the case of Ne-like Ar [2]. The region of gain is a narrow plasma column of about 300 μm in diameter surrounded by lower density plasma containing sulfur ions of a lower degree of ionization, which in turn is surrounded by material ablated from the capillary walls. The deceleration of the plasma column near the axis prior to stagnation results in a velocity gradient that, due to motional Doppler broadening, considerably facilitates the radial escape of the lower laser level radiation. At the time of lasing, which for our discharge condition is observed to occur near the time of maximum current, the electron density and temperature in the gain region are computed to be about $(2-3)\ 10^{18}$ cm^{-3} and 60-80 eV, respectively. As described below, this temperature corresponds to a plasma that is overheated with respect to the temperature range Te=20-40 eV for maximum Ne-like sulfur abundance in a steady-state plasma.

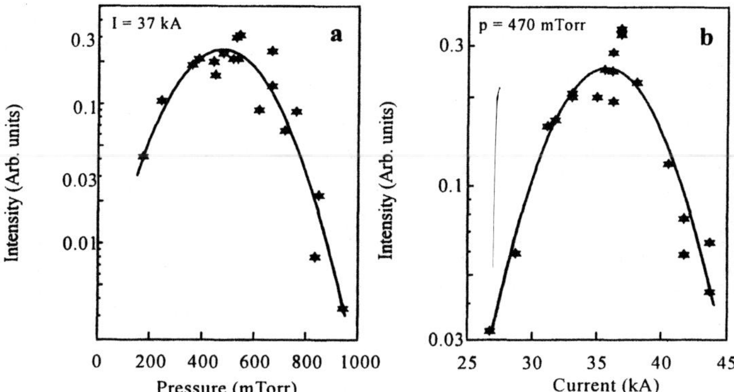

Figure 9. a) Integrated intensity of the 60.84 nm lasing line as a function of filling sulfur vapor pressure. The data corresponding to 37 kA discharges through 4-mm diameter. 16.8 cm long capillary channels. b) Variation of the integrated intensity of the J= 0-1 line of Ne-like sulfur as a function of the discharge current on the same capillary. The pressure was maintained constant at about 470 mTorr in all the shots.

The change in the integrated laser line intensity was also measured as a function of the amplitude of the fast current pulse and the sulfur filling pressure. Figure 9a shows the dependence of the intensity of the lasing line as a function of filling pressure for discharge shots with a current of 37±1 kA. The J=0-1 line was observed to lase strongly for a broad range of pressures, from 300 to 700 mTorr. Figure 9b shows the variation of the integrated laser line intensity as a function of the peak current, for a filling pressure of 470±10 mTorr. It is observed that large amplification occurs in discharge shots driven by peak currents between 33 and 38 kA.

According to the approximate scaling laws of atomic kinetics valid for lasing in Ne-like ions in steady-state conditions the gain scales for ions of charge Z approximately as $G \sim Z^{4.5}$.[29] Consequently for sulfur, a gain ~ 3 times smaller than that for the J=0-1 line of argon could be estimated. As mentioned above in the case of the J=0-1 line of argon, the effective gain was measured to reach $1.16 \ cm^{-1}$, a value that is smaller than the maximum computed gain of $1.5-1.8 \ cm^{-1}$ due to refraction.[2] According to the above scaling law the effective gain in Ne-like sulfur would be expected to be rather small, less than $0.3 \ cm^{-1}$. Our computations using the code RADEX[6,12] indicate important contributions to the generation of the population inversion by plasma overheating and transient population effects. Due to the exponential dependence of the excitation rates on the electron temperature, a larger population inversion and, consequently a larger gain, arises from overheating. Such overheating of the plasma respect to steady-state ionization conditions can be more easily achieved in low-Z elements like sulfur, due to a large decrease of the ionization time with ion charge. In addition, in this sulfur laser transient population effects are found to play a more important role than in the argon laser. A transient increase in the population inversion can arise when the characteristic time of the rise of the excitation is of the order of the effective lifetime of the laser upper level. While in these relatively long-lived discharge plasmas transient effects are not nearly as dramatic as in plasmas produced by subpicosecond lasers,[3] their contributions to the gain can be noticeable. The computations indicate that in the case of sulfur transient population effects can increase the gain by 20-40%. As a result of plasma overheating and transient population effects the maximum gain in the J=0-1 line of Ne-like sulfur is computed to approach $1 \ cm^{-1}$, a value which taking into account refraction losses is in satisfactory agreement with the measured effective gain of $0.45 \ cm^{-1}$. This result is to our knowledge the first observation of gain enhancement due to transient effects in discharge-created plasma.

ACKNOWLEDGEMENTS: This work was supported by the National Science Foundation. We also acknowledge the support of the Colorado Advanced Technology Institute for a collaboration with Hyperfine Inc.

REFERENCES

1. J.J. Rocca, V.N. Shlyaptsev, F.G. Tomasel, O.D. Cortazar, D. Hartshorn and J.L.A. Chilla, Demostration of a discharge pumped table top soft x-ray laser, *Phys. Rev. Lett.* **73**:2192 (1994).
2. J.J. Rocca, D.P. Clark, J.L.A. Chilla, and V.N. Shlyaptsev, Energy extraction and achievement of the saturation limit in a discharge pumped table-top soft X-ray amplifier, *Phys. Rev. Lett*, 77:1476, (1996).
3. P.V. Nickles, M.P. Kalachnikov, M. Schnürer, W. Sandner, V.N. Shlyaptsev, C. Danson, D. Neely, E. Wolfrum, M. Key, A. Behjat, A. Demir, G. Tallents, G.J. Pert, J. Warwick and C. Lewis, Low energy pumped X-ray lasers with saturated transient gain. In these *Proceedings*.
4. M.H. Key, D.H. Kalantar, J. Nilson, B.A. Remington, S.V. Weber, E. Wolfrum, D. Neely, S.J. Rose, J. Zhang,

N.S. Kim, J.S. Wark, C.L. Lewis, A.G. Mac Phee, J. Warwick, A. Demir, J. Lin, R. Smith and G.J. Tallents. Developments in XUV laser radiography of laser driven targets. In these *Proceedings*.

5. H. Daido, Y. Kato, T. Imani, S. Sezaki, S. Hirose, G.Y. Yoon, T. Jitsuno, Y. Takagi, K. Mima, S. Wang, Z. Lin, Y. Gu, G. Huang, H. Tang, D. Ximing, G. Zhang, K. Murai and H. Takenaka. Generation of intense X-ray laser radiation at 8 nm in Ni-like Nd ions. In these *Proceedings*.

6. V.N. Shlyaptsev, J.J. Rocca, P.V. Nickles, M.P. Kalashnikov, W. Sadner and A.L. Osterheld. Modeling of table-top capillary and transient inversion x-ray lasers. Soft x-ray lasers and applications II. JJ. Rocca and L.B. Da Silva, eds, SPIE vol 3156 (1997).

7. J.J. Rocca, F.G. Tomasel, M.C. Marconi, V.N. Shlyaptsev, J.L.A. Chilla, B.T. Szapiro, and G. Giudice, Discharge-pumped soft x-ray laser in Ne-like Ar, *Physics of Plasmas* , 2:2547 (1995).

8. F.G. Tomasel, J.J. Rocca and V. Shlyaptsev. Dynamics of the plasma column of a discharge pumped soft x-ray laser in cylindrical plasmas, *IEEE Trans. on Plasma Science*, 49:24 (1996).

9. J.J. Rocca, M.C. Marconi, J.L.A. Chilla, F. Tomasel and V. Shlyaptsev. Discharge driven 46.9 nm amplifier with gain-length approaching saturation. *IEEE Select. Topics in Quantum Electronics*, 1:945 (1995).

10. A. Hildebrand, A. Ruhrmann, S. Maurmann, and H.J. Kunze, Amplified spontaneous emission on the J=2 to 1, 3p-3s transition of neonlike argon in a capillary discharge, *Phys. Lett. A.* 221:335 (1996).

11. G. J. Linford, E. R. Peressini, W. R. Sooy and M. L. Spaeth, Very long lasers, Appl. Opt, 13:379 (1974).

12. V. Shlyaptsev, A.V. Gerusov, A.V. Vinogradov, J.J. Rocca, O.D. Cortazar, F.G.Tomasel, and B.T. Szapiro. Modeling of fast capillary discharges for collisionally excited soft x-ray lasers: comparison with experiments, S. Suckewer, ed. , SPIE Vol.2012, Ultrashort Wavelength Lasers II (1993).

13. F.G. Tomasel, V.N. Shlyaptsev and J.J. Rocca, Enhanced beam characteristics of a discharge pumped soft x-ray amplifier by an axial magnetic field. *Phys. Rev. A.* 54:2474 (1996).

14. B.R. Benware, C.H Moreno, D. Burd and J.J. Rocca. Operation and output pulse characteristics of an extremely compact capillary discharge table-top soft x-ray laser. *Opt. Lett.* 22:796 (1997).

15. J.C. Moreno, J. Nilsen, Y. Li, P. Lu and E.E. Fill, Two-dimensional near-field images of the neonlike germanium soft-x-ray laser, *Opt. Lett.* 21:866 (1996).

16. J. Nielsen, J.C. Moreno, L.B. Da Silva and T.W. Barbee Jr, Two-dimensional spatial imaging of the multiple-pulse-driven 196-Å neonlike germanium x-ray laser, *Phys Rev A.* 55:827 (1997).

17. J.E. Trebes, K.A. Nugent, S. Mrowka, R.A. London, T.W. Barbee, Jr., M.R. Carter, J.A. Koch, B.J. MacGowan, D.L. Matthews, L.B. Da Silva, G.F. Stone and M.D. Feit, Measurement of the saptial coherence of a soft X-ray laser, *Phys. Rev. Lett.* 68:588 (1992).

18. R.E. Burge, G.E. Slark, X. Cheng, M.T. Browne, D. Neely, C.L.S. Lewis and A. MacPhee, Studies of the spatial coherence of an x-ray laser, Soft X-ray Lasers and Applications, SPIE vol 2520, J.J. Rocca and P.L. Hagelstein, eds., SPIE, Bellingham, WA, 257, (1995).

19. F. Albert, B. Rus, Ph. Zeitoun, A. Carillon, P. Jaeglé, G. Jamelot, A. Klisnick, D. Ros and S. Sebban, New approach for measurement of the X-ray laser transverse coherence, X-Ray Lasers 1996, S. Svanverg and C-G Wahlström, eds., IoP, Bristol, 427, (1996).

20. R.A. London, M. Strauss and M.D. Rosen, Modal analysis of x-ray laser coherence, *Phys. Rev. Lett.* 65:563 (1990).

21. R.P. Ratowsky and R.A. London, Propagation of mutual coherence in refractive x-ray laser using a WKB method, *Phys. Rev. A.* 51:2361 (1995).

22. G. Hazak and A. Bar-Shalom, Mode-selecting effects and coherence in hot-plasma x-ray, *Phys. Rev. A.* 40:7055 (1989).

23. M.D. Feit and J.A. Fleck, Jr., Spatial coherence of laboratory soft-x-ray lasers, *Opt. Lett.* 16:76, (1991).

24. M.D. Feit and J.A. Fleck, Jr., Wave-optics description of laboratory soft-x-ray lasers, *J. Opt. Soc. Am. B.* 7:2048 (1990).

25. M.C. Marconi, J.L.A. Chilla, C.H. Moreno, B.R. Benware and J.J. Rocca, Measurement of the spatial coherence build up in a discharge-pumped table-top soft x-ray laser. *Phys. Rev. Lett.* 79:2799 (1997).

26. J.J. Rocca, O.D. Cortazar, B. Szapiro, K. Floyd, F.G. Tomasel, Fast discharge excitation of hot capillary plasmas for soft x-ray amplifiers, *Phys. Rev. E*, 47:1299 (1993).

27. R.R. Gayazov, A.E. Kramida, L.I. Podobedova, E.N. Ragozin and V.A. Chirkov, X-Ray Plasma Spectroscopy and the Properties of Multiply-Charged Ions. Nova Science, NY, (1988)

27. Y Li, P. Lu, G. Pretzler, and E.E Fill, Lasing in neonlike sulphur and silicon, *Opt Commun.* 133:196 (1997).

29. A.V. Vinogradov and V.N Shlyaptsev, Amplification of UV radiation in a laser plasma, *Sov. J of Quantum Electronics.* 13:1511 (1983).

PHASE-LOCKING OF HIGH-ORDER HARMONICS TO THE FUNDAMENTAL FIELD

M. B. Gaarde[1,2], C. Altucci[1], M. Bellini[3], T. W. Hänsch[3,4], A. L'Huillier[1], C. Lyngå[1], C.-G. Wahlström[1], and R. Zerne[1]

[1]Physics Department, Lund Inst. of Technology, S-221 00 Lund, Sweden
[2]Niels Bohr Institute, Universitetsparken 5, 2100 Copenhagen, Denmark
[3]L.E.N.S., Largo E. Fermi, 2, I-50125 Florence, Italy
[4]M.P.Q., Hans-Kopfermann-Str. 1, D-85748 Garching, Germany

INTRODUCTION

High harmonic generation in rare gases is by now a well understood and well characterized phenomenon (L'Huillier et al., 1995). The applicational aspect of the generated radiation attracts more and more interest, and the harmonic radiation is currently being both optimized and characterized. Recently the harmonic radiation has been extended into the water window (Chang et al., 1997), and both the temporal (Schins et al. 1996, Glover et al 1996), and spatial (Tisch et al., 1994, Salières et al., 1994) characteristics have been measured. For applicational purposes, the coherence properties of the harmonic radiation are of course very important. The spatial coherence has recently been investigated both experimentally (Ditmire et al., 1996) and theoretically (Salières et al., 1997), and found to be dependent on the focusing conditions as well as the intensity.

The property we want to examine in the present work is whether we can create two harmonic sources which are phase locked. We present an experiment in which we generate two harmonic sources by splitting a laser beam in two (Zerne et al. 1997). Then we study if the harmonic beams are coherent enough to generate an interference fringe pattern in the farfield.

We also study the influence of the dipole phase on this property. Theoretical studies of high-order harmonic generation by a single atom exposed to an intense laser field, has shown that the dipole phase varies rapidly as function of the laser intensity (Lewenstein et al., 1995). Since the harmonics are generated with a pulse, and the dipole phase is strongly dependent on the instantaneous intensity in the pulse, one could expect that even the slightest difference in the intensities driving the two sources would destroy the mutual coherence.

The basic configuration, which is reminiscent of a Young's double slit experiment, and some notation is presented in Figure 1. Two harmonic sources, with driving field

Applications of High Field and Short Wavelength Sources
Edited by DiMauro *et al.*, Plenum Press, New York, 1998

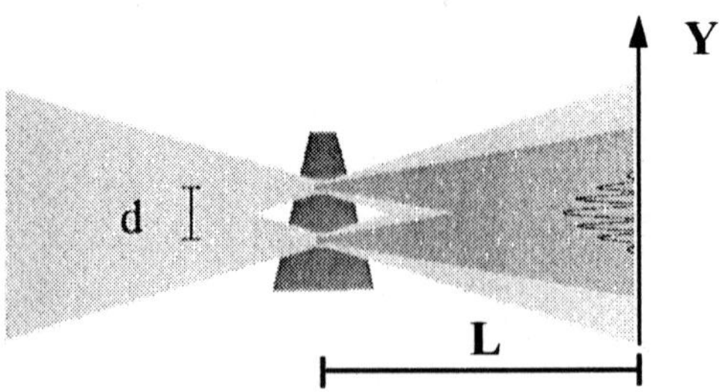

Figure 1: We show the basic configuration along with some notation.

intensities $I_1(t)$ and $I_2(t)$, are separated by the distance d. The farfield interference pattern of the qth harmonic is observed at a distance L from the source. Y is the distance from the axis in the observation plane. The instantaneous fringe pattern is determined by the following formula:

$$I(t) = I_1(t) + I_2(t) + 2\sqrt{I_1(t)I_2(t)} \cos(\frac{2\pi}{q\lambda}\frac{dY}{L} + \phi_1(t) - \phi_2(t)) \tag{1}$$

where λ is the laser wavelength. One recognizes the first part of the formula as the usual Young's double slit expression - the fringe separation is determined by the separation of the sources, and by the wavelength. However, in this case, there is in addition the dipole phase terms $\phi_1(t)$ and $\phi_2(t)$ of the two fields which are time dependent through their dependence on the intensities of the driving fields. This means that for each time, it is the phase difference $\phi_1(t) - \phi_2(t)$ that determines the position of the fringes. If this phase difference varies by more than $\approx \pi$ the fringe pattern will be smeared out by time integration over the entire pulse. The contrast, which is determined as $(I_{max} - I_{min})/(I_{max} + I_{min})$, is therefore determined by the variation of the *phase*, as well as of the variation of the *strength* of the two fields.

As mentioned above, the dipole phase varies rapidly as function of the intensity. From the above equation, one would therefore expect that, having high intensities in the two focii, these would have to be matched very carefully during the entire pulse, not to smear out the interference fringe pattern.

EXPERIMENT

In Figure 2 we show the experimental setup used in the interference experiment. We use the picosecond laser system of the Lund High Power Laser Facility (Svanberg et al., 1994). This is a mode-locked Nd:YAG laser, giving 35 ps pulses at 1064 nm, with pulse energies of up to 80 mJ. The laser light is focused by a 25.5 cm focal length lens. After the lens the beam passes through a bi-refringent BBO crystal, with the axes at 45 degrees to the laser polarization, and is thereby split up into two spatially displaced beams with orthogonal polarization. A polarizer which is placed after the crystal selects a common polarization component. Note that by rotating the polarizer, the relative intensities in the two beams can be varied. One ends up with two focii in a gas jet of xenon, vertically separated by ≈ 150 μm. The FWHM of each of the (Gaussian) focii is estimated to be 50 μm and the two focii are therefore well separated and independent of each other. The intensity in each of the focii is estimated to be approximately 10^{13} W/cm^2. The desired wavelength is selected by a normal incidence

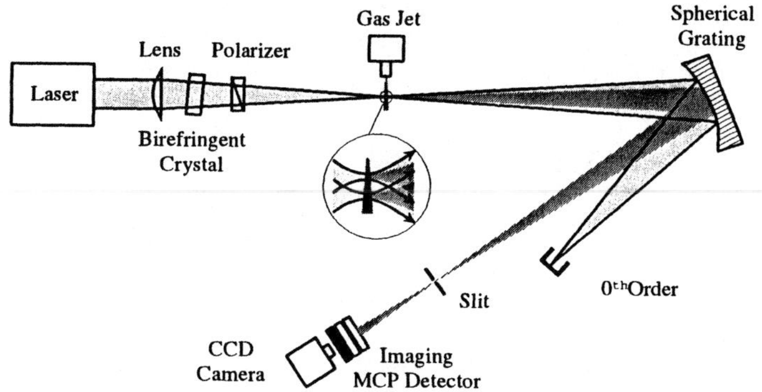

Figure 2: Experimental setup.

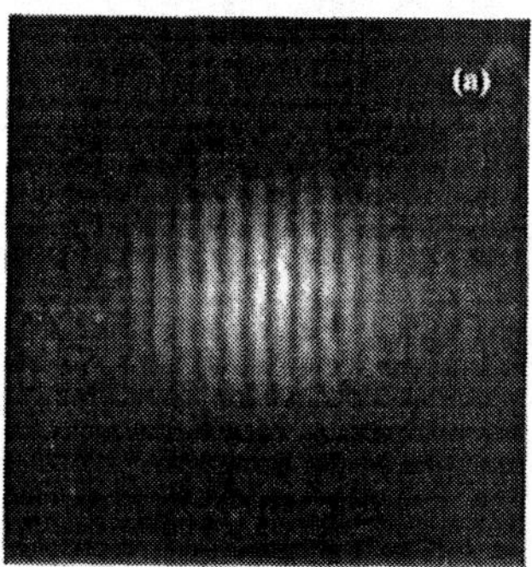

Figure 3: Farfield pattern of the 13th harmonic in xenon.

spherical grating, and imaged through a (wide) slit, onto a micro channel plate, coupled to a CCD camera. The observation distance, which is the distance from the focal plane of the spectrometer to the detector, is 75 cm.

In Figure 3 we show a typical farfield pattern of the 13th harmonic in xenon, obtained when the intensities in the two focii are equal. The interference fringes are clearly visible in the figure, with a contrast of about 30%. The picture shown is an average over 30 laser shots, but the fringes are visible even on a shot to shot basis, with a similar contrast. The fringe separation is 0.44 mm, in good agreement with the expectation $\frac{\lambda L}{qd} = 0.41$ mm. We have obtained similar results for the harmonics 9-17, of course with different fringe separations for the different wavelengths.

We observed a very robust fringe contrast - even towards changes in the relative intensities in the two focii by almost a factor of 2.

CALCULATIONS

To understand this behaviour, we have performed calculations simulating in detail the experimental conditions, including both the response of individual atoms to the strong field as well as the response of the macroscopic collection of atoms in the gas jet and the propagation of the generated radiation through the gaseous medium. The single atom calculations have been performed within the strong field approximation as described by Lewenstein et al. (1994). The propagation equations have been solved within the paraxial and slowly varying envelope approximations, for a number of times spanning the laser pulse duration, as described by Antoine et al (1995). The farfield of the qth harmonic in (r', z') for each of the two fields has been calculated from the electric field $E_q(r, z)$ at the exit of the medium (the nearfield), by means of a Hankel transformation (L'Huillier et al., 1992). We have calculated the interference pattern by numerically superposing the farfields of the two harmonic beams, separated by the distance d[1].

An explanation for the robust fringe contrast can be found in Figure 4, where we show the phase variation of the 13th harmonic as function of time. We show the phase at the exit of the medium, on axis, for two different intensities $I_1 = 1.5 \times 10^{13}$ W/cm^2, and $I_2 = 0.75 \times 10^{13}$ W/cm^2. Also the time profile of the 13th harmonic is shown. We see that during the entire harmonic pulse, the phase *difference* changes only by $\approx$ 0.8 radians, which means that the fringe pattern will not be totally smeared out. The reason that the phase change is so small is that the 13th harmonic in xenon is generated at a relatively low intensity, below 2×10^{13} W/cm^2.

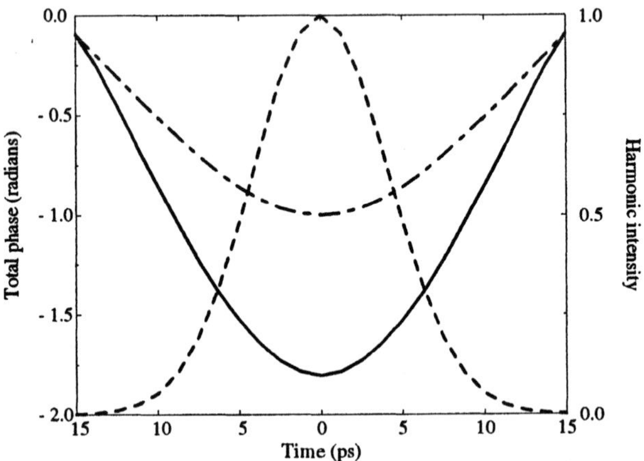

Figure 4: The total phase of the 13th harmonic in xenon at the exit of the medium, for intensities of $I_1 = 1.5 \times 10^{13}$ W/cm^2 (solid line), and $I_2 = 0.75 \times 10^{13}$ W/cm^2 (dot-dashed line). In dashed line is shown the time profile of the 13th harmonic.

We have seen above how the robustness of the fringes is due to the slow phase variation at this low intensity. To investigate the influence of the phase variation, we have performed similar calculations for a higher order harmonic, namely the 43rd harmonic in neon. To simulate realistic conditions for the generation of the 43rd harmonic, we have chosen the field parameters to mimic a femtosecond Ti:Sapphire laser system. The wavelength has thus been chosen to be 790 nm, the pulselength 150 fs, and the peak intensity a few times 10^{14} W/cm^2.

In Figure 5 we show the total phase of the 43rd harmonic, after propagation through a 1 mm gas jet. We show again two different intensities, $I_1 = 5 \times 10^{14}$ W/cm^2 and $I_2 = 4 \times 10^{14}$ W/cm^2. We see that the phase variation is now much more rapid, and

---

[1]Note that we actually do not probe the spatial coherence of the harmonic fields, since we essentially assume that the separation of the two farfields is so small that the *spatial* variation of the fringe pattern is determined by the $\frac{2\pi}{q\lambda}\frac{dY}{L}$ term (c.f. Eq.(1)).

42

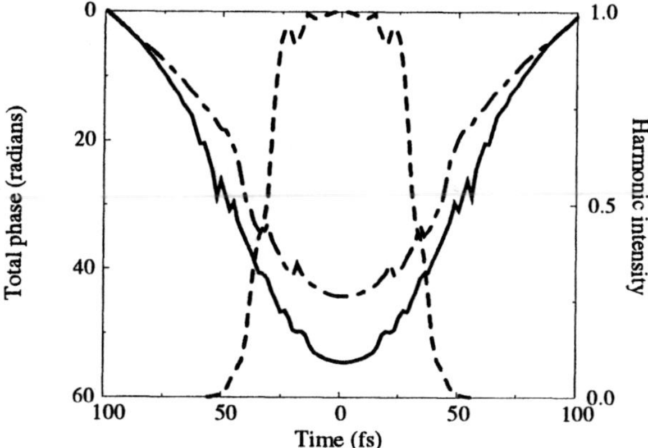

Figure 5: The total phase variation of the 43rd harmonic in neon at the exit of the medium, for an intensity of $I_1 = 5 \times 10^{14}$ W/cm^2 (solid line) and $I_2 = 4 \times 10^{14}$ W/cm^2 (dot-dashed line). The time profile of the 43rd harmonic for an intensity of 4×10^{14} W/cm^2 is shown in dashed line.

that the phase difference changes by several π over the duration of the harmonic pulse, even though the intensity ratio is higher than for the 13th harmonic.

Finally, we compare the results of the 13th harmonic in xenon and the 43rd harmonic in neon, in Figure 6, where we show the contrast as function of the intensity ratio of the driving fields. For the 13th harmonic the intensity I_1 has been held fixed at 1.5×10^{13} W/cm^2, whereas for the 43rd harmonic $I_1 = 5 \times 10^{14}$ W/cm^2. As expected from the

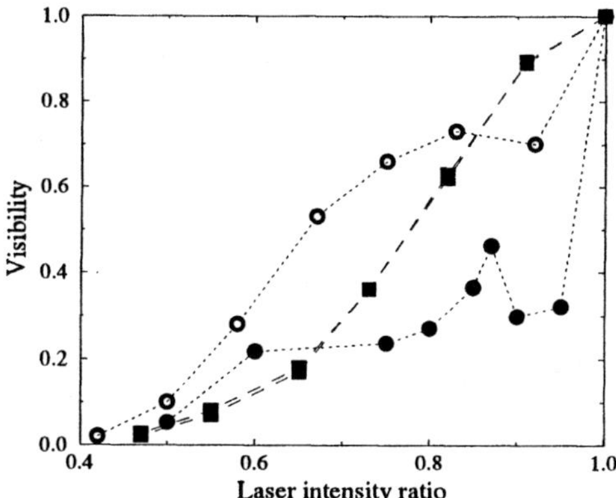

Figure 6: The contrast as function of the intensity ratio, for the 13th harmonic in xenon (filled squares), and the 43rd harmonic in neon (filled circles). The empty symbols result from calculations where the time dependent phase has been omitted.

above, the contrast decreases much faster for the higher order harmonic, generated at high intensity, than for the harmonic generated at lower intensity. In the figure is also shown with open symbols the corresponding results obtained when the temporal variation of the phase of the field has been removed, i.e. by setting $\phi_1(t) = \phi_2(t)$ for all values of t in Eq.(1). The results show that for the 13th harmonic, the decrease of the contrast when the two intensities are different is due to the difference in *amplitudes*, whereas it is the variation of the *phase* which is the dominant contribution for the 43rd harmonic.

CONCLUSION

We have presented an experiment investigating the farfield interference of two harmonic fields, generated by two sources originating in the same laser pulse. We have shown experimentally, that high order harmonics are locked in phase with the fundamental field.

This is very interesting from an applicational point of view, since it enables XUV interferometry *without* the use of XUV beamsplitters. In the experiment, we use a beamsplitter for near infrared radiation, and to perform interferometry experiments, one would only have to insert a disturbing medium into one of the harmonic beams and monitor the change in the fringe pattern.

We observed that the fringe contrast is very robust towards changes in the relative intensities in the two focii. The calculations showed that this is due to the low intensity necessary to generate the 13th harmonic in xenon, making the dipole phase variation negligible.

The calculations for a higher order harmonic, generated at much higher intensity, showed the influence of the phase variation: The contrast falls off much more rapidly as function of the intensity ratio. It is an open question how this sensitivity would influence the applicability of the highest order harmonics. However, as long as the two intensities are exactly matched, there is no fundamental difference with the lower order harmonics.

We acknowledge the support of the Swedish Natural Research Council and the EC 'Access to Large Scale Facilities' Programme (Contract No. ERBFMGEC-T950020). Fruitful discussions with D. Lappas and M. Ivanov are gratefully acknowledged.

REFERENCES

Antoine, Ph., L'Huillier, A., Lewenstein, M., Salières, P., and Carré, B., 1995, PRA **53**:1725.

Chang, Z., Rundquist, A., Wang, H., Murnane, M. M., and Kapteyn, H. C., 1997, PRL **79**: 2967.

Ditmire, T., et al., 1996, PRL **77**:4756.

Glover, T. E. , Schoenlein, R. W., Chin, A. H., and Shank, D. V., 1996, PRL **76**:2468.

Lewenstein, M., Balcou, Ph., Ivanov, M. Yu., L'Huillier, A., and Corkum, P., 1994, PRA **49**:2117.

Lewenstein, M., Salières, P., and L'Huillier, A., 1995, PRA **52**:4747.

L'Huillier, A. , Lompré, L.-A., Mainfrey, G., and Manus, C., 1992, in *Atoms in Intense Laser Fields*, Ed. M. Gavrila (Academic Press, New York).

L'Huillier, A., et al., 1995, J. of Nonl. Opt. Phys. and Mat. **4**:647.

Salières, P., Ditmire, T., Budil, K. S., Perry, M. D., and L'Huillier, A., 1994, J. Phys. B **27**:L217.

Salières, P., L'Huillier, A., Antoine, P., and Lewenstein, M., 1997, Adv. in At., Molec. and Opt. Phys.

Schins, J. M., Breger, P., Agostini, P., Constantinescu, R. C., Muller, H. G., Bouhal, A., Grillon, G., Antonetti, A., and Mysyrowicz, A., 1996, JOSA B **13**:197.

Svanberg, S., Larsson, J., Persson, A., and Wahlström, C.-G., 1994, Physica Scripta **49**:187.

Tisch, J. W. G., Smith, R. A., Muffet, J. E., Ciarocca, M., Marangos, J. P., and Hutchinson, M. H. R., 1994, PRA **49**:R28.

Zerne, R., Altucci, C., Bellini, M., Gaarde, M. B., Hänsch, T. W., L'Huillier, A., Lyngå, C., and Wahlström, C.-G., 1997, PRL **79**:1006.

Soft-X-Ray Harmonics in the Water Window

Andy Rundquist, Zenghu Chang, Haiwen Wang, Erik Zeek, Margaret Murnane, and Henry Kapteyn

The Center for Ultrafast Optical Science
The University of Michigan
Ann Arbor, MI 48109-2099
rundqust@eecs.umich.edu

INTRODUCTION

Over the last decade, much attention has been devoted to the generation and understanding of the process of high-order harmonic generation (HHG).[1-5] The potential applications for extremely short-duration (~fs), coherent, VUV and soft-x-ray radiation are numerous.[6-9] For in-vitro biological imaging in particular, there has been a need for a compact source of radiation in the so called "water window" region, between the carbon and oxygen K-edges, where water is less absorbing than carbon. In this work we report on our efforts to extend the excellent qualities of high-harmonic generation as a radiation source into the water window region.

The simplest way to understand the harmonic generation process is to consider the semi-classical three-step method proposed by Kulander and Corkum.[10,11] As the strong electric field of the laser pulse suppresses the coulomb barrier, the electron first tunnels out of the resulting barrier, then evolves in the field as a free particle, and finally returns to possibly recombine with the core, giving up both the kinetic energy it acquired during its orbit and its ionization potential energy. A simple calculation assuming sinusoidal orbits shows that the maximum kinetic energy an electron can acquire and still return to the nucleus is 3.2 times the pondermotive or "wriggle" energy of the laser field. Thus the simple three-step model predicts that the highest possible harmonic photon energy is given by $h\nu_c = I_p + 3.2U_p$, where I_p is the ionization potential, and U_p is the pondermotive energy (Note: $U_p \propto I\lambda^2$, where I is the laser intensity at the point of ionization). The final step required is then to calculate the intensity of the laser (I) at the point of ionization, when the electron tunnels through the barrier. This does not correspond to the peak intensity of the laser, since the atom is usually ionized on the leading edge of the laser pulse. However, using quantum mechanical ADK tunneling rates,[12] the ionizing intensity can be calculated,

Applications of High Field and Short Wavelength Sources
Edited by DiMauro *et al.*, Plenum Press, New York, 1998

and as a result, the three-step model accurately predicted the cutoff harmonic order for all experiments to date.

In our efforts to extend the highest achievable harmonic orders into the water window region, we have concentrated on increasing both terms in the simple three-step equation for hv_c. To this end we used helium as a target, because it has the largest ionization potential for a neutral gas. While ionic species have even higher ionization potentials, the free electrons present due to ionization would severely disrupt phase matching of the harmonic generation process, leading to very low output. To increase U_p, we systematically studied the effect of using shorter pulses on the threshold intensity required for ionization. We determined both experimentally and numerically that using shorter pulses allows the atom to survive to higher intensities prior to ionizing. This can be understood, to a first approximation, by recognizing that the ionization rate is finite, and therefore the ionization probability will be less for a short pulse with the same peak intensity as a longer pulse.[3,7,13,14] Shorter pulses also increase the efficiency of harmonic generation, since it takes less energy to reach a given intensity. It should be noted that the choice of gas also determines U_p because higher ionization potentials allow higher intensities to be reached for a given degree of ionization. Thus, by exciting helium with ultrashort pulses, we can attain harmonic photon energies well beyond those of previous work which used longer pulses and other gases.

EXPERIMENTAL SET-UP

The laser system we used for these experiments is a 2-stage, 10Hz, Ti:sapphire chirped-pulse amplifier which produces 26fs pulses with 100mJ of energy, corresponding to a peak power of 4TW. For our experiments using He, we typically used up to 10mJ of energy, with 30fs pulsewidths, and an f50 focusing geometry, corresponding to a peak intensity $\approx 2 \times 10^{16} W/cm^2$. The experimental setup is shown in Fig 1. The gas target is a piezo-driven, sub-sonic, pulsed jet which provides a 1mm, 8 torr interaction area. Similar to past experiments,[15] the interaction length is much shorter than the confocal parameter of the laser beam. In contrast to a free-flowing gas jet, the gas load generated by a pulsed nozzle is lower and the lighter noble gases are better controlled, allowing us to observe our first measurable harmonics from He. The harmonics were analyzed using a grazing-incidence, flat-field, imaging VUV/soft-x-ray spectrometer (Hettric Scientific). The signal was measured using an area detector consisting of two micro-channel plates (MCP's) and a phosphor screen coupled to a CCD. This set-up dramatically improved upon our previous scanning geometry, because it allowed us to observe multiple harmonics in a single shot, allowing real-time optimization.

THEORY

There have been numerous semi-classical and quantum models developed that have had varying degrees of success in explaining experimental data to date.[1-3,13,14,16-23] Since we have concentrated in this work on the extension of the harmonic plateau to
higher energies, we have investigated the validity of the three-step model combined with the ADK rate[12] since these give a simple prediction of the highest harmonic photon energy achievable. Our theoretical approach has been to integrate the ADK rate assuming a given pulseshape, while monitoring the total ionization probability until it reaches a given value

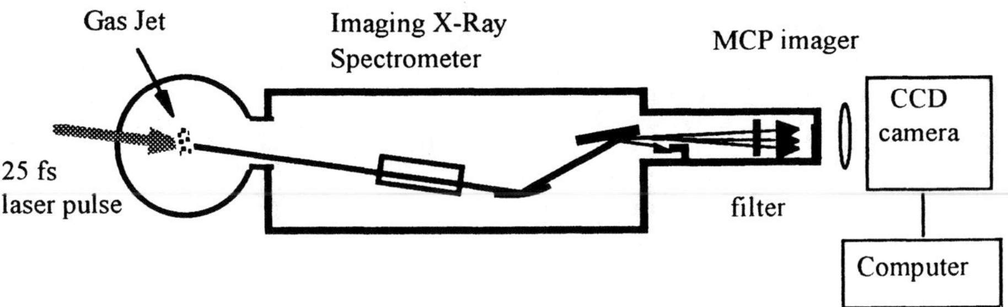

Figure 1. High harmonic generation set-up.

(typically 98%) where the intensity is high and there are still enough neutral atoms left to provide a measurable signal. At this point, the intensity of the laser is used to calculate U_p and thus the highest harmonic photon energy. This approach can be used using our experimentally measured pulse shapes (through the use of the FROG technique[24]) or using an assumed analytic pulse shape. Recently, we derived an analytic expression for the cutoff photon energy as a function of pulse duration, gas species, and ionization state:[13]

$$h\upsilon_c = I_p + \frac{0.5 I_p^{(3+a)} \lambda^2}{\left(\ln\left(0.86\tau 3^{2n^*-1} G_{lm} C_{n^*l^*} I_p\right) / \left(-\ln(1-p)\right)\right)^2}$$

where $h\upsilon_c$ and I_p are in eV, a = 0.5, λ is the laser wavelength in μm, τ is the FWHM of the pulse in fs, and p is set to 98%. It can be seen that the harmonic plateau cut-off scales as the inverse of the log of the pulse duration, and that the gas species (which enters as the ionization potential and the electron polarization state) is a highly non-linear parameter. Due to the highly nonlinear process of harmonic generation, a pulse which can be described analytically over only a few orders of magnitude can still be described by this equation, but we point out that any pulse shape can be used in the more general application of this technique.

In addition to modeling our current experiments, we used the simple 3-step model to test different pulse shapes to optimize the harmonic generation process, in an effort to capitalize on advances in temporal pulse shaping,.[25] Since harmonic generation ceases when the atom is fully ionized, we only need to shorten the rise time of the pulse for example, in order to increase the cutoff photon energy. One way to accomplish this is through interferometric pulse shaping.[26] By using a Michelson interferometer to destructively interfere two pulses, it is possible to produce a pulse with a faster rise time than the original pulse. This method does produce a small pre-pulse, but the intensity is low, resulting in little pre-ionization of the gas. There are several experimental problems with using a Michelson interferometer, primarily one of stability, since the arm lengths must be stable to a fraction of a wavelength. Therefore, more recently we used a multiple-order wave plate which allows two pulses to form along the slow and fast axis of the wave plate and interfere in a polarizer. Tests of this form of pulse shaping have shown decreases in rise-time of 10% for a 20fs pulse.

EXPERIMENTAL RESULTS

Using our current set-up, we have been able to extend harmonic generation in both neon and helium. The single-shot capabilities of the area detector have greatly enhanced

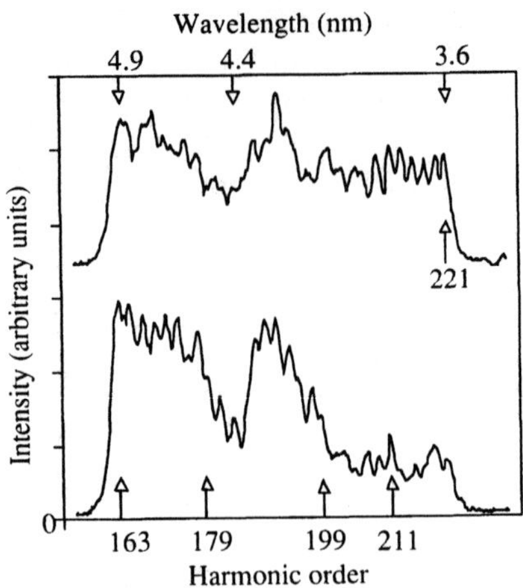

Figure 2(a). Harmonic emission from He, up to order 221. The cut-off due to our spectrometer at short wavelengths is artificial

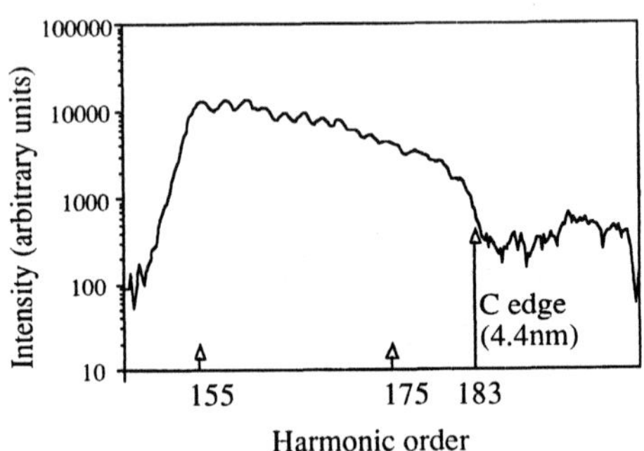

Figure 2(b). Harmonic emission from He observed through a 0.4 micron carbon filter on a log scale, showing the carbon K-edge.

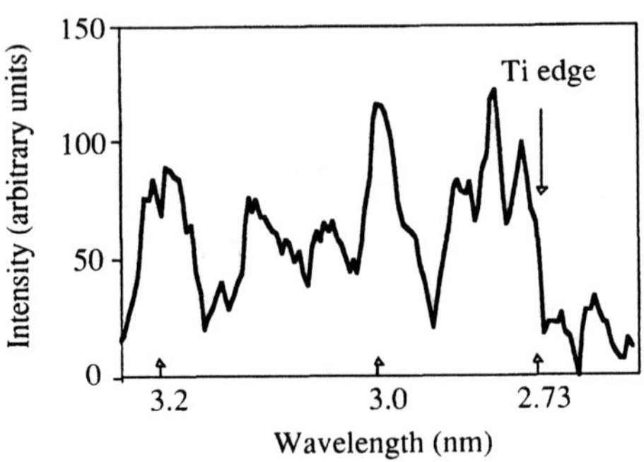

Figure 3. Harmonics from He using a different grating, showing emission up to order 299, down to the titanium K-edge.

our ability to optimize the harmonic signal. In Ne, we can generate up to the 155th harmonic, which agrees very well with our three-step model prediction. In He, we observe harmonics up to the 299th order, corresponding to a wavelength of 2.7nm, or a photon energy of 460eV. The wavelength range we observe fom He is so large that we cannot

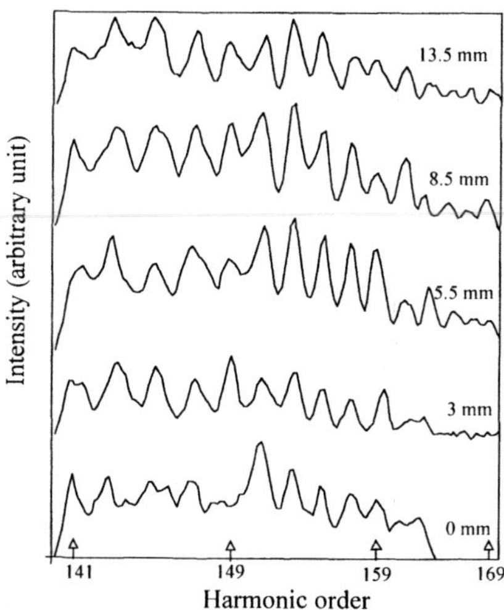

Figure 4. The effect of the pulse-jet position relative to the focus on high harmonic emission from He.

acquire the entire spectrum at once. This is primarily because in order to protect our MCPs from the zero-order radiation from the spectrometer grating, we were forced to limit the range over which we observed harmonics at any one time. Fig. 2(a) shows the harmonic spectrum from He up to order 221, while Fig. 2(b) shows the result of adding a 0.4 micron carbon filter between the x-ray source and the detector. The carbon edge is clearly visible on a log scale, verifying that the harmonic radiation is real. By using a more dispersive grating, we were able to observe harmonics in He down to the titanium K-edge at 2.7nm, which corresponds to the 299th harmonic, as shown in Fig. 3. This grating is not as efficient as the one used in Fig. 2, and therefore the signal-to-noise ratio is not as good. We estimate that there are $\approx$ 300 photons per harmonic peak per pulse in the 2.7nm spectral region. It should be noted that the contrast ratio for biological imaging is greatest for wavelengths just below the carbon edge where we clearly have resolved peaks. Finally, in Fig. 4 we show the dependence of the harmonic spectra on the jet position relative to the focus. These results will be considered in the next section.

DISCUSSION

We have shown that extremely short wavelength, coherent, radiation can be generated via high harmonic emission. Our new data for Ne and He, combined with our previous data for the heavier noble gases, is presented in Table 1. We compare the experimentally observed highest harmonic with the prediction of the simple three-step model.

There is excellent agreement between theory and experiment for all the nobel gases except He, where the experimentally observed cut-off falls short of the prediction by 10%. This disagreement may be due to the fact that the ADK rate is not expected to be valid for helium.[27] Also, since we cannot remove the Ti filter from our set-up, because it is needed to prevent damage to our detector from the zer-order scattered light, we may be generating harmonics above 299 from He. It should also be noted that for even shorter pulses, the ADK rate will not be valid, because non-adiabatic effects start to play a role.[28] These effects are linked to an intrinsic intensity-dependent phase of the harmonic field which comes about because the sinusoidal laser field increases its intensity appreciably within a single cycle for such short pulses. For every cycle in an increasing field, the phase at which

Table 1. Comparison of predicted and experimental highest harmonics

Gas	Xe	Kr	Ar	Ne	He
3-step prediction	31	43	61	165	331
measured harmonic	29	41	61	155	299

the intensity is high enough to ionize slips back to an earlier point. Also, the electron trajectories in such a field are only approximately those of an electron in a pure sinusoidal field, allowing trajectories which can return to the nucleus with kinetic energies higher than $3.2\ U_p$, which is calculated for pure sinusoidal fields only. Both of these descriptions depend strongly on the slope of the envelope of the field, which in turn depends on the intensity, thus giving an intensity-dependent phase of the harmonics. Both of these effects have been modeled by analytically and numerically integrating the Schrödinger equation.[9,29,30] The intensity-dependent phase manifests itself both temporally and spatially. In Fig. 4 we show an example of the spatial effect on the harmonic peaks as the jet is positioned at different points along the focusing laser beam. The qualitative features shown agree well with those of the numeric simulation in reference.[30]

CONCLUSION

We have demonstrated that high harmonic emission can be extended to wavelengths as short as 2.7nm, which spans the water window region of x-ray transmission. This radiation is coherent, with pulse durations as short as 3fs.[7,8] We have successfully applied a simple model to predict the highest possible photon energy attainable, and we observe excellent agreement between our predictions and experiment. The highest photon energy we observe thus far is approaching 500eV. Finally, we have shown that the intrinsic intensity-dependent phase effects play a role in determining the output spectra of the harmonics.

We would like to acknowledge Ivan Christov for useful discussions. We gratefully acknowledge support from the National Science Foundation. H. Kapteyn acknowledges support from a Sloan Foundation Fellowship.

BIBLIOGRAPHY

1. A. McPherson, G. Gibson, H. Jara, U. Johann, T. S. Luk, I. A. McIntyre, K. Boyer, C. K. Rhodes, JOSA B **4**, 595 (1987).
2. A. L'Huillier, P. Balcou, Phys. Rev. Lett. **70**, 774 (1993).
3. J. Zhou, J. Peatross, M. M. Murnane, H. C. Kapteyn, I. P. Christov, Phys. Rev. Lett. **76**, 752 (1996).
4. S. Backus, J. Peatross, M. Murnane, H. Kapteyn, Opt. Lett. **21**, 665 (1996).
5. A. L'Huillier, K. J. Schafer, K. C. Kulander, Journal of Physics B **24**, 3315 (1991).
6. J. H. Glownia, D. R. Gnass, P. P. Sorokin, JOSA B **11**, 2427 (1994).
7. I. P. Christov, M. M. Murnane, H. C. Kapteyn, Phys. Rev. Lett. **78**, 1251 (1997).

8. P. B. Corkum, N. H. Burnett, M. Y. Ivanov, Opt. Lett. **19**, 1870 (1994).

9. K. J. Schafer, K. C. Kulander, Phys. Rev. Lett. **78**, 638 (1997).

10. P. B. Corkum, Phys. Rev. Lett. **71**, 1994 (1993).

11. K. C. Kulander, K. J. Schafer, J. L. Krause, at the *NATO 3rd Conference on Super Intense Laser-Atom Physics* (Han-sur-Lesse, Belgium, 1993).

12. M. V. Ammosov, N. B. Delone, V. P. Krainov, Soviet Physics JETP **64**, 1191 (1986).

13. Z. Chang, A. Rundquist, H. Wang, H. C. Kapteyn, M. M. Murnane, Phys. Rev. Lett. **79**, 2967 (1997).

14. I. P. Christov, J. P. Zhou, J. Peatross, A. Rundquist, M. M. Murnane, H. C. Kapteyn, Physical Review Letter 77, 1743 (1996).

15. J. Zhou, J. Peatross, M. M. Murnane, H. C. Kapteyn, at the *QELS* (Baltimore, MD, 1995).

16. W. Becker, S. Long, J. K. McIver, Phys. Rev. A **41**, 4112 (1990).

17. J. L. Krause, K. J. Schafer, K. C. Kulander, Phys. Rev. Lett. **68**, 3535 (1992).

18. M. Lewenstein, P. Balcou, M. Y. Ivanov, P. B. Corkum, Phys. Rev. A **49**, 2117 (1993).

19. I. P. Christov, M. M. Murnane, H. C. Kapteyn, Phys. Rev. Lett. **78**, 1251 (1997).

20. A. L'Huillier, K. J. Schafer, K. C. Kulander, Phys. Rev. Lett. **66**, 2200 (1991).

21. J. J. Macklin, J. D. Kmetec, C. L. Gordon, III, Phys. Rev. Lett. **70**, 766 (1993).

22. S. Preston, A. Sanpera, M. Zept, W. Blyth, C. Smith, J. Wark, M. Key, K. Burnett, M. Nakai, D. Neely, A. Offenberger, Phys. Rev. A **53**, 31 (1996).

23. N. Sarukura, K. Hata, R. N. T. Adachi, M. Watanabe, S. Watanabe, Phys. Rev. A **43**, 1669 (1991).

24. K. W. DeLong, R. Trebino, D. J. Kane, JOSA B **11**, 1595 (1994).

25. A. Weiner, Progress in Quantum Electronics **19**, 161 (1995).

26. A. Giles, J. Posthumus, M. Thompson, L. Frasinski, K. Codling, A. Langley, W. Fhaikh, P. Taday, Opt. Commun. **118**, 537 (1995).

27. S. Augst, D. D. Meyerhofer, D. Strickland, S. L. Chin, JOSA B **8**, 858 (1991).

28. I. P. Christov, J. Zhou, J. Peatross, A. Rundquist, M. M. Murnane, H. C. Kapteyn, Physical Review Lettters 77, 1743 (1996).

29. M. Lewenstein, P. Salieres, A. L'Huillier, Phys. Rev. A **52**, 4747 (1995).

30. P. Salieres, A. L'Huillier, M. Lewenstein, Phys. Rev. Lett. **74**, 3776 (1995).

HARMONIC GENERATION IN PRESENCE OF IONIZATION

A. Bouhal[1], G. Hamoniaux[1], A. Mysyrowicz[1], A. Antonetti[1],
P. Salières[2], P. Breger[2], P. Agostini[2],
R. C. Constantinescu[3], H. G. Muller[3]
[1]Laboratoire d'Optique Appliquée, ENSTA-Ecole-Polytechnique,
91125 Palaiseau, France
[2]Service des Photons, Atomes et Molécules
DSM-DRECAM CEA Saclay, 91191 Gif-sur-Yvette France
[3]FOM-Institute for Atomic and Molecular Physics Kruislaan 407, 1098 SJ
Amsterdam, The Netherlands

INTRODUCTION

Most experiments and theoretical studies of High-Order Harmonics have been limited
to situations of low levels of ionization. It is the purpose of this work to investigate
experimentally and theoretically the situation of intensities around the saturation intensity
since it is under such conditions that harmonic generation is most efficient. In case of strong
ionization, both the spatial and temporal behavior of harmonics are expected to be affected.
This study focuses on the temporal aspect, with special attention of the respective roles of
the intensity-dependent atomic phase and the ionization. Since direct autocorrelation of the
XUV pulses is not yet possible, a cross-correlation method is the only available technique
to approach short pulses measurements in this wavelength range[1-4]. Experimentally, the
cross-correlation of the 19th harmonic pulses generated in argon by 800 nm pulses of
various lengths from 120 fs to 1 ps and a 120 fs 800 nm probe pulse were measured at
intensities above 10^{14}W.cm^{-2}. Theoretically, the corresponding profiles were calculated
using numerical values of the atomic dipole and the ionization rate in a propagation code
solving the coupled wave equations by a finite- difference method. The main result is that
the harmonic pulse is strongly shortened by the combined effect of the intensity-dependent
phase mismatch and the dipole reduction due to ionization.

EXPERIMENT

The cross-correlation method is based on two-color Above-Threshold Ionization
(ATI): The high-frequency, weak-intensity (harmonic) beam lifts the electron into the
continuum while a high-intensity, low-frequency (fundamental) pulse induces multiphoton

Applications of High Field and Short Wavelength Sources
Edited by DiMauro *et al.*, Plenum Press, New York, 1998

free-free transitions. In the electron energy spectrum, such free-free transitions produce "sidebands" of energy E_s on both sides of the peak resulting from the VUV ionization with photon energy separated from it by the low frequency photon energy. Experimentally, I_s is measured as a function of the time delay between pump and probe pulses. The harmonic duration is then obtained by deconvolution, the fundamental pulse being characterized by either autocorrelation measurements or more sophisticated methods like cross-correlation with its third harmonic or nonlinear pulse retrieval. We have generated the harmonics with pulses of various duration in the range .12 to 1 ps and analyzed them with the shortest probe pulse available (120 fs) using two separate compressors in the laser system (Fig. 1).

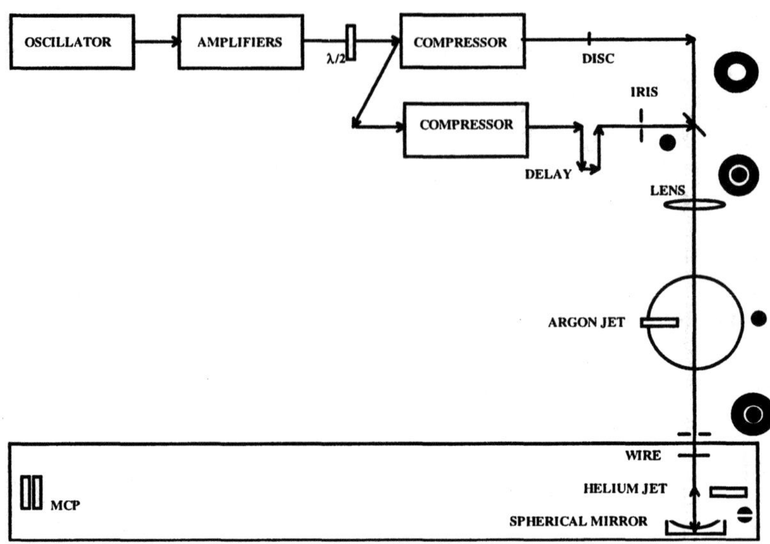

Figure 1. Experimental setup.
Cross-sections of the pump and probe beams at various points are shown along the beams paths.

The output of a self-mode-locked Ti:S laser is stretched to 400 ps and amplified in two stages[5]. After recompression, one obtains 50 mJ pulses of 120 fs at 800 nm with a repetition rate of 10 Hz. The input grating of the compressor was used as a beam splitter: its zero order sent to a second pair of gratings where it could be compressed to an independently adjustable value and used as a probe beam. A half-wave plate placed upstream of the first compressor is rotated to control the energy balance between the pump and probe beams. The central part of the pump beam is blocked by an 8 mm diameter circular disk in order to create an annular beam. The resulting annular beam is focused by a 1m spherical lens inside a pulsed argon jet where the harmonics are generated. The outer diameter of the ring is adjusted to optimize harmonic conversion and depends of the laser energy (between 16 and 20 mm). The effective f-number for the beam is then between 62 and 50 (neglecting the influence of the hole). The focus is diffraction-limited to a diameter of 50 to 63 μm with corresponding Rayleigh ranges of 2.5 to 4 mm. The pulsed jet, backed by a pressure of 350 torr, is collimated by a 5 mm long and .8 mm diameter nozzle and the beam is focused about .5 mm below it. The density in the jet is estimated to be of the order of 10^{18} cm^{-3}. The harmonics generated in the jet propagate along the system axis as if generated by a TEM_{00} mode while the fundamental quickly recovers the annular shape it had at the focusing lens. It can therefore be blocked by an adequate aperture. This device provides a simple way to separate with minimum losses the harmonic beam from the fundamental. The IR probe beam energy is limited to about 10 μJ. The two beams (probe and pump) are then recombined by a beam splitter after which they propagate colinearly through the focusing lens. They are both focused by a spherical (R=7cm) mirror. The delay between the probe and XUV beams is controlled by an optical delay line inserted in probe arm. Ionization of the target gas by the harmonic beam crossing the interaction region before reaching the mirror is prevented by a 1mm diameter movable wire shadowing the sensitive part of the spectrometer while letting about half of the beam section reach the mirror and be focused back into the same zone. The mirror coating[6] is a multilayer Mo/Si centered on the wavelength of the 19th harmonic (29 eV). The dressing and the harmonic beams have approximately the same diameter on the mirror (2.5 mm). The mirror reflects both the probe beam and the 19th and 21st harmonics, as well as the low-order harmonics. The dressing and the harmonic beams have approximately the same diameter on the mirror ($\approx$2.5 mm). The electron spectrometer (TOF) is a magnetic bottle, time-of-flight spectrometer collecting the electrons over 2π sr solid angle. Helium gas is injected in the sensitive zone of the spectrometer through a pulsed valve synchronized with the laser pulse. helium ionization potential (24.6 eV) is such that the harmonics below 17 are not seen in single-photon ionization. The filtering out of the fundamental beam is not perfect though: because of the residual misalignment of the disk image with respect to the pinhole and the scattering from the gas jet it is difficult to reduce the effective transmission below 10^{-6}. The two beams (probe and pump) are focused by a spherical (R=7cm) mirror. about 2 mm beyond the center of the spectrometer sensitivity zone to keep the IR intensity low enough. The actual intensity is determined from the red-shift of the photoelectrons energies due to the ponderomotive potential.

RESULTS

Harmonic 19 and 21 produce 4.9 and 8 eV respectively. The high energy sideband of Harmonic 19, has been used for the measurements. For each position of the stepping motor the electron energy spectra are collected over a few hundreds to a few thousands laser shots. The cross correlation is then obtained by plotting the electron counts summed over the sideband normalized to the electron counts summed over the harmonic peak to compensate for the slow variations of the laser intensity. The measurements were taken for four adjustments of the compressor gratings distance yielding pumping pulse durations of 1 ps, 600 fs, 400 fs, and 120 fs as measured by ω - 3ω cross-correlation. Fig. 2a-d show typical cross-correlations for a 1 ps, $(9.5 \pm 30\%)10^{13}$W.cm^{-2}, 600 fs $(1.6 \pm 30\%)10^{14}$W.cm^{-2}, 400 fs $(2.5 \pm 30\%)10^{14}$W.cm^{-2}, 120fs $(5\pm30\%)10^{14}$W.cm^{-2}) pump pulses respectively when the focus is close to the center of the jet (z=0).

The single-atom response is handled through the model of Lewenstein et al[7]: the atom is described by its ground state wavefunction, (in the hydrogenoid approximation). The influence of the atomic potential is neglected in the continuum where the wavefunctions are taken as plane waves. The harmonic spectrum is then computed as the Fourier spectrum of the atomic dipole calculated from the solution of the time-dependent Schrödinger equation. The harmonic spectrum is computed from the Fourier transform of the dipole. From the single-atom reponse only, one obtains rather long harmonic pulses (500 fs, typically for 1ps laser pulses), determined by the low index of the nonlinear response of the dipole to the pump intensity near saturation. The phase rate of change with intensity, and the ionization rate however, strongly modify the time dependence of the harmonic pulse.

The propagation is handled through a finite-difference method code described in details elsewhere[7-11]. The roles of phase mismatch and ionization are easily separated in the simulation. Without ionization, as the intensity in the rising edge of the laser pulse, the harmonic production grows rapidly but, at the same time, due to the rapid dephasing of the dipoles, it reaches a maximum and falls off rapidly. Close to the maximum of the laser pulse, the increase of the polarization compensates to some extent the dephasing and a secondary maximum appears in the time profile which is obviously symmetric with respect to t=0. The

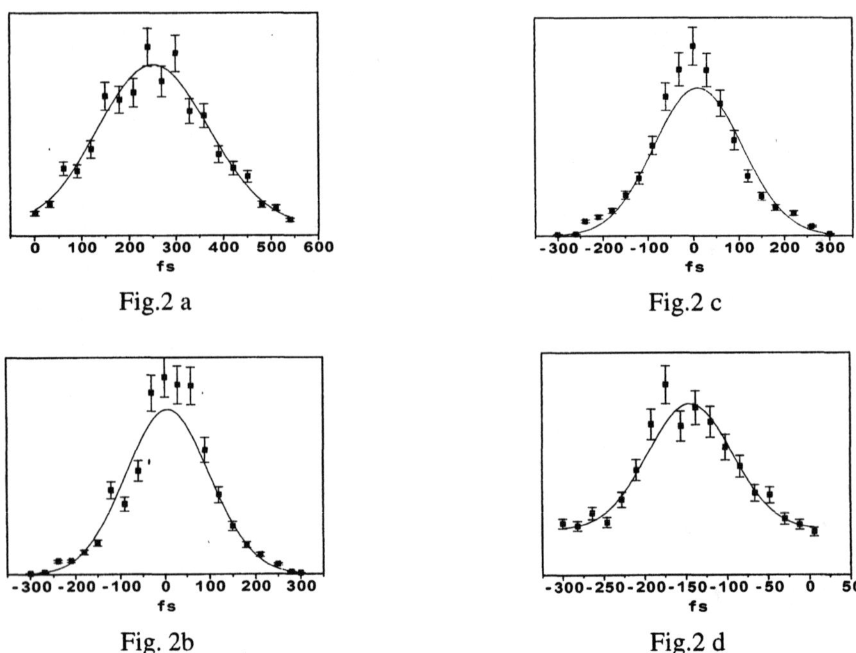

Fig.2 a

Fig.2 c

Fig. 2b

Fig.2 d

effect of ionization adds to the effect of the phase by essentially removing the atomic polarization as time increases and suppresses the revival of emission due to the re-phasing of dipoles as intensity decreases on the falling edge of the pulse. However, the effective duration is determined by the phase: taking the intensity dependence out from the simulation yields much broader pulses.

The trend derived from the simulations is illustrated in Fig.3 and 4. For a given intensity, and if the intensity-dependent phase is removed from the calculation, the profile width is determined by the nonlinearity of the dipole and the ionization rate (Fig. 3 dashed line). Including ionization, but neglecting the phase yields the dotted line curve. Finally, including both produces the solid line profile which is essentially half the profile due to the phase alone, the second half being removed by the ionization. Note that the resulting width is therefore due to these combined effects. The theoretical profiles including phase and ionization effects for different intensities are shown for z=0 in Fig.4. The data is plotted against the laser pulse width in Fig.5 in the four cases at z=-.48 mm The simulation is in good agreement with the data. For comparison, the simulations with the ionization turned off are also displayed. The effect of reducing the dipole lifetime by ionization is clearly seen even at the 1ps point taken at relatively low intensity. The convolution of course limits the resolution of the measurement. In both cases the theoretical curves including or not ionization are rather different. It appears that (i) at 1 ps pump pulse, the harmonic duration is close to 220 fs, i.e. more than twice shorter than expected on the basis of the effective order on nonlinearity (3.8) around 10^{14} W.cm^{-2} (512 fs). At 120 fs the accuracy of the measurement is poor since the convolution has a width close to that of the probe pulse. However, the shortest duration deducted from the comparison (13 fs) is even more drastically shorter than the 60 fs expected on the basis of the effective slope. (ii) this shortening results mainly from the ionization as clear from inspecting the predictions in Fig.3 and 4. (iii) In reference to the work of Glover et al.[3], the simulation does predict the trends reported there. However, harmonic durations longer than the fundamental one, as reported in the experiment are not found in the simulation, even at intensities twice the saturation intensity.

Let us point out that (i) the simulations have been run with both annular and gaussian beams on a few test cases. The results are (for the other parameters kept constants) approximately identical. This is not unexpected since the annular beam focuses practically as a gaussian. (ii) The theoretical profiles, convoluted with the probe profile (Gaussian, 120 fs FWHM) are then fitted by a Gaussian function whose width is compared to the width of the Gaussian fit to the experimental data. (iii) Even at the lowest intensity used in the experiment (9.5×10^{13} W.cm^{-2}) the ionization of the gas jet is important ($\gamma\tau \approx .8$). However, it appears that, apart from a slight defocusing of the beams (of course incorporated in the simulation of the harmonic generation itself), the consequenses on the both temporal profiles are negligible. (iv) Given the irregular shapes of the simulated temporal profiles, it was more reliable to compare the experiment to the convolution of these profiles and the probe pulse profile.

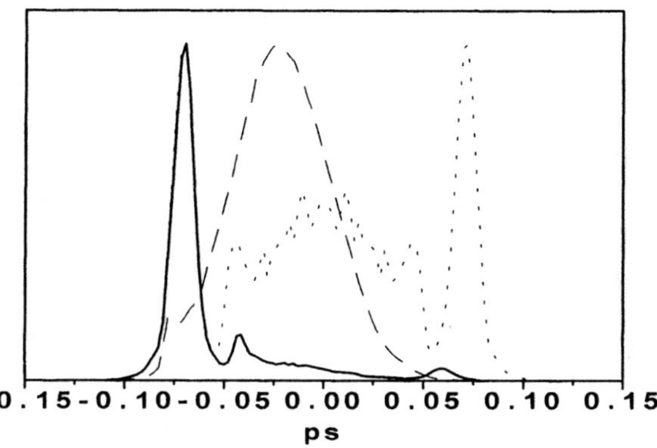

Fig.3

Theoretical profiles: ionization/no phase (dashed); phase/no ionization (dotted); phase and ionization (solid)

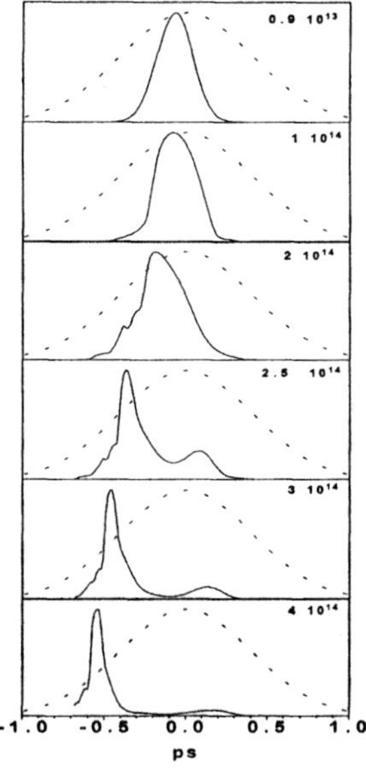

Fig.4

Theoretical profiles vs intensity

CONCLUSION

We have applied the cross-correlation technique to study the temporal behavior of high-order harmonics generation when the laser pulse is intense enough to saturate ionization. The main conclusion is that the harmonic pulses are more than twice shorter than expected on the basis of the intensity dependence of the atomic dipoles alone. The theory reproduces the experimental measurements. However the details of the predicted temporal profiles are lost in the convolution process, owing to the limitations imposed by the probe pulse. Given the extremely short pump pulses (few fs) currently available, it can be expected that subfemtosecond harmonic pulses will be reached in the near future, even without relying on "attosecond" schemes[12,13].

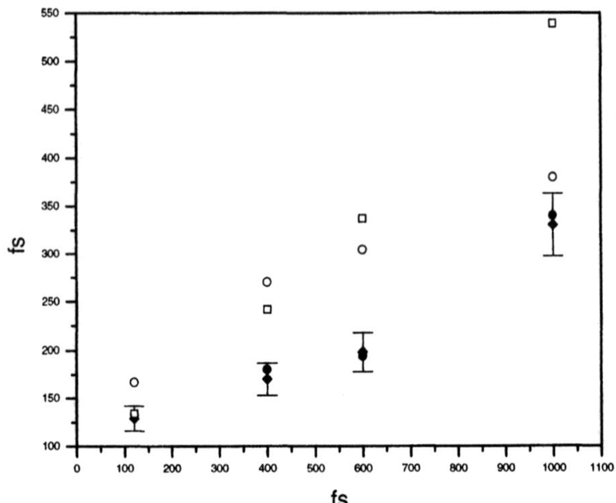

Fig.5

Experimental widths vs theory: ionization/no phase (squares); no ionization/phase; both ionization and phase (diamonds)

REFERENCES

1 J. M. Schins, P. Breger, P. Agostini, R. C. Constantinescu, H. G. Muller, G. Grillon,A. Antonetti, A. Mysyrowicz, Phys. Rev. Lett. **74**, 2180 (1994).

2. J. M. Schins, A. Bouhal, P. Breger, P. Agostini, R. C. Constantinescu,H. G. Muller, G. Grillon, A. Antonetti, and A. Mysyrowicz, JOSA B, **13**, 197 (1996).

3. C. Glover et al, Phys. Rev. Lett. **76**, 2468 (1996).

4. A. Bouhal et al. JOSA B, **4**, 950 (1997).

5 C. Le Blanc, G.Grillon, J. P. Chambaret, A. Migus,and A. Antonetti, Opt. Lett. **18**, 140 (1993).

6. J. P. Chauvineau, Rev. Phys. Appl. **23**, 1645 (1994).

7. M. Lewenstein, Ph. Balcou, M. Yu. Ivanov, A. L'Huillier, and P. Cockum, Phys. Rev. A **49**, 2117 (1994).

8. P. Balcou, P. Salières, A. L'Huillier and M. Lewenstein Phys. Rev. A **55**, 3204, (1997).

9. P. Salières, A. L'Huillier and M. Lewenstein, Phys. Rev. Lett. **75**, 3776 (1995).

10. A. L'Huillier and P. Balcou Laser Physics **3**, 654 (1993).

11. P. Salières et al. Adv. Atom. Mol. Opt. Physics to be published.

12. P. Antoine et al. Phys. Rev. Lett, **77**, 1234 (1996).

13. P. B. Corkum et al., Opt. Lett, **19**, 1870 (1994).

THE OPTIMISATION OF SOFT X-RAY LASER OUTPUT

G J Tallents[1], J Y Lin[1], J Zhang[2], A Behjat[1*], A Demir[1+], M M Güzelgöz[1+], C L S Lewis[3], A MacPhee[3], D Neely[4], G J Pert[5], R Smith[1], J S Wark[2], P J Warwick[3•] and E Wolfrum[2]

[1]Department of Physics, University of Essex, Colchester CO4 3SQ, England.
[4]Clarendon Laboratory, University of Oxford, Oxford OX1 3 PU, England.
[3]Department of Pure and Applied Physics, Queen's University Belfast, Belfast BT7 1NN, Northern Ireland.
[4]Central Laser Facility, Rutherford Appleton Laboratory, Chilton, Didcot, OX11 0QX, England.
[5]Department of Physics, University of York, York YO1 5DD, England.

INTRODUCTION

The efficiency of producing soft X-ray laser output from nanosecond duration optical pumping pulses has been found (see, for example, Carillon et al[1]) to be low ($< 10^{-6}$). This has meant that moderately large optical pumping lasers were required (typically producing 100 - 1000 joules in nanosecond pulses) and that because of the rapid scaling of efficiency with X-ray laser wavelength, the X-ray wavelengths able to be pumped with such medium-sized laser facilites were limited to typically > 15 nm.

Recent experiments and simulations have shown that soft X-ray laser output can be increased by orders-of-magnitude by pumping with multiple-pulse, short duration (< 100 ps) pulses[2,3]. At the Rutherford Appleton Laboratory, lasing on nickel-like transitions has produced saturated output at 14 nm in Ni-like Ag[4] and at 7 nm in Ni-like Sm[5] with double pulse pumping. The lasing at 7 nm is currently the shortest wavelength saturated output ever achieved. Earlier multi-pulse work with neon-like germanium[6] produced saturated output on the J = 0 - 1 line at 19.6 nm and only a small output on the J = 2 - 1 lines at 23 nm. With single pulse irradiation, the J = 2 - 1 lines had previously exhibited the most X-ray laser output, despite calculations which predicted that there should be higher gain for the J = 0 - 1 lines[1].

In this paper, recent experiments undertaken at the Central Laser Facility, Rutherford Appleton Laboratory with double pulse pumping will be briefly reviewed. Experiments and simulations with muliple pulse 75 ps duration pulses and with short duration (2 ps) pulses superimposed onto long pulses (300 ps) will be discussed. Lasing on Ne- and Ni-like ions has been observed.

---

* Now at University of Yazd, PO Box 89195-741, Yazd, Iran.

+ Now at University of Kocaeli, Department of Physics, Kocaeli, 41000, Turkey.

• Now at Max-Born Institüt, Rudower Chaussee 6, 12489 Berlin, Germany.

Applications of High Field and Short Wavelength Sources
Edited by DiMauro *et al.*, Plenum Press, New York, 1998

EXPERIMENTS

Double short (75 ps) pulse pumping

Three beams of the VULCAN glass laser at the Rutherford Appleton Laboratory were focused to line foci of 100 μm width and 16 - 25 mm length using three refracting f/2.5 lenses and off-axis spherical mirrors[7].   The foci were overlapped to produce an approximately uniform irradiation $\approx 4 \times 10^{13}$ W cm^{-2} on target with approximately 75 J of laser energy at 1.06 μm wavelength and pulse lengths of 75 ps.   Deploying another three of the VULCAN beams 180° opposed in a second line focus was used for some of the results presented here to produce a plasma with an opposed density gradient that compensates for refractive deviation of the X-ray laser beam from the first plasma[1].

A pre-pulse was introduced by splitting the oscillator pulse into two pulses before amplification in the VULCAN laser.   A variable pre-pulse energy of typically 20% of the total energy in the second pulse was produced typically 2.2 ns before the second pulse.   The pre-pulse level and pulse separation were monitored by a fast photodiode system and an optical streak camera.

The effectively slab targets used in the experiment were 16-18 mm long stripes of width 100 μm and depth 0.6 μm made up of the element (germanium, silver, indium, tin or samurium) under investigation.   The principal diagnostic employed was a grazing incidence flat field spectrometer with either (i) a grazing incidence cylindrical focusing optic to image the exit aperture of the X-ray laser in the horizontal plane along the spectrometer entrance aperture or (ii) with the spectrometer entrance slit aligned perpendicular to the target normal so as to resolve the angular variation of the X-ray laser emission.   In addition, a space resolving time integrating KAP crystal spectrometer incorporating another 16 bit CCD detector was used to monitor the resonance line emission in the wavelength range of 7 - 10.1 Å and to measure the uniformity of the line focus plasmas.

Pumping with a long ($\approx$ 300 ps) and a superimposed ~ 2 ps pulse

Two beams of the VULCAN laser were employed to produce a pulse of 300 ps duration with irradiance $\approx 10^{12}$ W cm^{-2} and a superimposed 2 - 4 ps duration pulse of irradiance $\approx 8 \times 10^{14}$ W cm^{-2}.   A 100 mm diameter beam of 55 J energy was focused into a line using a standard f/2.5 refraction lens and off axis spherical mirror, while an off axis parabola and spherical mirror system was used with another beam of 20 J energy and chirped pulse amplification (CPA) to produce the short pulse line focus.   The long pulse focus was of length 17 mm and width 100 μm, while the short pulse was of length 11 mm and similar 100 μm width.   Both foci were superimposed onto stripe targets (as described above) of length in the range 1.5 - 10 mm.   The relative timing of the long and short pulse was varied so as to optimise the X-ray laser output.   Experiments were undertaken with both travelling wave pumping and without a travelling wave for the short pulse beam.   The travelling wave was produced by introducing an additional grating into the CPA beam[8].   With germanium targets, a long (300 ps) prepulse of irradiance $\approx 10^{11}$ W cm^{-2} was incident 3.3 ns before the main long pulse.   The plasma diagnostics employed were identical to those used with the double 75 ps pulse pumping.

RESULTS

Results for 75 ps double pulse pumping with 2.2 ns separation between the pre-pulse and main pulse have been published for Ne-like Ge (atomic number 32)[6] and for Ni-like Ag (atomic number 47)[5] and Sm (atomic number 62)[4].   Elements of intermediate atomic number were also found to exhibit saturated lasing on the Ni-like ionisation stage.   The spectrum of Ni-like Sn (atomic number 50) is shown in figure 1.   The spectrum is dominated by the lasing line at 12 nm.   Figure 2 shows a target length scan for Ni-like In (atomic number 49).

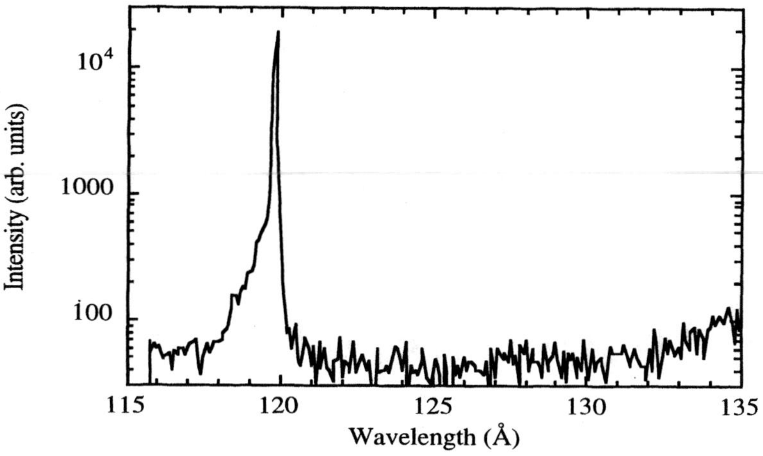

Figure 1. Spectrum of the Ni-like Sn X-ray laser pumped by double 75 ps pulse irradiation.

A gain coefficient of 9 cm^{-1} can be deduced from the exponential X-ray laser output for short target lengths up to ≈ 1.8 cm with saturation occuring at a gain length product of ≈ 16.

The 2 ps duration travelling wave pumping of Ni-like Sn produced a gain coefficient of ≈ 32 cm^{-1} with saturation at ≈ 4.5 mm length of target corresponding to a gain length product of ≈ 14 - 15 (see figures 3). The maximum gain was produced with the 2 ps pulse incident 500 ps after the peak of the ≈ 300 ps duration long pulse. Removing the travelling wave pumping so that the irradiation travelled close to instantaneously along the direction of travel of the X-ray laser beam rather than at the speed of light results in a ≈ 20 fold drop in X-ray laser output for a target length of 10 mm (figure 3). This illustrates the importance of using the travelling wave with this short duration pumping. The short and long pulse

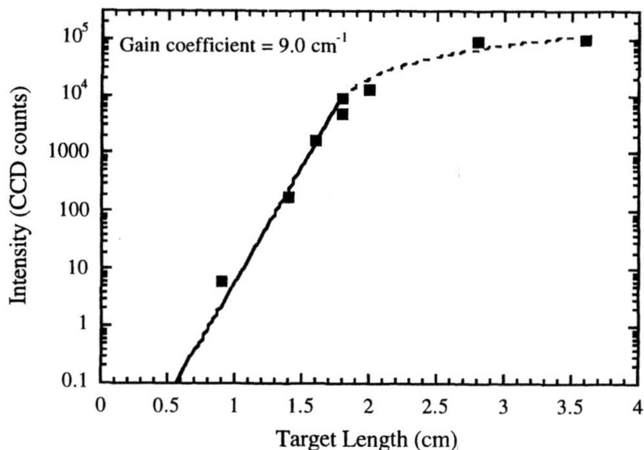

Figure 2. Ni-like In X-ray laser output at 126 Å as a function of target length for double 75 ps pulse irradiation. The light squares are for single targets, while the dark squares are for double targets of total length as plotted.

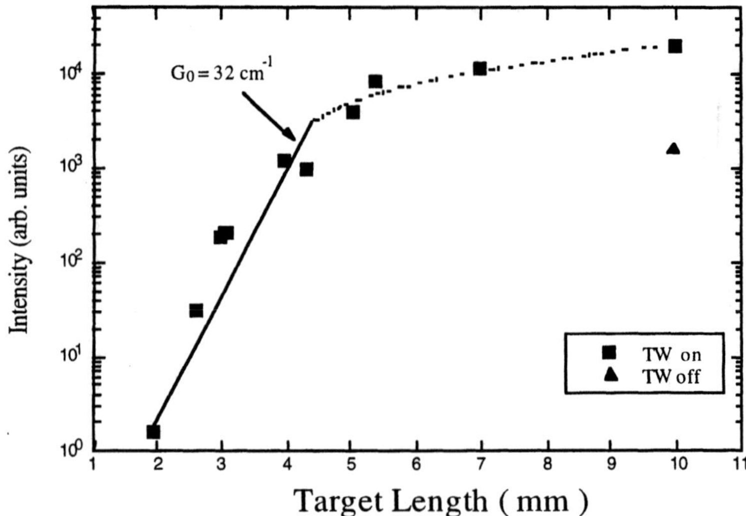

Figure 3. X-ray laser output for Ni-like Sn as a function of target length for pumping by a 2ps laser pulse at irradiance 10^{15} Wcm^{-2} incident 500 ps after of the peak irradiance (at 10^{12} Wcm^{-2}) of a 300 ps duration pulse. The ■ data points are for travelling wave pumping and show a gain coefficient of 32 cm^{-1} for short target lengths with saturation at a gain length product of 14-15. The ▲ data point is for the same irradiance conditions, but with the travelling wave pumping turned-off.

pumping configuration produced lasing on Ni-like ions of atomic number up to samarium (Z = 62), corresponding to a wavelength of 7.3 nm (see figure 4).

The 2 ps duration travelling wave pumping of Ne-like Ge has been modelled using the hydrodynamic and atomic physics code EHYBRID[9] and a ray tracing post-processor of the X-ray laser beam propagation[10] (see figure 5). The simulated X-ray laser outputs as a function of target length for various timing positions of the short 2 ps pulse relative to the peak of the long 300 ps pulse are shown. The experimental data points coincide with the 2ps pulse 150 ps early relative to the long pulse. It seems that a significantly higher experimental X-ray laser output could be obtained by optimising the relative timing between the 2 ps and 300 ps pumping pulses.

Simulations with EHYBRID and the ray tracing post-processor have elucidated the processes producing the enhanced X-ray laser output with multiple-pulse pumping[3,11]. A prepulse ensures that the gain during the main pumping pulse occurs over a large spatial region and that the density gradients in the gain region are flat. This enables the X-ray laser beam to propagate in the gain region for a long length of target and hence the X-ray laser output is maximised. It seems that the plasma degree of ionisation is not significantly enhanced with a prepulse[3].

CONCLUSIONS

Recent enhancements in X-ray laser output produced by multiple short pulse pumping of neon- and nickel-like ions at the Rutherford Appleton Laboratory have been reviewed. Pulses of 75 ps duration with a 20% prepulse, 2.2 ns earlier have produced saturated lasing on nickel-like ions down to 7 nm wavelength (in samarium). Short duration 2 ps pulses synchronised with long 600 ps pulses have produced saturated lasing down to 12 nm in nickel-like tin and we have observed a clear enhancement of X-ray laser output with travelling wave pumping where the 2 ps pulse is incident at different times along the lasing medium length to coincide with the amplifying X-ray laser pulse travelling at the speed of light. Simulation studies of the output of Ne-like Ge X-ray lasers agree well with experimental results.

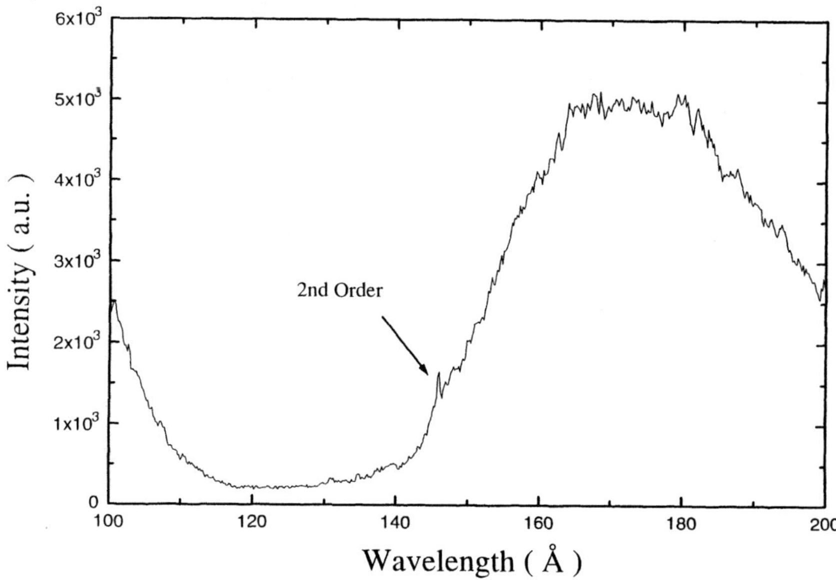

Figure 4. Spectrum of Ni-like Sm viewing axially along the target surface for 300 ps and superimposed 2 ps irradiation. The X-ray lasing line at 7.3 nm can be seen in the second order diffraction from the grating.

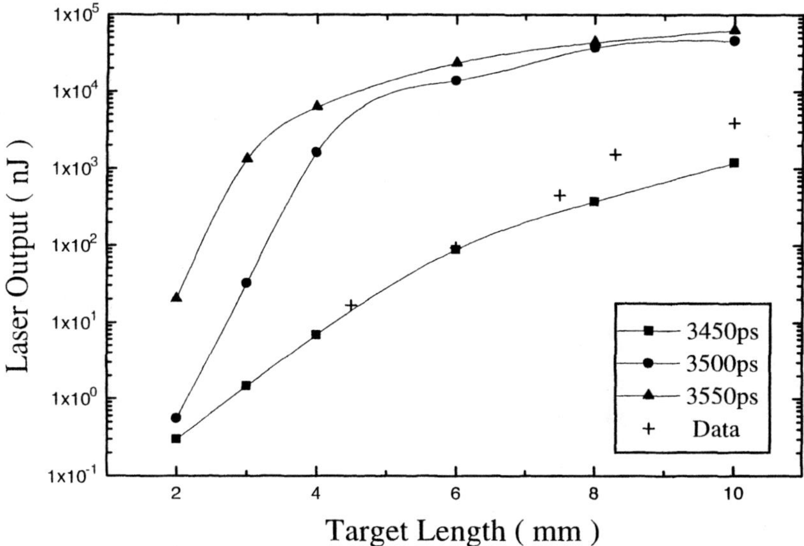

Figure 5. Simulation of the Ne-like Ge X-ray laser output as a function of target length for various times of short 2 ps pulse irradiance ($\approx 10^{15}$ Wcm^{-2}) relative to the peak irradiance ($\approx 10^{12}$ Wcm^{-2}) at 3600 ps of the 300 ps pulse. There is a 300 ps duration prepulse (of irradiance $\approx 10^{11}$ Wcm^{-2}) incident 3300 ps before the main 300 ps pulse Experimental data points (+) are also shown.

ACKNOWLEDGEMENTS

The support of United Kingdom Engineering and Physical Sciences Research Council grants GR/L 11540, 11946, 11779 and 11809 is gratefully acknowledged. We thank the laser and other staff at the Central Laser Facility, Rutherford Appleton Laboratory for their support of the research program.

REFERENCES

1.   A Carillon *et al Phys. Rev. Lett.* **68**, 2917 (1992).
2.   J Nilsen and J C Moreno *Phys. Rev. Lett.* **74**, 3376 (1995).
3.   A Behjat, J Lin, G J Tallents, A Demir, M Kurkcuoglu, C L S Lewis, A G MacPhee, S P McCabe, P J Warwick, D Neely, E Wolfrum, S B Healy and G J Pert 1996 *Optics Comm.* **135**, 49 (1997).
4.   J Zhang, A G MacPhee, J Nilsen, J Lin, T W Barbee, C Danson, M H Key, C L S Lewis, D Neely, R M N O'Rourke, G J Pert, R Smith, G J Tallents, J S Wark and E Wolfrum *Phys. Rev. Lett.* **78**, 3856 (1997).
5.   J Zhang, A G MacPhee, J Lin, E Wolfrum, R Smith, C Danson, M H Key, C L S Lewis, D Neely, J Nilsen, G J Pert, G J Tallents and J S Wark *Science* **276**, 1097 (1997).
6.   J Zhang, P J Warwick, E Wolfrum, M H Key, C Danson, A Demir, S Healy, D H Kalantar, N S Kim, C L S Lewis, J Lin, A G MacPhee, D Neely, J Nilsen, G J Pert, R Smith, G J Tallents and J S Wark *Phys. Rev. A* **54**, R4653 (1996).
7.   I N Ross *et al Appl. Optics* **26**, 1584 (1987).
8.   J Collier, C Johnson and C Mistry *Central Laser Facility report TR-96-066* , Rutherford Appleton Laboratory, Didcot (1997) p88.
9.   G J Pert 1983 *J Fluid Mech.* **131**, 401.
10.   P B Holden, S B Healy, M T M Lightbody, G J Pert, J A Plowes, A E Kingston, E Robertson, C L S Lewis and D Neely 1994 *J Phys B* **27**, 341.
11.   J Y Lin, G J Tallents, A Demir, S B Healy and G J Pert 1997 *J Appl Phys* submitted for publicaton.

SPECTROSCOPIC INVESTIGATIONS OF AN OPTICAL-FIELD-IONIZED X-RAY LASER WITH A MICROCAPILLARY TARGET

Katsumi Midorikawa and Yutaka Nagata

Laser Technology Laboratory,
The Institute of Physical and Chemical Research, RIKEN
2-1 Hirosawa, Wako-shi, Saitama 351-01, Japan

INTRODUCTION

The use of optical field ionization (OFI) for the production of plasma media of re-combination-pumped x-ray lasers introduces some favorable characteristics. The first advantage is that it is possible to produce population inversion with respect to the ground state of an ion,[1,2] because the OFI and the following three-body recombination processes are faster than radiative decay rates of the relevant resonance lines. This leads to a laser transition with a much shorter wavelength compared with that between excited states. The second advantage is that x-ray lasers by OFI could be operated more efficiently than ones pumped by conventional electron collisional ionization because of their high quantum efficiency and the controllability of plasma characteristics.

We previously reported the observation of gain on the 13.5-nm Lyman-α transition of H-like Li ions by OFI of a preformed Li^+ plasma.[3] The Li^+ preformed plasma was produced by irradiation of a 20-ns KrF laser onto a plane Li target. In this experiment, we obtained a small-signal gain of 20 cm^{-1}. The gain-length (GL) product of 4 with an interaction length of 2 mm was attained. However, further increase of the GL product was interrupted by propagation problem of the pump beam.

It is difficult to keep high intensity in high density plasma medium.[4,5] Since typical laser pulses have a Gaussian profile in time and in space, the electron density distribution produced by OFI becomes high at a center of the focused pulse and low at both edges. This electron density gradient in the transverse direction causes defocusing of the ionizing pulse. The ionization-induced self-defocusing is unavoidable for OFI x-ray lasers using the axial pumping geometry.

Recently, Suckewer et al.[7] proposed the use of a microcapillary filled with a preformed plasma to increase the gain length of the 13.5-nm Lyman-α laser. They reported that the microcapillary made it possible to create a lasing medium up to 5 mm long with a small signals gain of 11 cm^{-1}. Although the result indicated the possibility of producing a saturated output of the 13.5-nm OFI x-ray laser, the missing of some resonance lines in the observed spectra and a lack of details of the plasma condition in the microcapillary made it difficult to fully understand the results reported.

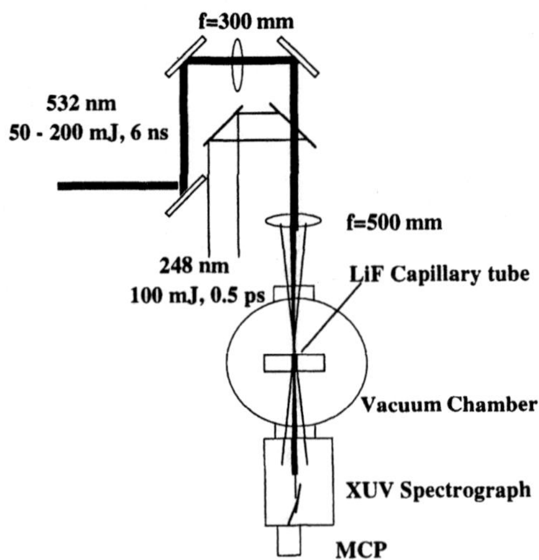

Fig. 1. Schematic diagram of the experimental setup.

In this paper, we report the results of spectroscopic investigation of a 13.5-nm OFI x-ray laser with a microcapillary target. The plasma condition in the microcapillary is discussed.

EXPERIMENTAL

Figure 1 shows the experimental setup for the measurement of time-integrated spectra. Capillaries used in this experiment are made of LiF and have a bore diameter of 300 μm. A 6-ns second harmonic pulse of a low intensity Q-switched Nd:YAG laser was focused at a few mm front of the microcapillary entrance with an f=500 mm lens. After a certain delay with respect to the Nd:YAG laser pulse, a linearly-polarized, high intensity KrF laser pulse with a pulse width of 500 fs was focused inside the microcapillary with the same lens. The focused laser intensity of the subpicosecond KrF laser was 10^{17} W/cm^2. The Nd:YAG laser was operated at 10 Hz, while the subpicosecond KrF was operated at 2.5 Hz.

The spectra from a microcapillary after irradiation of the subpicosecond KrF laser pulses were detected by using a flat-field grazing-incidence XUV spectrograph equipped with combination of a microchannel plate detector having a time-gate duration of 80 ns and a linear diode array. The time-integrated spectra were obtained by accumulating two shots.

TIME-INTEGRATED SPECTRA

We first optimized a delay time between the Nd:YAG and subpicosecond KrF laser pulses by monitoring the time-integrated spectral intensity at 13.5 nm. When the delay was between 500 and 1000 ns, the strong 13.5-nm line signal was observed with a 2-mm microcapillary. The 13.5-nm line intensity reached its maximum at 800 ns.

Time integrated spectra along an optical axis of a subpicosecond KrF laser were observed with 2-mm and 4-mm long microcapillary targets. A 2ω-YAG laser energy was 150 mJ. Observed spectra are shown in Figs. 2(a) and (b). For comparison, observed spectra with a LiF disk target for a plasma length of 2 mm is shown in Fig. 2(c). We observed the Lyman-α line at 13.5 nm, the Lyman-β at 11.4 nm and He-α at 19.9 nm for H-like and He-like lithium ions, respectively. In addition to these lithium-ion spectra, we also clearly

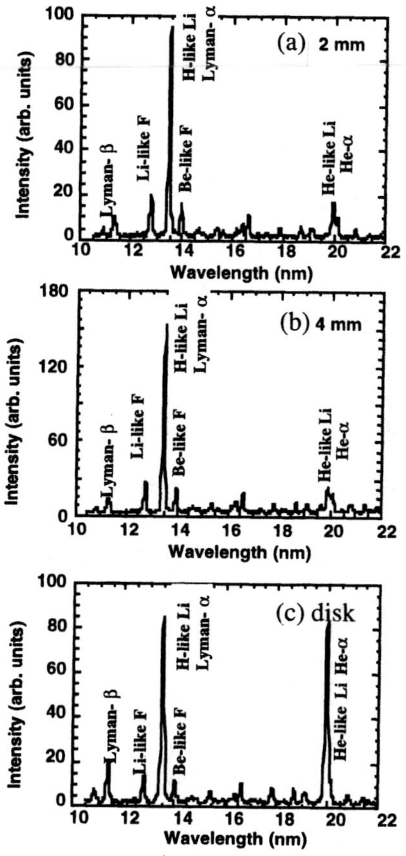

Fig. 2 Time integrated spectra from LiF targets irradiated with subpicosecond KrF pulses: (a) 2-mm microcapillary, (b) 4-mm microcapillary, and (c) flat disk.

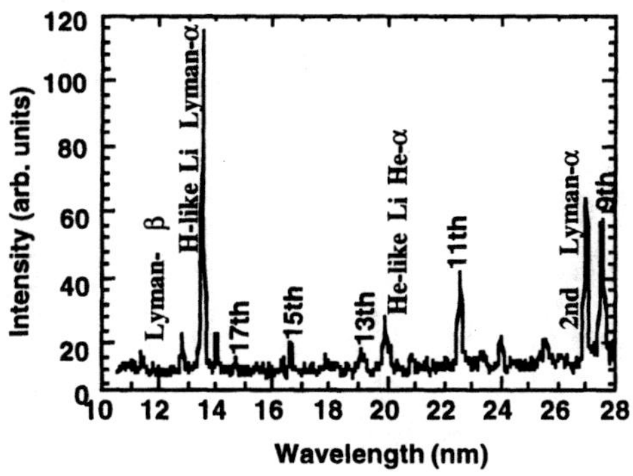

Fig. 3. Observed harmonic and line spectra with a 5-mm microcapillary target.

obtained the Li-like fluorine line (3p-2d) at 12.8 nm and the Be-like fluorine line (2s3d-2s2p) at 14.0 nm. This observation is reasonable, because the LiF microcapillary contains the same amount of lithium and fluorine atoms. However, these fluorine lines were not observed in Ref. 7. When we increased the microcapillary length from 2 mm to 4 mm, only the Lyman-α line intensity increased by a factor of more than two. Large intensity ratios of the Lyman-α to the Lyman-β and the He-α suggested the appearance of gain for the microcapillary targets. However, further increase of the microcapillary length led the 13.5-nm signal to the saturation. Similar intensity ratio to the 4-m microcapillary spectra was observed in the original experiment using a Li disk target,[3] in which a small signal gain of 20 cm^{-1} and a GL product of 4 were obtained.

OBSERVATION OF HIGH-ORDER HARMONICS

With a 5-mm microcapillary, we sometimes observed high-order harmonics as shown in Fig. 3, notwithstanding a 600 ns delay of a subpicosecond KrF laser pulse from a YAG laser pulse. When we used a flat LiF target, we could not observed high-order harmonic signals with a delay longer than 200 ns.[8] These results indicates that microcapillary structure plays a role in confining the preformed plasma, because the harmonics intensity is proportional to a square of plasma density. The observed intensity distribution of the harmonics is flatter than those reported previously.[8] Although the waveguiding effect was not confirmed for optical field ionization x-ray laser scheme, nonuniform electron density distribution of the confined plasma in the radial direction may change the propagation of the harmonics or phase matching condition.

DISCUSSION

In order to guide the pumping laser pulse, the preformed plasma created in a microcapillary has a minimum electron density on the center of axis with parabolic electron density distribution.[9] Such an electron distribution was achieved in a microcapillary discharge plasma.[10] Jackel et al.[11] also reported the waveguiding of ultrashort high intensity pulses by microcapillary tubes. As a practical aspect, however, it was too difficult to reproduce the OFI plasma spectra after the change of capillaries. This is partly due to the difficulty in beam alignment and also partly due to the slight difference of the inner surface condition of the targets employed.

When we slightly tilted the view angle of the spectrograph with a 5-mm microcapillary, only the 13.5-nm Lyman-α line was observed as shown in Fig. 4. In this experiment, no

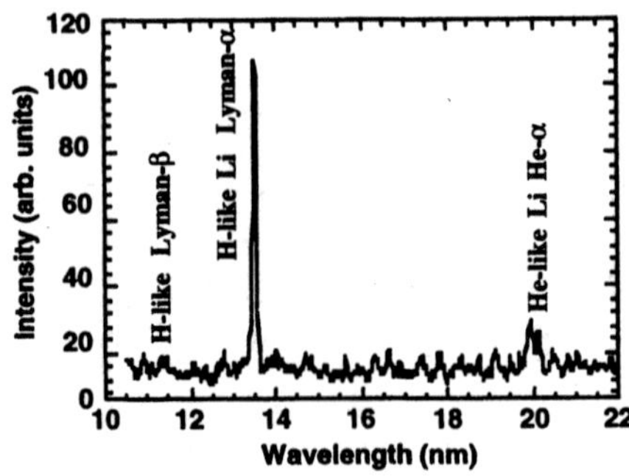

Fig. 4. Observed spectra with a 5-mm microcapillary target. Observed angle was slightly tilted.

fluorine-ion lines were observed. Although the observed spectra were very promising, we could not confirm the appearance of gain by changing the microcapillary length, because of a lack of the reproducibility. Therefore, we could not obtain the waveguiding effects of the microcapillary target.

REFERENCES

1. N. H. Burnett and P. B. Corkum, Cold-plasma production for recombination extreme-ultraviolet lasers by optical-field-induced ionization, J. *Opt. Soc. Am B* 6: 1195(1989).
2. N. H. Burnett and G. D. Enright, Population inversion in the recombination of optically-ionized plasmas, *IEEE J. Quantum Electron.* 26: 1797(1990).
3 Y. Nagata, K. Midorikawa, S. Kubodera, M. Obara, H. Tashiro, and K. Toyoda, Soft-x-ray amplification of the Lyman-α transition by optical-field-induced ionization, *Phys. Rev. Lett.* 71: 3774(1993).
4 P. Monot, T. Auguste, L. A. Lompre, G. Mainfray, and C. Manus, Focusing limits of a terawatt laser in an underdense plasma, *J. Opt. Soc. Am. B* 9: 1579(1992).
5 Y. M. Li, J. N. Broughton, R. Fedosejevs, and T. Tomie, Formation of plasma columns in atmospheric pressure gases by picosecond KrF laser pulses, *Opt. Commun.* 93: 366(1992).
6. C. D. Decker, D. C. Eder, and R. A. London, Ionization-induced refraction in recombination x-ray lasers, *Phys. Plasmas* 3: 414(1996).
7 D. V. Korobkin, C. H. Nam, and S. Suckewer, Demonstration of soft x-ray lasing to ground state in Li *HI, Phys. Rev. Lett.* 77: 5206(1996).
8 S. Kubodera, Y. Nagata, Y. Akiyama, K. Midorikawa, M. Obara, H. Tashiro, and K. Toyoda, High-order harmonic generation in laser-produced ions, *Phys. Rev. A* 48: 4576(1993).
9. P. Sprangle and E. Esarey, Interaction of ultrahigh laser fields with beams and plasmas, *Phys. Fluids B* 4: 2241(1992).
10. Y. Ehrlich, C. Cohen, A. Zigler, J. Krall. S. Sprangle, and E. Esarey, Guiding of high intensity laser pulses in straight and curved plasma channel experiments, *Phy. Rev. Lett.* 77: 4186(1996).
11. S. Jackel, R. Burris, J. Grun, A. Ting, C. Manka, K. Evans, and J. Kosakowskii, Channeling of terawatt laser pulses by use of hollow waveguides, *Opt. Lett.* 10: 1086(1995).

GUIDED-WAVE OPTICAL PARAMETRIC AMPLIFICATION IN GASES: A NOVEL PHASE-MATCHING TECHNIQUE FOR ULTRAFAST PULSES

Charles G. Durfee III, Sterling Backus, Margaret M. Murnane and
Henry C. Kapteyn

Center for Ultrafast Optical Science
University of Michigan
Ann Arbor, MI 48109-2099

INTRODUCTION

Despite much recent progress in developing ultrafast laser systems with pulse duration less than 25 fs in the near-infrared[1], many applications in solid state physics, surface science, biophysics, and chemistry[2] require ultrafast pulses with energies $>\mu J$ in the ultraviolet (UV). In the past, third harmonic generation (THG) by sum-frequency mixing in nonlinear crystals has been used to generate light in the deep ultraviolet[3]. While this technique can result in conversion efficiencies of several percent, it suffers from pulse broadening due to group velocity walkoff in the medium. The resulting pulse duration of 200 fs is too long to monitor reaction dynamics in many cases, particularly in small molecules and organic species. THG by focusing high-intensity infrared pulses in a gas[4,5] does not suffer from severe temporal walkoff, and thus can be used to generate very short-duration pulses. In previous work[4] we demonstrated that by focusing 20 fs pulses in air or noble gases, third-harmonic light at 270 nm with 14 fs pulse duration can be generated. However, in this case, the conversion efficiency ($\sim 0.1\%$) is relatively low due to the short interaction length and poor phase matching.

Recently, we presented the first experimental demonstration of guided-wave nonresonant phase-matched parametric amplification in gases[6]. This technique preserves high conversion efficiency and good output beam quality, while maintaining a short pulsewidth. Broad-bandwidth ultrafast pulses, tunable around 270 nm, were generated from an ultrafast Ti:sapphire amplifier system using $2\omega + 2\omega - \omega$ parametric wave mixing in a noble-gas filled capillary waveguide. For a beam propagating in a gas-filled capillary waveguide, the mode-dependent contribution to the propagation phase is decoupled from the material dispersion, allowing the two to be balanced. Both the fundamental and the second-harmonic light were coupled into the near-Gaussian, lowest-order (EH_{11}) mode. Phase-matching for signal in the lowest-order mode was then achieved by tuning the gas pressure. The conversion efficiency from the pump is greater than 13%, giving output pulses with an energy >4 µJ at a 1 kHz repetition rate. Since the process is nonresonantly phase-matched, it is suitable for the wide bandwidth of femtosecond laser pulses, and can be applied to a wide range of input and output wavelengths to provide a tunable source throughout the visible, uv and vacuum ultraviolet (VUV). At high pump intensity, the output spectral bandwidth is increased by cross-phase modulation, which should allow compression to sub-5 fs in the UV. Moreover, the technique may be applied to other parametric combinations for higher-order frequency conversion[7] in the VUV and extreme-ultraviolet (EUV).

Previous efforts to phase-match frequency conversion in gases have made use of narrow spectral regions in vapors[8] or gases[9,10] that are negatively dispersive to compensate for the positive dispersion of the active gas species as well as the focusing phase shift[11,12]. Near-resonant phase matching with a capillary waveguide has been proposed[13] and tested[14], but molecular absorption limited the efficiency ($<10^{-3}$). More important, the intrinsic bandwidth limitations of this technique limit its usefulness for frequency conversion of ultrashort pulses. In other work, difference frequency mixing in a plasma waveguide has been proposed[7], in which the modal properties of the plasma waveguide facilitate phase matching. However, a second laser is required to make the plasma channel. In this work, we guide the beams with a glass capillary waveguides, which have been used extensively for Raman generation[15] (a self-phase matched process) and have recently proven very useful in the compression of femtosecond optical pulses[16,17]. Propagation in an optical waveguide offers several options for phase matching parametric processes, such as mixing of light propagating at different frequencies and/or in different spatial modes[7,18,19]. Although the importance of the waveguide phase in high-order Raman Stokes generation has been demonstrated[20], the scheme presented here is the first demonstration that the gas pressure may be used to tune the phase matching condition for optical parametric amplification. As will be shown, the use of a gas-filled capillary for frequency conversion has many advantages: atomic gases with high ionization potential may be used, and the length and gas pressure may be controlled easily, making the scheme easy to scale and implement.

PHASE MATCHING IN CAPILLARY WAVEGUIDES

Since it counters diffraction of the wave, a waveguide adds a geometrical component to the wavevector. The propagation constant for a capillary filled with a medium of index n_g is given by -

$$k = \frac{2\pi \, n_g(\lambda)}{\lambda}\left[1 - \frac{1}{2}\left(\frac{u_{nm} \, \lambda}{2\pi \, a}\right)^2\right] \tag{1}$$

where a is the core radius, and u_{nm} is the modal constant[21]. Although the core/cladding index ratio is less than unity, Fresnel reflections at the gas-glass interface allow lossy guiding of optical beams. Consider the general process of difference frequency mixing between two pump beams in an optical fiber at central frequencies ω_1 and ω_2. The phase mismatch for difference frequency mixing, $\omega_3 = N\omega_2 - M\omega_1$, where $\omega_2 = R\omega_1$, $\Delta k = Nk_2 - Mk_1 - k_3$. Δk may be written as (where the index $n_g = 1 + p\delta_g$, and p is the gas pressure) -

$$\Delta k = \Delta k_{mode} - \Delta k_{mat}$$

$$= \frac{\lambda_1}{4\pi \, a^2}\left(\frac{u_3^2}{RN-M}+Mu_1^2 - \frac{Nu_2^2}{R}\right) - \frac{2\pi \, p}{\lambda_1}\left((RN-M)\delta_3 + M\delta_1 - N\delta_2\right) \quad (2)$$

The phase mismatch results from a modal dispersion term ($\Delta k_{mode} \propto 1/pa^2$) minus a material dispersion term ($\Delta k_{mat} \propto p$). In gases (or plasmas) with normal dispersion, $\Delta k_{mat} > 0$; if the modes of the pump and signal are chosen such that $\Delta k_{mode} > 0$, there will exist an optimum pressure, p_{opt}, for which $\Delta k = 0$. This is also possible for harmonic generation ($R = 1, M = 0$), providing the output is in a higher-order spatial mode such that $u_3/N > u_1$. The phase matching inherently takes place over a wide bandwidth, since the modal dispersion can balance the positive gas dispersion in a spectral region where they are both slowly varying.

EXPERIMENTAL SETUP AND RESULTS

In the first experiment, the fundamental was coupled to the capillary and phase-matched third harmonic light ($N = 3$) was generated in high-order modes. With the fundamental in the lowest order EH_{11} mode ($u_{11} = 2.405$), the lowest order mode that satisfies the phase matching condition is the EH_{13} mode ($u_{13} = 8.654$). Output to this mode should be phase matched at an optimum pressure (using Eqn 2 and the the dispersion formulae given by Dalgarno and Kingston[22]) of $p_{opt}= 61$ Torr. This pressure tuning was accomplished by mounting the capillary in a cell with 1.6 mm thick fused silica windows that could be evacuated and pressurized. The output beam of a kHz Ti:sapphire multipass amplifier system[23] capable of producing 4 µJ in 20 fs was focused at the entrance of a capillary (153 µm core diameter, 30 cm long) with 46% throughput. The throughput of a HeNe laser beam was measured to be 80%, very close to the calculated value of 83%. Thus, imperfect mode matching of the fundamental is responsible for the lower than expected throughput. The UV output of the fiber was separated from the fundamental using dichroic mirrors centered at 270 nm, with a bandwidth of 30 nm. The beam was then directed into a CCD spectrograph or, after passing through a fused silica prism, a calibrated photodiode power meter (Newport 818-UV detector).

Figure 1 shows the 3ω signal vs. the pressure of krypton in the cell. The output energy had a clear maximum at 55 Torr, in good agreement with the value calculated for phase

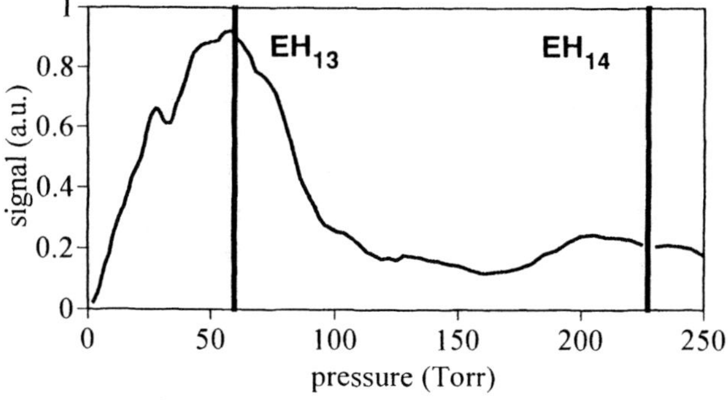

Figure 1: Third harmonic energy versus krypton gas pressure. The reference line shows the calculated optimum pressures for output into the EH_{13} and EH_{14} modes

matching to the EH$_{13}$ mode. Perturbation theory for Gaussian pulses predicts a spectral width of $\Delta\lambda_3 = \Delta\lambda_1/n^{3/2}$; for $\Delta\lambda_1 = 40$ nm, $\Delta\lambda_3 = 8$ nm. The measured bandwidth was also 8 nm, indicating that the process is clearly phase matched over the whole bandwidth. For 250 μJ input, the output UV pulse was measured to be 0.28 μJ, giving an efficiency of 0.2% (assuming 145 μJ actually enters the waveguide), which compares favorably with the case of a focus in air[4].

Unlike harmonic conversion in unbounded media, the overlap of the signal mode in a waveguide with the induced nonlinear polarization is an important factor in the efficiency. Despite good phase matching, the above harmonic conversion experiment was limited by the poor overlap of the EH$_{13}$ mode with the cube of the EH$_{11}$ mode profile. Conversion to the EH$_{11}$ would result in more than a factor of 15 in gain. The restriction on the output mode may be removed by performing sum-difference frequency generation. For the particular case of $N = 2$, $M = 1$, and $u_1 = u_2$, $\Delta k_{mode} > 0$ for any signal mode. In the present experiment, second-harmonic light is used as the pump and the fundamental as the idler ($3\omega = 2\omega + 2\omega - \omega$). With these beams coupled to the lowest order mode, the output is also in the lowest order mode. For this experiment, a 100 μm LBO crystal was used to frequency double the 800 nm light, with 20% conversion efficiency. The two colors were then separated, and later recombined with a relative time delay, before being focused into the capillary (127 μm core diameter, 60 cm long). The divergence of the input telescope was adjusted to optimize the coupling of the 400 nm light (35% throughput); the fundamental beam was not as well optimized (<10% throughput).

Figures 2(a) and (b) shows the 3ω output energy versus argon and krypton gas pressure. The positions of the peaks are in excellent agreement with the values calculated

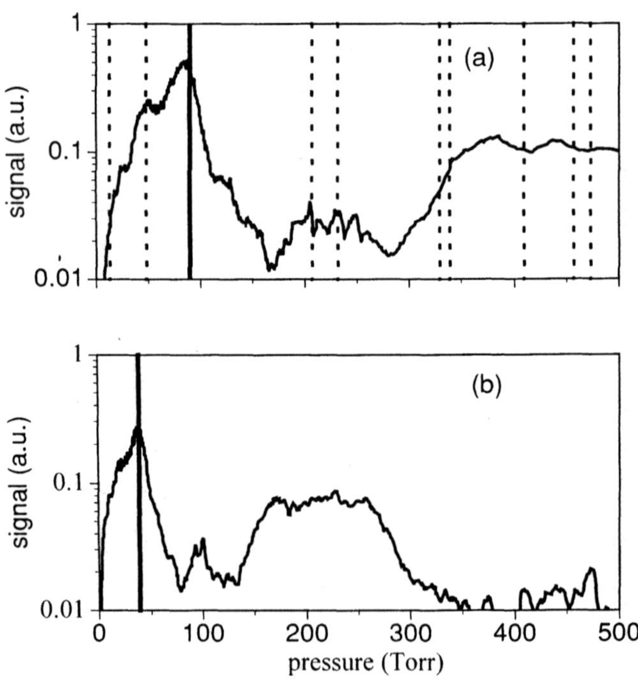

Figure 2: Strength of signal beam (near third-harmonic) versus argon (a) and krypton (b) gas pressure. The reference lines show the calculated optimum pressures for coupling among lowest-order modes (solid) and among various low-order modes (dashed).

from Eqn. 3. The measured (calculated) output energy shows a clear maximum at 85.1 (89.7) Torr for argon and 38.9 (39.2) Torr for krypton, where the phase matching was optimized for conversion to the lowest order (EH_{11}) mode. The output mode is very nearly Gaussian. The pressure tuning curves also show low secondary peaks that correspond to conversion with other combinations of spatial modes which are not well overlapped. The structure of the peaks for krypton is similar to argon, scaled according to the respective dispersion characteristics of the gases. With 30 µJ at 2ω and 64 µJ at ω leaving the fiber, the output energy at 3ω was 4 µJ, giving 13% conversion from 2ω. The larger conversion efficiency in this case allowed us to use a standard thermal power meter rather than a photodiode. It was found that when the laser compressor was positioned for maximum 2ω yield there was a positive chirp on the fundamental and the second harmonic pulses. Dispersion in the optics leading to the capillary adds to this positive chirp. The 2ω beam passed through a 1.6 mm window, resulting in broadening to ~32 fs. In addition to the input window, the fundamental passed through a waveplate, lenses and two dichroic mirrors, resulting in a broadening to ~70 fs. A more sophisticated setup could pre-compensate for the pulse broadening in the input optics, and give improved output efficiency. When the pump beam precedes the idler, the short wavelength part of the blue spectrum is mixed with the long wavelength part of the IR spectrum, resulting in UV output that is shifted to short wavelengths. As the delay is reduced, the situation is reversed, and the UV output is tuned in this manner over 8nm.

Figure 3 shows the UV output spectrum with the compressor adjusted for maximum SHG (as above) and also for shorter pump pulses. The FWHM of the spectra are 8.5 nm and 26 nm, respectively. The short pump pulses induce both self- and cross-phase modulation, resulting in an ultrabroadband UV signal. The transform limit of the spectra shown (assuming flat spectral phase) are 9.5 fs and 4.5 fs, respectively. These pulses should be compressible because of the small value of the group velocity walkoff (5 fs).

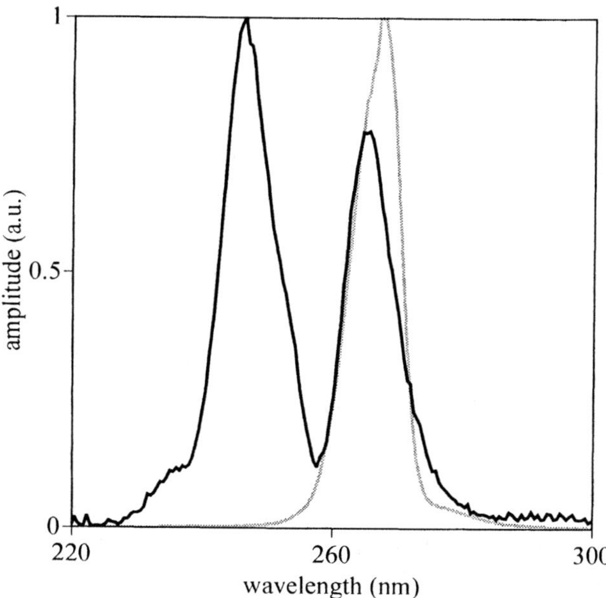

Figure 3. Spectrum of the signal with chirped (gray) and short (black) input pulses.

DISCUSSION

With several straightforward improvements, it should be possible to increase the conversion efficiency and output energy of this OPG technique. As discussed above, the conversion efficiency is governed primarily by the intensities of the pump and idler in the waveguide, rather than by the waveguide dimensions. Thus improvements in the coupling efficiency, fiber transmission, and dispersion compensation of the pump and idler will result in greatly improved efficiency and yield. In conclusion, we have demonstrated that nonresonant phase-matched frequency conversion in gases can generate tunable ultrafast pulses with several μJ of energy at 270 nm, in a near-Gaussian (EH_{11}) spatial mode. The output bandwidths should support a sub-5 fs pulse in the UV. With several straightforward improvements in the coupling efficiency, fiber transmission, and dispersion compensation of the pump and idler, it should be possible to increase the conversion efficiency and output energy. Preliminary results indicate that the output of a crystal OPA used as the idler in the gas OPA allows the signal to be tuned throughout the deep UV (217-305 nm). Finally, while in press, we have extended this technique to higher-order frequency conversion to provide a new source of ultrashort pulses in the VUV and EUV[24].

ACKNOWLEDGEMENTS

The authors gratefully acknowledge support for this work from the Department of Energy and the National Science Foundation. H. Kapteyn acknowledges generous support from an Alfred P. Sloan Foundation Fellowship.

REFERENCES

1.   S. Backus, C. Durfee, M. M. Murnane, H. C. Kapteyn, Rev. Sci. Instrum. **69**, 1207 (1998).

2.   P. Barbara, J. Fujimoto, W. Knox, W. Zinth, *Ultrafast Phenomena X* (Springer-Verlag, Berlin, 1996).

3.   J. Ringling, O. Kittelmann, F. Noack, G. Korn, J. Squier, Opt. Lett. **18**, 2035 (1993).

4.   S. Backus, J. Peatross, M. Murnane, H. Kapteyn, Opt. Lett. **21**, 665 (1996).

5.   C. W. Siders, N. C. Turner. III, M. C. Downer,, *et al*, JOSA B **12**, 330 (1996).

6.   C. G. Durfee, S. Backus, M. M. Murnane, H. C. Kapteyn, Opt. Lett. **22**, 1565 (1997).

7.   H. Milchberg, C. G. D. III, T. McIlrath, Phys. Rev. Lett. **75**, 2494 (1995).

8.   T. Jong, G. Bjorkland, A. Kung, R. Miles, S. E. Harris, Phys. Rev. Lett. **27**, 1551 (1971).

9.   J. F. Reintjes, *Nonlinear Optical Parametric Processes in Liquids and Gases* (Academic Press, 1984).

10. T. Chu, A. Bouvier, A. J. Bouvier, R. Fischer, Journal de Physique France **49**, 1725 (1988).

11. J. F. Ward, G. H. C. New, Physical Review **185**, 57 (1969).

12. G. C. Bjorklund, IEEE J. Quantum Electron. **QE-11**, 287 (1975).

13. V. G. Arkhipkin, Y. I. Heller, A. K. Popov, A. S. Provorov, Applied Physics B **37**, 93 (1985).

14. M. Castillejo, J. Y. Zhou, M. H. R. Hutchinson, Applied Physics B **45**, 293 (1988).

15. P. Rabinowitz, A. Kaldor, R. Brickman, W. Schmidt, Applied Optics **15**, 2005 (1976).

16. M. Nisoli, S. De Silvestri, O. Svelto, Appl. Phys. Lett. **68**, 2793 (1996).

17. M. Nisoli, S. D. Silvestri, O. Svelto, *et al*, Opt. Lett. **22**, 522 (1997).

18. R. H. Stolen, J. E. Bjorkholm, IEEE J. Quantum Electron. **QE-18**, 1062 (1982).

19. G. P. Agrawal, *Nonlinear Fiber Optics* (Academic Press, 2nd Ed., San Diego, 1995).

20. W. Heuer, H. Zacharias, IEEE J. Quantum Electron. **24**, 2087 (1988).

21. E. A. J. Marcateli, R. A. Schmeltzer, Bell Systems Technical Journal **43**, 1783 (1964).

22. A. Dalgarno, A. E. Kingston, Proc. Roy. Soc. A **259**, 424 (1966).

23. S. Backus, C. Durfee, G. Mourou, H. C. Kapteyn, M. M. Murnane, Opt. Lett. **22**, 1256 (1997).

24. A. Rundquist, C. Durfee, Z. Chang, C. Herne, S. Backus, M.M. Murnane, H. C. Kapteyn, Science, **280,** 1412 (1998).

INCREASED COHERENCE LENGTH IN HIGH-ORDER HARMONIC GENERATION BY A SELF-GUIDED BEAM

H.R. Lange,[1] A.A. Chiron,[2] A. Bouhal,[1] J.-F. Ripoche,[1] P. Breger,[3] P. Agostini,[3] and A. Mysyrowicz[1]

[1]Laboratoire d'Optique Appliquée, Centre National de la Recherche Scientifique URA 1406, ENSTA-Ecole Polytechnique, F-91761 Palaiseau
[2]Dipartimento di Fisica, Universita di Pisa, Piazza Torrecelli 2, I - 56100 Pisa, Italy
[3]Servie des Photons, Atomes et Molecules, Centre d'Etudes de Saclay, F- 91191 Gif sur Yvette

INTRODUCTION: OPTIMISATION OF HIGHER HARMONIC GENERATION

The increase of the yield of high-order harmonic generation in gases is a topic of current interest in the search of high-brightness UVX sources. The yield depends on the pump intensity, the jet density, the nonlinear polarizability of the target and the coherence length. The first three parameters have been thoroughly investigated and optimized : the intensity is limited by the saturation of ionisation[1], the jet density is limited by the defocusing effects of the free electrons[2] and the rare gases offer, to the present knowledge, the most efficient targets[3] . The coherence length is limited by three different phase-mismatch factors: the medium dispersion, the intensity-dependent atomic phase and the phase-shift due to the focusing geometry (the Goüy phase) normally used to achieve the required intensity. While little can be gained from the weak dispersion of relatively low density gases, the suppression of the focusing geometry is potentially an important source of improvement of harmonic generation that would remove the Goüy phase itself, and also cancel the atomic phase longitudinal gradient. The ideal plane-wave configuration can be approached by different ways. In this contribution we investigate harmonic generation by self-guided beams which can propagate over distances of several tenth of meters [4-7].

In a self-focusing beam, the radial variation of the intensity in combination with the intensity dependent refractive index introduces a phase front curvature in the medium which is equivalent to a lens. As a consequence, the laser pulse focuses itself on the axis of rotational symmetry. This focusing process is eventually counterbalanced by a negative refractive index change introduced by ionisation of the propagation medium.

The compensation of positive Kerr index and negative ionisation index leads to quasi-planar wave fronts and eliminates ideally the phase mismatch due to the Goüy phase.
We present here preliminary results from both numerical simulations and experiment of harmonic generation by a self-guided beam. These include generation of up to the fifteenth harmonic in Xenon, as well as the yield dependence on the gas pressure and the length of the gas medium.

NUMERICAL SIMULATIONS

In order to characterise the properties of the laser pulses we have studied the self-guiding mode numerically and experimentally. A 3-dimensional numerical propagation code has been developed for the simulation of the nonlinear propagation of femtosecond laser pulses. The algorithm is based on the non-linear Schrödinger equation for the normalised scalar complex electrical field propagating in the paraxial wave approximation for slowly varying envelopes. A cylindrical transverse geometry was used in the calculations. The nonlinearity of the medium consists of the time-dependent Kerr effect as well as multiphoton ionisation. Above $I = 10^{14}$ W/cm^2 multiphoton ionisation is substituted by a tunnelling regime. The propagation equation was solved through a finite difference scheme based on the method of Crank-Nicholson.

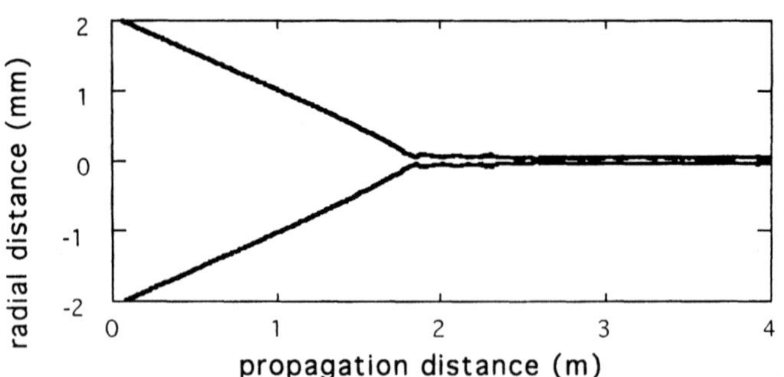

Figure 1. Simulated diameter at full width at half maximum as function of the propagation distance z for a f = 2 m lens.

We have chosen the nonlinearity as an instantaneous response n_2=2.8 10^{-23} W/cm^2 and a retarded response proportional to $n_{2,t}$ = 10^{-23} W/cm^2 in the form of a single-sided exponentially decaying function with a decay time of 70 fs corresponding to the nonlinear refractive index in air. The initial pulse parameters are 120 fs, 1.5 mJ initial pulse energy and a super-gaussian beam profile (power N=6) in order to account for the experimental top hat spatial structure introduced by a diaphragm placed in the laser beam. The simulation yields a stable filamentary mode of 160 mm diameter at full width at half maximum. Figure 1 shows the resulting simulation of the self-guided mode. The laser beam coming from the right is focused with a f=2 m lens. The Kerr nonlinearity leads to a self-focusing prior to the linear focal spot at 2m. The energy channelled into the self-guided mode is typically 10 percent of the initial laser energy. The simulations give a peak intensity of 4x10^{13} W/cm^2 on axis.

HIGH-ORDER HARMONIC GENERATION SET-UP

In the experiment, the initial pulses stem from a Ti:Sapphire chirped pulse amplification laser chain. The pulse energy W_0 after a diaphragm of 7 mm diameter is approximately 1.5 mJ, the pulse duration about 120 fs and the centre wavelength 800 nm. The measured diameter of the self-guiding mode is 380 μm - we attribute the larger diameter compared to the simulations to inhomogeneities in the spatial intensity distribution. Values for the coherence lengths for typical parameters of the experiment presented in this article - 100 Torr pressure of Xenon for the fifth harmonic assuming an electron density of $n_e = 3.0 \cdot 10^{14} cm^{-3}$ (estimated for Xe at an intensity of $4.\times10^{13} W/cm^2$) - give the coherence length $L_{coh} \cong 1m$ due to electronic dispersion, $L_{coh} \cong 7.6mm$ due to atomic dispersion and $L_{coh} = 30mm$ due to a geometric phase mismatch based on the assumption that the filament propagates as a Gaussian beam in the low pressure gas cell.

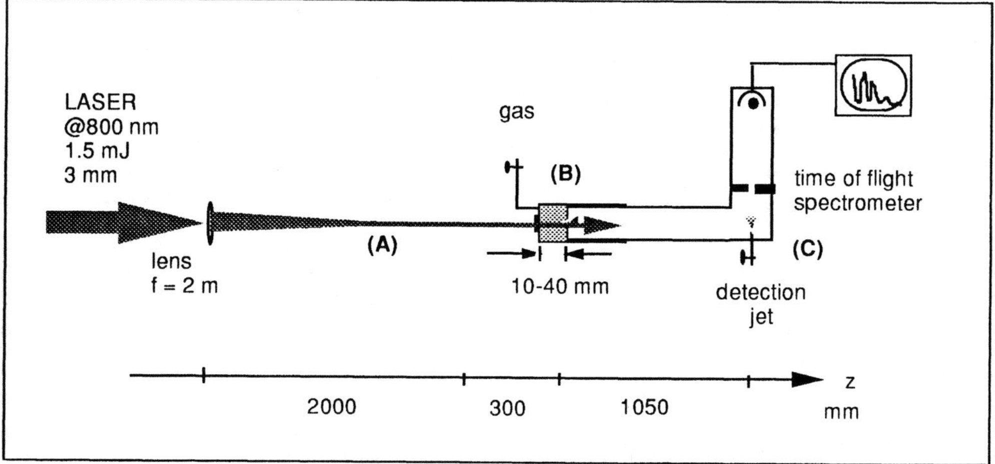

Figure 2. Experimental set up of the experiment for higher harmonic generation with a filament.

Figure 2 shows the experimental set up. A diaphragm in the laser beam reduces the shot to shot fluctuations in the position of the self-guided beam. The self-guided mode is generated in air with a f=2m lens and passes 30 cm after the geometric focal spot through a thin window into the gas cell. The thin glass window leads to a small reduction in diameter of the filament, but does not destroy the stable self-guided propagation. The gas cell is of variable length (10-45 mm). The pressure can be varied between a few mbar and normal pressure. After the passage through the gas cell, the light enters through a 500 μm aperture in the vacuum cell which is at a pressure inferior to 10^{-3} mbar. After a propagation over 1050 mm in vacuum the light pulse hits a gas jet. The ionized electrons are detected and resolved with respect to their kinetic energy in a time of flight spectrometer of a length of 2m. The distance of 1050 mm propagation in vacuum is sufficient to diffract the fundamental - in absence of gas in the first gas cell no higher harmonic signal could be detected. The higher harmonic signal diffracts less due to the shorter wavelength.

The detected photo-electron energy spectrum has been accumulated over 1000 shots. Figure 3 (a) displays the higher harmonic spectrum generated and detected with Xenon.

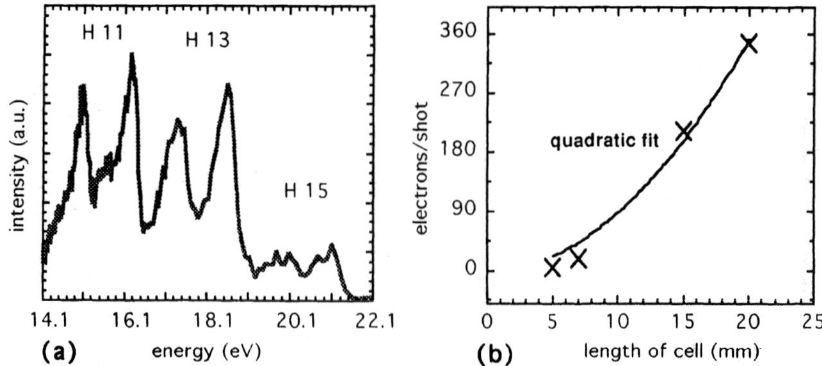

Figure 3. Higher harmonics in Xenon: (a) cut off in the energy spectrum of photo-electrons and (b) signal intensity as a function of cell length.

One clearly sees the eleventh, thirteenth and fifteenth order. The double peak structure is due to the presence of the two spin states S=1/2 and S=3/2. The photo-electron spectra from higher harmonic generation can be used to estimate the peak intensity of a pulse which can be related to the cut off energy of the highest order harmonic observed in the spectrum. The highest order of the harmonics is q=15.

The estimate of intensity has been obtained from calculations of dipole strengths below the tunnel ionisation regime[8] made at a slightly different wavelength of 1064 nm (in our experiment λ=800 nm). The intensity estimate yields $I = (2.5 \pm 0.5)10^{13}\, W/cm^2$. As the laser intensity is just below the tunnel ionisation regime, one can obtain a second estimate within the tunnel ionisation regime. The harmonic of highest order is related to the laser intensity by[9]: $I_{laser} = (qh\nu - I_p)/(3.2\lambda^2)$ where q is the highest observed harmonic order, hν the single photon energy of the fundamental, I_p the ionisation potential of Xenon in our experiment and λ the wavelength of the fundamental. This results in an estimate of the peak intensity in the filament of 6×10^{13} W/cm^2. Figure 3 (b) shows the variation of the signal with the length of the cell and a quadratic fit to the experimental data. These first results show a super-linear growth in the integrated signal of all harmonics indicating that the measured coherence length of 20 mm has been successfully increased with respect to the coherence length in the lens case in the order of a one mm.

CONCLUSION

We have shown to the first time to our knowledge the generation of higher harmonics in a self-guided beam. The experiments show the successful elimination of the geometric phase matching limit. In the chosen configuration the generation of higher harmonics is limited by the atomic dispersion.

NOTE ADDED IN PROOF

Recently we have observed further evidence of higher harmonic generation with filaments for the case of lower order harmonics generated in a filament. We interpret the observed periodic oscillations with cell length and pressure as Maker fringes with a coherence length of 10 mm for the fifth harmonic. This gives a further indication of plane wave phase fronts in the self-guiding mode of intense femtosecond laser beams.

ACKNOWLEDGEMENT

We thank Philippe Balcou and Eric Constant for illuminating discussions. H. R. Lange gratefully acknowledges the support by a Marie-Curie-Fellowship of the European Community (grant ERBFMBICT950065).

REFERENCES

[1]   A. l'Huillier, L.-A. Lompré, G. Mainfray and C. Manus: *Higher Harmonic Generation in Rare Gases*, in: Milhai Gavrila (ed): Atoms in Intense Laser Fields, Academic Press Inc., New York 1992, p.187

[2]   C. Altucci, T. Starczewski, E. Mevel, and C.-G. Wahlström: *Influence of atomic density in high-order harmonic generation*, J. Opt. Soc. Am. B 13, 148 (1996)

[3]   C. Lynga, A. l'Huillier and C.-G. Wahlström: *High-order Harmonic Generation in Molecular Gases*, J. Phys.B 29, 3293 (1996)

[4]   J. H. Marburger: *Self-Focusing: Theory*, Prog. Quantum Electron. 4, 35 (1975).

[5]   A. Braun, G. Korn, X. Liu, D. Du, J. Squier, and G. Mourou: *Self-channelling of high-peak-power femtosecond laser pulses in air*, Opt. Lett. 20, 73 (1995).

[6]   E. T. J. Nibbering, P. F. Curley, G. Grillon, B. S. Prade, M. A. Franco, F. Salin, and A. Mysyrowicz: *Conical Emission from self-guided femtosecond laser pulses in air*, Opt. Lett. 21, 62 (1996).

[7]   H.R.Lange, A.Chiron, E.T.J.Nibbering, G.Grillon, J.-F.Ripoche, M.A.Franco, B.Lamouroux, B.S.Prade and A.Mysyrowicz: *Anomalous long range propagation of femtosecond laser pulses through air: moving focus or pulse selfguiding?*, accepted for publication in Opt. Lett.

[8]   A. l'Huillier, PH. Balcou, S. Candel, K.J. Schafer and K.J. Kulander: *Calculation of high-order harmonic generation processes in Xenon at 1064 nm*, Phys. Rev. A. 46 2778 1992

[9]   P.B. Corkum: *Plasma Perspectives on Strong-Field Multiphoton Ionisation*, Phys.Rev.Lett. 73, 1995 (1993)

ACCELERATOR BASED SOURCE DEVELOPMENT: HIGHER, WIDER AND SHORTER

Erik D. Johnson
Brookhaven National Laboratory, National Synchrotron Light Source
Upton, Long Island New York, 11973-5000

ABSTRACT

A host of new sources are currently being pursued throughout the world which push the performance boundaries **higher, wider** and **shorter** than thought possible only a few years ago for accelerator based technologies. Various free electron laser configurations open the window to high power, sub-picosecond synchrotron radiation sources over a wide range of wavelengths. Storage ring sources still have some room for improvement within reason, and intense pulsed x-ray sources by Thomson scattering are now being experimentally investigated. This paper outlines the scope of these developments as they relate to the production, and possible utilization, of novel sources of radiation based on accelerator technology.

INTRODUCTION

The range of photon sources available to the researcher today represents a spectacular span of parameter space in wavelength and pulse energy. However, the primary tools available, namely lasers and synchrotron facilities leave gaps in the coverage of particular attributes. For example, solid state lasers can provide very high peak power, while the synchrotrons tend to be high average power sources. The tunability and wavelength range at a synchrotron are enormous (IR to gamma radiation) but they typically require monochromators to provide wavelength selection. Lasers usually operate with much narrower bandwidth, but through a comparatively limited wavelength range with limited tunability. Lasers are also known for their capability to provide variable pulse lengths down to sub-picosecond, while synchorotrons operate with essentially fixed pulse width ranging from tens of picoseconds to a few nanoseconds in existing machines. Pressing scientific needs have been identified that fall within these gaps.

Demands exist for sources with; **Higher-** Pulse energy, Average Power, and Repetition Rate, with **Wider-** Bandwidth, and Tuning Range, while operating at **Shorter-** Wavelengths and Pulse lengths. Many technologies are being pursued in an effort to address the requirements of these experiments. Recent developments make it seem likely that accelerator technology will have a significant role in satisfying these needs, particularly at shorter wavelengths. Within the field of these sources there are a variety of technological approaches. However, from the standpoint of understanding their behaviour, one can take

Applications of High Field and Short Wavelength Sources
Edited by DiMauro *et al.*, Plenum Press, New York, 1998

the position that essentially all accelerator based sources of radiation can be thought of as synchrotrons.

While there are obviously many differences in the details of these machines they share the common feature of producing their radiation by changing the direction of motion of charged particles (usually electrons) moving at nearly the speed of light. The character of this motion as either transverse or longitudinal is used to broadly classify the particular machine. Another fashionable way to classify these machines has become to think of them by 'generations'. Depending on the venue, the generations are couched either in terms of size of facility, energy of the machine, or degree of coherence or 'spectral brightness' they produce in their radiation.

In this view, first generation machines were available for synchrotron radiation experiments only as a parasitic use of machines developed for other applications, usually high energy physics. The second generation machines were dedicated for synchrotron research, and third generation machines added specialized magnetic devices to produce higher power or more coherence. The move from first to second generation machines made synchrotron radiation more broadly available since an individual storage ring based machine may support many users simultaneously. For instance, the two storage rings at the National Synchrotron Light Source (NSLS) support over 80 beamlines.

With the specialization of third generation machines around insertion devices, a given machine supports fewer simultaneous users, but at more powerful sources. It appears that this trend may continue with the development of so called 'fourth generation' light sources. Possible avenues to these higher performance instruments include further optimization of storage rings, Compton or Thomson scattering sources, and various free electron laser schemes.

While any one of these topics provide ample scope for an extensive review article, the purpose of this paper is to provide a broad brush overview of the general subject. The readership of these proceedings is presumed to be generally conversant with laser technology and perhaps less familiar with the development of synchrotron radiation sources. Hence a brief overview of the development of these accelerators and their objectives is given. In addition, since the field of advanced accelerator development continues to evolve, world wide web resources for further study are included in the reference list.

SYNCHROTRON SOURCE DEVELOPMENT

It is interesting to consider the history of this field, now almost exactly fifty years after the first visual observation of synchrotron radiation in 1947. The presence of synchrotron radiation was actually experimentally observed as an energy loss in a 100 MeV betatron machine at the General Electric Laboratory one year earlier by Blewett. He describes the early history of synchrotron radiation in a delightful paper presented at the 1987 synchrotron radiation instrumentation meeting[1].

In essence, the basis for synchrotron radiation can be traced back to Maxwell's equations on "Electricity and Magnetism" which show that a changing electric current will emit electromagnetic radiation. By 1897 the existence of the electron had been established by J.J. Thomson, and it was shown that currents could be carried by electrons. Indeed, elegant (but computationally intractable) theory describing synchrotron radiation energy losses had been developed by Schott[2] by 1908. This work was buried in the literature, and only revived in the 1940's when the topic became relevant to problems in contemporary accelerator physics. The work of Schwinger[3] published in 1949 is much more widely known and used because he recognized the similarity between the derivation of the synchrotron radiation spectral distribution function and of Airy integrals. This insight allowed him to reduce the formulae to forms that used tabulated Bessel functions. Hence quantitative analysis of the properties of synchrotron radiation became readily accessible.

Qualitatively, the basic properties of synchrotron radiation from an idealized dipole source are shown in figure 1. In most storage rings used as synchrotron radiation sources, the electrons (or positrons) are circulated at an energy ranging from a few hundred MeV to several GeV. In the equations describing the properties of the radiation, the electron beam energy is referenced to the rest mass of the electron; $\gamma=E/mc^2$ or in practical units $\gamma=1957E[\text{GeV}]$. The emitted radiation has a broad spectrum at low energies, and falls off exponentially above a critical energy ε_c (frequency ω_c) described by $\varepsilon_c = \hbar\omega_c = 3\hbar c\gamma^3/2\rho$, where ρ is the radius of the electron trajectory.

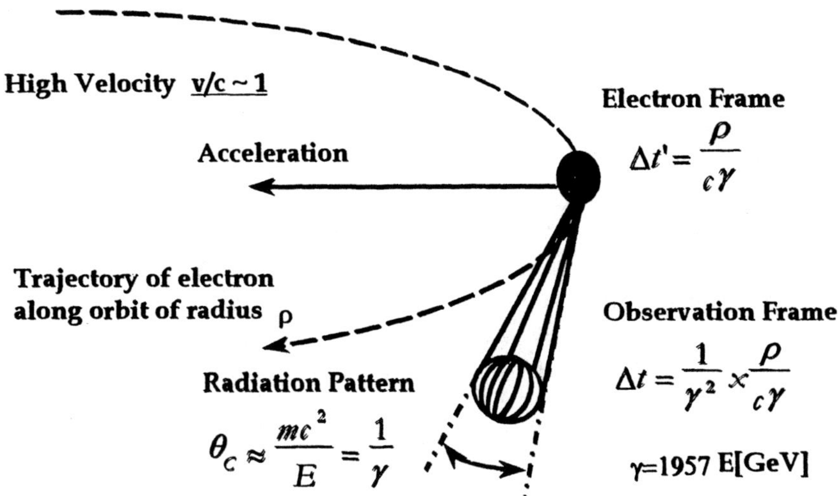

Figure 1. Schematic of radiation pattern produced by a relativistic electron following a circular trajectory.

Half of the total power is radiated above and half below the critical energy. As shown in figure 1, since the electron is traveling at nearly the speed of light, the radiation pattern is strongly peaked (or blue shifted) in the forward direction. At the critical energy, the opening angle of the radiation (Θ_c) is approximately $1/\gamma$ for an ideal accelerator. In practice several other factors may increase this apparent angle.

At a bending magnet source, the beam sweeps across the observers view in one direction, so in this direction (usually the horizontal) the apparent angle is just the collection angle. In addition, the actual size of the electron bunch (as opposed to a single electron in the ideal case) will smear the vertical angle. To keep the electron beam in the machine, energy must be supplied (usually by a radio frequency cavity) to make up for the beam energy lost producing synchrotron radiation. This introduces some spread of electron energies, increasing the effective size of the electron beam. Focusing optics must also be used to keep the electron beam on a closed orbit which, depending on the location of the observer, will introduce further angular spread to the beam and apparent source size. The product of the actual electron beam size and angular divergence is known as the emittance of the storage ring.

Minimizing this parameter for a given storage ring puts its radiated power into a smaller phase space yielding a "higher brightness" machine. A high brightness source can have decided benefits for the photon optical system used to deliver the synchrotron radiation to the experiment[4] yielding a higher useful power in a narrower spectral bandwidth. The choice of beam energy, lattice, current and emittance are all factors in defining the range of science that can be served by a particular machine. Murphy[5] gives a compendium of machine designs and their governing equations that illustrate the diversity of approaches to the optimization of storage rings. A broad range of machines and their subsystems are also described in a recent volume edited by Winick[6] .

Another way to increase the radiation power available to an experiment is to look at multiple sources. One can achieve this by passing the electron beam through a series of alternating dipole magnets arranged so the average trajectory is a straight line. These periodic magnetic structures are known as insertion devices since they are placed in the straight sections between the discrete dipoles used in most storage rings.

If the deviation angle of the trajectory from one period to the next is large compared to the opening angle of the radiation, the device is known as a wiggler. For this type of device, the power and brightness of the source are enhanced by a factor approximately equal to the number of magnetic poles in the structure. Essentially this is a linear superposition of a string of dipole magnets. The trajectory and spectra for this type of device are provided in figure 2.

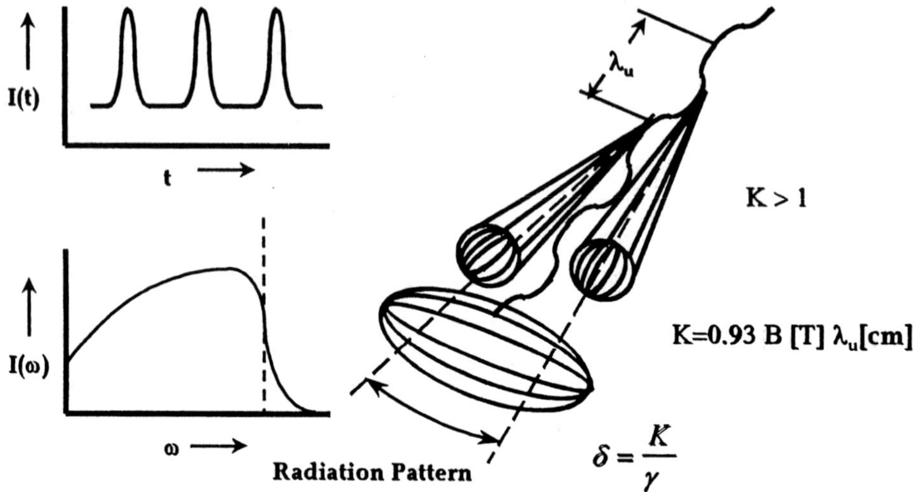

Figure 2. Schematic of radiation pattern produced in a wiggler. The intensity as a function of time and emitted radiation frequency are also shown.

If one reduces the amplitude of the angular deviation to be comparable to the emission angle of the radiation the device is known as an undulator, shown schematically in figure 3. Operated in this mode, the intrinsic brightness of the source is preserved. In addition coherent effects are possible from the interaction of radiation of one period with the electron beam in a subsequent period of the device. This interaction produces a spectrum that can be substantially enhanced at certain characteristic energies, and it is the coupling between the electron beam and the photon beam that forms the basis for the Free Electron Laser (FEL)[7]. It is worth noting that the highest average brightness machines currently available at energies beyond the vacuum ultra-violet are based on undulators in storage rings. Although their coherence properties are far removed from those of conventional lasers, prototype experiments have been performed that require coherent illumination[8]. This early exposure has whetted the appetite of the synchrotron community for FELs operating down to x-ray wavelengths.

FREE-ELECTRON LASERS

In 1971, Madey[9] realized that an undulator operating as an amplifier in an optical

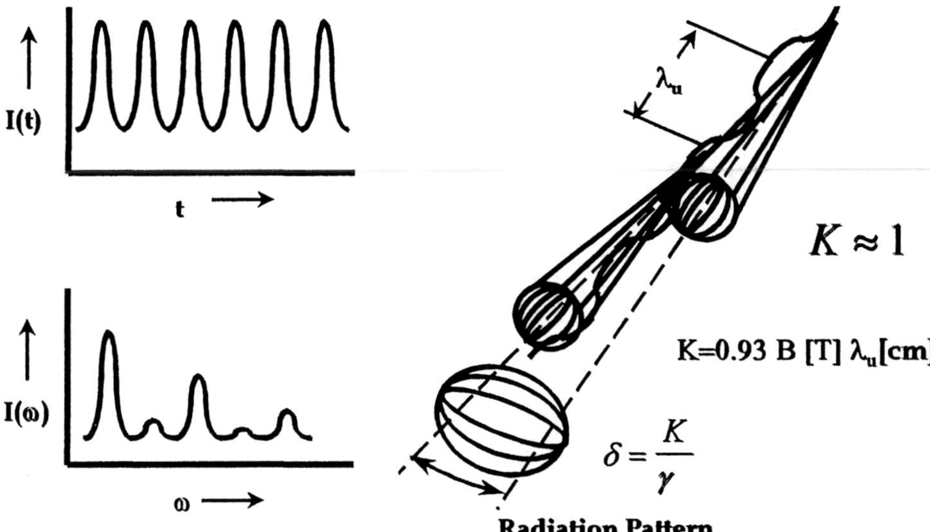

Figure 3. Schematic of radiation pattern produced in an undulator. The intensity as a function of time and emitted radiation frequency are also shown.

cavity could replace the gain medium used in conventional lasers up to that time. He in fact coined the name Free Electron Laser (FEL) to describe this device. Deacon, Elias, Madey, Schwettman and Smith[10] succeeded in constructing the first operational FEL at 3.4μm in 1977. The history leading up to these developments is described in a monograph by Luchini and Motz[7], which is in itself quite interesting reading. Dattoli and Torre[11] also provide a good tutorial article on the physics of FELs. Current information on FEL projects around the world is available on the world wide web[12]. This site has links to 25 operational facilities and many more at the proposal stage or under active development.

Within the context of these proceedings, figure 4 perhaps best indicates why FEL's have become an area of current interest. It illustrates the peak power from representative sources of short wavelength radiation. Lasers and synchrotrons are represented by existing sources, the FEL's by proposed instruments operating at wavelengths substantially shorter than any of the existing oscillator based FEL's. The principle difference between the FEL's and the storage ring sources is that the FEL's are nearly fully coherent.

Qualitatively, this is due to a strong interaction between the photon and electron beams in machines configured as FEL's which results in density modulations in the electron beam that are on a length scale comparable to the wavelength of the undulator radiation. In essence, these microbunches emit like a single particle with the charge of the entire microbunch. Until recently, technological problems have confined this mode of operation to relatively long wavelengths, since multiple passes through optical oscillators were required to produce adequate bunching. Some governing equations provide clues as to the barriers to be overcome.

For an FEL to operate at a photon wavelength λ_{ph} on the central axis, one must satisfy the resonance condition:

$$\lambda_{ph} = \frac{\lambda_u}{2\gamma^2}\left(1 + \frac{K^2}{2}\right) \qquad \text{where} \qquad K = \frac{eB_0\lambda_u}{2\pi mc} = 0.934\lambda_u[cm]B_0[T]$$

The γ is the scaled electron beam energy as before, λ_u is the undulator magnetic period, and K is known as the undulator parameter. It essentially scales the deviation angle of the radiation from the central axis of the undulator δ to the energy of the beam γ ($\delta=K/\gamma$), as illustrated in figures 2 and 3. For the photon beam to interact coherently with the electron

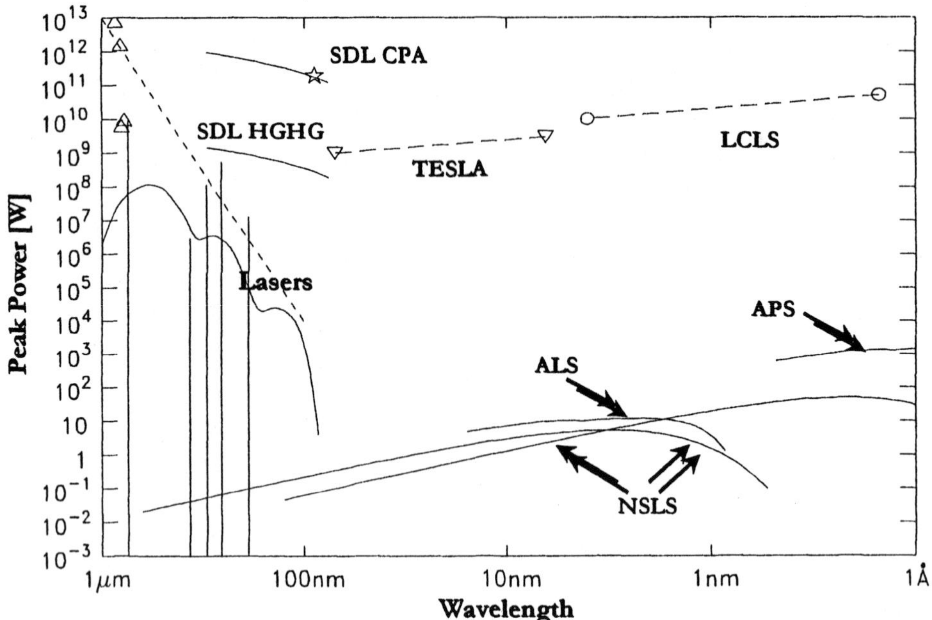

Figure 4. Comparison of peak power from some existing and proposed short wavelength sources. The curves to the left are representative 'conventional' lasers. NSLS, ALS, and APS are undulator sources at existing synchrotron facilities. SDL-HGHG and SDL-CPA are the Source Development Laboratory FEL operated in high gain harmonic generation and chirped pulse amplifier mode. TESLA and LCLS are the DESY and SLAC SASE FEL's respectively. See references 12 through 16 for more information.

beam, K must be the order of 1. To obtain resonance at short wavelengths, one must then increase the energy of the beam, and/or reduce the period of the undulator. Practical values for the undulator period are currently about 2 cm and longer for magnetic devices. This means that beam energies from a few hundred MeV to tens of GeV are required to span the wavelength range shown in figure 4. While these are prodigious machines, they are possible; for example the SLAC linac runs up to 45 GeV.

The optical field of a laser can however present what is effectively a much shorter period device, as in the recent Thomson Scattering experiments at Berkeley[17]. This technique has the interesting feature of providing a very high-energy x-ray beam from a modest energy accelerator with a pulse as short as the drive laser. The pulse can be a factor of 10 to 100 times shorter than from a synchrotron, but for existing systems, it turns out that the peak brightness is only just comparable to existing third generation storage rings. For some types of experiments this will however be an interesting source, and a number of projects are underway that utilize Thomson or Compton scattering.

For an FEL to function as a coherent source, in addition to maintaining the resonance condition, the phase space of the electron beam must be matched to the diffraction limit for the radiation. This clearly becomes much more difficult as the target wavelength of the FEL is reduced. To achieve an interesting level of power, an FEL must also provide a sufficient number of 'gain lengths' of undulator for the interaction to reach saturation. In an optical resonator, this can be achieved by feeding optical beam back onto a subsequent electron beam and essentially using the same undulator over again. This becomes much harder to achieve, particularly in the vacuum ultraviolet where high reflectivity optical systems are not readily available.

Increasing the peak current of the electron beam in the accelerator reduces the gain length for a particular FEL configuration, so the number of passes through the cavity could be reduced. Taken to its limit, the number of passes could be reduced to one. This single pass approach neatly sidesteps the demands of an optical resonator, but it places severe requirements on the electron beam. This has been an area of intense activity in the accelerator physics community over the last decade, and a number of radio-frequency photocathode gun systems for linear accelerators (linacs) have been developed that can provide electron beams with the requisite properties[18].

The most straightforward implementation of a single pass FEL is to feed a high quality electron beam through a very long undulator. The early part of the undulator is used to produce spontaneous emission as in conventional synchrotron insertion devices. The last part of the undulator is used to provide the amplification through the FEL process. This arrangement is known as a self amplified spontaneous emission or SASE FEL. While it is comparatively simple, it requires a very long undulator for any given wavelength, and is very sensitive to instabilities in the input electron beam. From the standpoint of the end user of such a machine, this could cause difficulties since one would need to monitor the photon beam pulse energy, wavelength, duration and bandwidth for every shot.

An alternative approach described by Yu[19] is to use the electron beam as a high quality non-linear gain medium. This would be accomplished by seeding the electron beam with a conventional laser in a short undulator which would introduce an energy modulation on the electron beam. Using a dispersive magnet, the energy modulation can be converted to a spatial modulation so the electron beam is prebunched as it enters a longer amplification wiggler. Yu realized that the second wiggler can have its resonance condition matched to a harmonic of the seed laser fundamental. In this way the desireable properties of the conventional seed laser such as the pulse length, well defined wavelength and bandwidth can be shifted to a shorter wavelength. The experimenter need in practice only monitor the FEL pulse energy from one shot to the next, since that is the only parameter that is influenced by instabilities in the accelerator. Chirped pulse amplification (CPA) techniques now common in solid state laser systems can even be applied to the FEL[20] so, in principle, mJ ultra-violet pulses 10 fs or shorter can be produced.

NEAR TERM PLANS, LONG TERM PROSPECTS

There is a great deal of excitement within the accelerator community regarding the development of short wavelength FEL's for research applications. Each of the approaches outlined in this paper (and several others not mentioned) has at least one project under development or construction at the time of writing with the aim of experimentally verifying the predictions of theory. These machines are qualitatively different in their output than existing synchrotrons. However if one regards them as the next generation of accelerator based radiation source, past experience might be used as a guide as to their likely appearance as routine research tools. If indeed this were to hold true, dedicated short wavelength FEL facilities should be with us between 2005 and 2010. Based on the projects already in progress, it seems likely that prototype user experiments should be underway on the R&D machines by the year 2000.

ACKNOWLEDGMENTS

This work was performed under the auspices of the U.S. Department of Energy, under contract DE-AC02-76CH00016. The author would like to thank his many colleagues in the SDL project, and at other light source projects around the world, for their many useful (and patiently presented) discussions of the various aspects of this extremely interesting field.

REFERENCES AND RESOURCES

1. John P. Blewett, Synchrotron radiation -1873 to 1947,*Nucl. Instrum. & Methods,* A266:1 (1988)
2. 2 G.A. Schott, Electromagnetic radiation and the mechanical reactions arising from it, (Cambridge University, 1912)
3. J.S. Schwinger, *Phys. Rev.,* 75:1912 (1949)
4. E.D. Johnson, S.L. Hulbert, L.E. Berman, Beam lines at synchrotron facilities:The link between the user and the machine, in:*The Physics of Particle Accelerators, AIP Conference Proceedings 249,* M. Month and Margaret Dienes ed. American Institute of Physics, New York (1992).
5. J.B. Murphy,*Synchrotron Light Source Data Book,*BNL Publication 42333, Version 4, Brookhaven National Laboratory, Upton Long Island NY (1996). Also available at www.nsls.bnl.gov//AccPhys/hlights/dbook/Dbook.Menu.html
6. H. Winick, ed. *Synchrotron Radiation Sources A Primer,* World Scientific Publishing Co, Singapore, River Edge NJ, London (1994)
7. P.Luchini, H. Motz, *Undulators and Free-electron Lasers,* Clarendon Press, Oxford (1990)
8. See for example C. Jacobsen *etal,* these proceedings or M. Sutton *etal, Nature,* 352:608 (1991)
9. J.M.J. Madey, Stimulated emission of bremsstrahlung in a periodic magnetic field, *J. Appl. Phys.* 42:1906 (1971)
10. D.A.G.Decon *etal,* First operation of a free-electron laser, *Phys. Rev. Lett.* 38:892 (1977)
11. G. Dattoli *etal,* An introduction to the theory of free electron lasers, in:Proceedings of the CERN accelerator school on Synchrotron Radiation and Free Electron Lasers, S. Turner ed. CERN Publication 90-03, CERN, Geneva (1990). Other articles in this volume are also valuable sources of basic information on topics in Synchrotron radiation and Free Electron Laser science.
12. The World-Wide Web Virtual Library:Free Electron Lasers, http://sbfel3.ucsb.edu/www/vl_fel.html, maintained by G. Ramian, UCSB.
13. For current information on the synchrotron facilities see for example http://www.nsls.bnl.gov/Intro/AllSynch.html. The cited acronyms are ALS (Advanced Light Source, Lawrence Berkeley National Laboratory), APS (Advanced Photon Source, Argonne National Laboratory), and NSLS (National Synchrotron Light Source, Brookhaven National Laboratory).
14. I.Ben-Zvi *etal,* The BNL source development laboratory, *Nucl. Instrum. & Methods,* A393:II-10 (1997) Information also available at www.nsls.bnl.gov/BeamRD/Erik/SDL.html
15. J. Rossbach, A VUV free electron laser at the TESLA test facility at DESY, *Nucl. Instrum. & Methods,* A375:269 (1996) Information also available at www.desy.de/~wroblewt/scifel/scifel.html
16. R. Tatchyn *etal,* Research and development toward a 4.5-1.5Å linac coherent light source (LCLS) at SLAC, *Nucl. Instrum. & Methods,* A375:274 (1996). Information also available at www-ssrl.slac.stanford.edu/lcls/lcls.html
17. W.P. Leemans *etal,* X-ray based subpicosecond electron bunch characterization using 90° Thomson scattering, *Phys. Rev. Letters* 77:4182 (1996) and Schoenlein *etal,Science* 274:236 (1996). See also the ALS web page www-als.lbl.gov.
18. X.J. Wang *etal,* Experimental characterization of the high-brightness electron photoinjector, *Nucl. Instrum. & Methods A,* 375:82 (1996). See also X. Qiu *etal,* Demonstration of emittance compensation through the measurement of the slice emittance of a 10 ps electron bunch, *Phys. Rev. Lett.* 76:3723 (1996)
19. L.H. Yu, Generation of intense UV radiation by subharmonically seeded single-pass free electron laser, *Phys. Rev A* 45:1163 (1992).
20. L.H. Yu *et al,* Femtosecond free electron laser by chirped pulse amplification, *Phys. Rev. E,* 49:4480 (1994)

A DEBRISLESS LASER-PLASMA SOURCE
FOR EUV AND XUV GENERATION

Christopher M. DePriest, David S. Torres, and Martin C. Richardson

CREOL, University of Central Florida
Laser Plasma Laboratory
4000 Central Florida Blvd.
Orlando, FL 32816

INTRODUCTION

Interest in debris-free sources of short-wavelength EUV radiation has risen in recent years due primarily to anticipated needs in the area of projection lithography, where smaller feature sizes and increased production rates are targeted for the near-future in the semiconductor industry. Consequently, laser-plasmas have come to the forefront as attractive source candidates for such lithography systems. As compact, modular, and high-rep-rate sources, laser-plasmas have already demonstrated the required minimum efficiency[1,2]. The water-droplet laser-plasma source[3] offers the additional advantages of low-cost (~\$10^{-6}/shot), continuous, debris-free operation that future lithography systems require. Broadband emission from solid-target laser-plasmas created from high-Z materials, which can lead to off-band heating by absorption of the primary collection optics[3], is another problem that is circumvented with the narrow-band droplet laser-plasma. In addition, solute-doped droplet targets are promising candidates for generating debris-free XUV radiation having, in general, the same advantages as the water droplet EUV source.

EUV DROPLET SOURCE

Debris-free operation of any laser-plasma source requires the implementation of "mass-limited" targets, which, by definition, are composed of just enough mass (typically ~10^{-6} g) to provide the minimum number of ionized radiators needed for a given set of laser conditions [4]. Liquid droplet technology lends itself well to the production of such targets due to the ease with which small, singular masses can be created. The operating parameters of the laser used in these experiments (a 1064 nm Nd:YAG with 10 Hz rep rate, 400 mJ/pulse, and 10 ns pulsewidth) led to the adoption of a piezoelectrically modulated water jet nozzle producing a well-defined, continuous stream of droplets ~20 μm in diameter with a nearest-neighbor separation of ~50 μm. The nozzle velocity of the water jet was 50 m/s when modulated at a frequency of 1 MHz. With precise synchronization of the arrival of the laser pulse with the arrival of the water droplet at the laser focus, the entire mass of the droplet can be utilized in producing the EUV emission, thereby leaving negligible debris.

The debrisless nature of the water droplet laser-plasma is shown vividly in Figure 1[5]. This data was obtained by exposing a Mo/Si multilayer mirror (reflectivity ~60% at 13 nm) to a large number of laser shots, with the mirror approximately 4 cm away from the target. An x-ray photodiode monitored the reflected throughput at an angle of 45 degrees for two multishot runs: one with a tin target and one with the droplet target. As is dramatically evident, the droplet laser-plasma target induced no reflectivity degradation (to within 1%) over more than 10^5 shots, whereas the tin target quickly reduced the reflectivity to a minimal level.

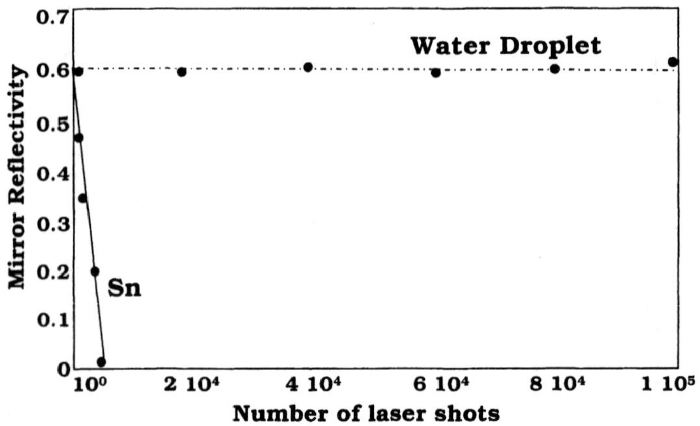

Figure 1: Reflectivity curves for Mo/Si multilayer with droplet vs. Tin targets.

The use of a Mo/Si multilayer in the reflectivity experiments was not coincidental. Mo/Si multilayers represent the highest reflectivity achieved to date with conventional optics in the EUV regime (~13 nm). Due to the narrow-band nature of these multilayers, it would be desirable for an EUV laser-plasma source to also emit in a narrow-band fashion, thereby avoiding the problem of off-band heating of the collection optics. The water droplet laser plasma satisfies this requirement. Two emission lines are produced by Li-like oxygen in the droplet laser plasma: 4d-2p (13 nm) and 4p-2s (11.6 nm). In fact, the emission of O^{5+} at these wavelengths is narrower than the bandwidth of current multilayers. Additionally, the laser conversion efficiency (within a 0.3 nm bandwidth) was ~0.6%, which is comparable to the best conversion efficiencies achieved with metal targets (~0.85%)[1].

Due to sensitive dependencies of the spectral intensity of emitted x-rays on the plasma conditions, a theoretical characterization of the water droplet emission spectrum was performed using three computer codes (in an effort to optimize irradiation and target conditions). A one-dimensional Lagrangian hydrodynamic code, Medusa[6], was used to temporally and spatially model (in linear and spherical geometries) the electron/ion temperature and density by division of the expanding laser-plasma into temporally evolving cells. In each cell, the principal plasma equations (Navier-Stokes, Maxwell's, continuity, energy conservation, and state equations) were solved based on initial target and irradiation parameters. The electron density and temperature (in each cell) generated by Medusa was then used as input to RATION[7], an atomic physics code which calculates the resulting steady-state emission spectra from H-like, He-like, and Li-like ion populations. Developed at the Laser Plasma Lab, a third code, SPIN, temporally and spatially integrates the spectral contributions from each cell, thereby coupling the output of Medusa and RATION to yield a synthetic emission spectrum for the entire droplet/laser-plasma interaction.

Experimental measurements of the emission spectrum obtained with a high-resolution, flat-field x-ray spectrograph along with a theoretical spectrum calculated in a one-dimensional spherical geometry are presented in Figure 2. As predicted by computer simulation, strong emission at 13 nm from the 4d-2p transition in Li-like oxygen is observed experimentally superimposed on the weaker broadband bremsstrahlung emission of the droplet plasma. The observed structure of the line emission confirms the predominant presence of Li-like oxygen ions, with additional contributions from Be-like and He-like ions[5].

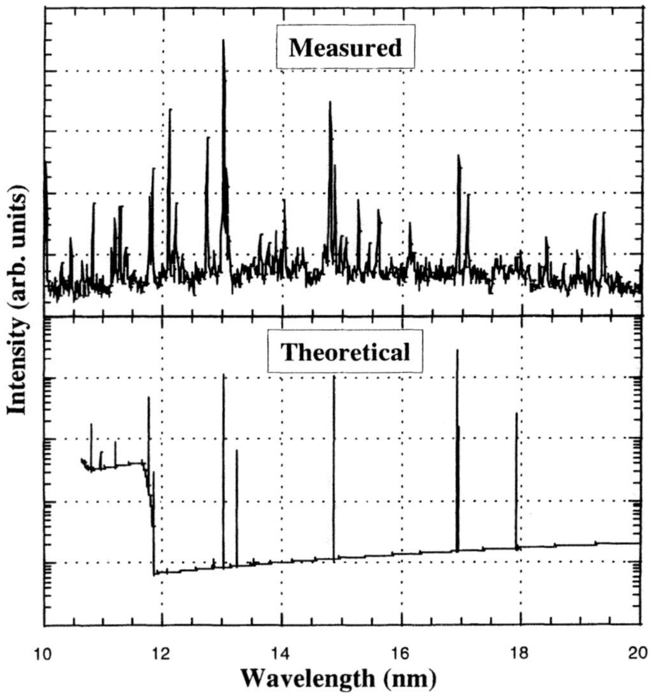

Figure 2: Experimental and synthetic spectra of water droplet.

XUV DROPLET SOURCES

The versatile nature of the water droplet laser plasma source, lends itself to EUV and XUV radiation generation at other wavelengths. A high-rep-rate source of radiation in this regime has many other applications. For biological microscopy, a high-power source operating in the water window (2.3-2.6 nm) would allow deeper probing of the specimen with much finer resolution of small details. The utilization of short pulses would have the added advantage of allowing in-vivo imaging. In addition, the droplet source renders such problems as image artifacts arising from expelled target debris inconsequential. Work by Rymell and Hertz on droplet sources has provided line-radiation at 3.37 nm and <1.7 nm radiation with the use of alcohols and other liquids.[8,9]

The nature of the droplet target also allows the incorporation of higher-Z dopants, suitable for nanometer and sub-nanometer lithography. Solutions and particulate suspensions are attractive methods of pushing droplet x-ray emission to shorter wavelengths. Zinc is a solute/particulate that is a promising candidate for use in a high-rep-rate XUV droplet system. Zinc chloride, for example, is highly water-soluble (4.32 g/cc), and the triatomic nature of the molecule is attractive due to the absence of a large host of extraneous ions. Fluorine-like Zinc (Zn^{+21}) also has strong XUV emission line in the region 1.0 nm. Utilizing the parameters of the current water droplet system (20 μm diameter), it is possible to dope a single droplet with $\sim 8_{10}^{13}$ $ZnCl_2$ molecules. For a 1 kHz laser system, under the assumption that 10% of the Zn atoms radiate within the desired band around 1.0 nm, it would be possible to achieve a steady-state x-ray power output of $\sim$1.6 W. This would require a 16 W laser system, assuming a 10% conversion efficiency[10].

In addition to solute-based droplet systems, the notion of nanoparticle suspensions will also be investigated. The incorporation of zinc nanoparticles (available in sizes upwards of 20 nm) into the droplet system is straightforward, with the added advantage of reducing the amount of extraneous solutes. Experiments characterizing the emission of such droplet targets are coming on line presently.

CONCLUSIONS

The EUV droplet source has proven to be a successful x-ray generation device with the much-needed advantages of debris-free, continuous, and high-rep-rate operation. Extension of the droplet source into the XUV spectral regime is now of primary importance. The XUV droplet source will be an attractive candidate for nanometer/sub-nanometer x-ray lithography and microscopy.

REFERENCES

1. R. L. Kauffmann, D. W. Phillion, and R. C. Spitzer, Appl. Opt., 32:6897 (1993).
2. G. D. Kubiak, K. W. Berger, S. J. Haney, P. D. Rockett, and J. A. Hunter, *OSA Proceedings on Soft X-ray Lithography*, A. M. Hawryluk and R. H. Stulen, eds. (Optical Society of America, Washington, DC 1993) 18, p. 127.
3. F. Jin, M. Richardson, G. Shimkaveg, and D. Torres, *Proceedings on the Applications of Laser Plasma Radiation II*, M. Richardson and G. Kyrala eds. (SPIE, Bellingham, WA., 1995), 2523, p. 81.: F. Jin, M. Richardson, G. Shimkaveg, and D. Torres, in *TOPS on EUV Lithography*, G. D. Kubiak and D. Kania eds. (Optical Society of America, Washington, DC, 1996), p.89.
4. M. Richardson, W. T. Silfvast, H. Bender, A. Hanzo, V. P. Yanovsky, F. Jin, and J. Thorpe, Appl. Opt., 32:6901 (1993).
5. M. Richardson, D. Torres, C. DePriest, F. Jin, and G. Shimkaveg, Opt. Comm., to be published.
6. J. P. Christansen, D. E. T. F. Ashby and K. V. Roberts, Com. Phys. Comm., 7:271 (1974).
7. R. Lee, *User Manual for RATION*, (University of California and LLNL, 1990).
8. L. Rymell & H. M. Hertz, Opt. Comm., 103:105 (1993).
9. L. Malmqvist, L. Rymell, M. Berglund, and H. M. Hertz, Rev. Sci. Inst., 67:4150 (1996).
10. Livermore reference on 10% conv. Eff in Cu...

ATTOSECOND PULSE GENERATION AT GAS IONIZATION
BY FEW-OPTICAL-CYCLE LASER PULSES

A.V. Kim, M.D.Chernobrovtseva, D.V. Kartashov, and A.M. Sergeev

Institute of Applied Physics
Russian Academy of Sciences
46 Ulyanov st.
603600 Nizhny Novgorod, Russia

INTRODUCTION

The application of laser pulses with supershort duration of the order of 10 fs has recently been proposed for enhancement of high order harmonic emission and subfemtosecond pulse production in the XUV range.[1-3] Such short driving pulses contain only a small number of optical cycles and therefore high energy photon bursts are generated due to atom ionization at rapidly increasing field amplitudes. It is obvious that the efficiency of this process becomes dependent on the concrete field distribution over the pulse or, in other words, on the absolute phase of the optical field. It should be also noted that several other schemes have been proposed for attosecond pulse generation. Corkum, Burnett, and Ivanov[4] have proposed to use the elliptically polarized light to take advantage of the strong polarization dependence of the harmonic emission process. To produce a train of attosecond pulses, one pulse each half cycle of the driving laser field, Antoine, L'Huillier and Lewenstein have suggested filtering and combining several plateau harmonics.[5] Schafer and Kulander[6] have suggested to generate single subfemtosecond pulse by using ultrashort pump laser and compression techniques for harmonics at the end of the plateau.

CLASSICAL APPROACH

The idea of an enhancement of energetic x-ray burst production with a decrease of the pulse duration consists in the following.[2] At a sharply increasing laser field amplitude, atoms may be ionized with high probability (or the main part of the electron Ψ-function may be detached from the intraatomic potential) for a fraction of an optical cycle when the field strength passes the ionization threshold value. For the subsequent laser period the electron wave packet including almost all particles will be accelerated by the laser field and may collide with the parent ion core at a velocity higher than that for the case of slowly varying field amplitude. By employing a 1D quantum model, consideration of this effect in detail was presented in the work,[7] where it was treated as nonadiabatic effect in high-harmonic generation with ultrashort pulses. It was observed that for shorter pulses, a redistribution of the energy between the different harmonics orders occurs, where the lower orders are weaker, while the higher ones are stronger, than for longer pulses.

The efficiency of this mechanism may be easily illustrated by considering the classical picture of the relevant trajectories of freed electrons that are responsible for the high energy photon emission. Let us assume that the leading front of the pulse is characterized by the exponential growth of the field amplitude with an increment β measured in the laser field frequencies. We will be interested in the phase φ_{col} and the energy U_{col} of return collisions of classical electrons with parent ions depending on the electron release phase φ. In the case $\beta=0$, considering the one-dimensional electron trajectory, precisely half of the electrons will collide with ions at least once. These are the electrons released from intraatomic potential during quarters of the field cycles after achieving the local maximum values. The electrons released within quarters of the increasing field cycles do not return to parent ions at all. The picture changes for $\beta>0$. It is seen in figure 1a that with increasing β the boundary of the "non-return phase" shifts to the negative domain, i.e. the electrons released before the field maximum may participate in the bremsstrahlung. In this case the maximum of the return collision energy (Fig.1b) broadens and shifts from $\approx 18°$ at $\beta=0$ to the earlier phases, so that it arrives at $\varphi=0$ when $\beta\approx 0.2$. In general, this fact indicates improved conditions for the soft X-ray burst generation and an increase in the maximum energy of emitted photons. However, it is clear that the process of soft X-ray burst generation is rather sensitive to a specific temporal profile of the pulse and one can imagine a situation when the non-return effect of the most released electrons will lead to a dramatic reduction of the bremsstrahlung efficiency.

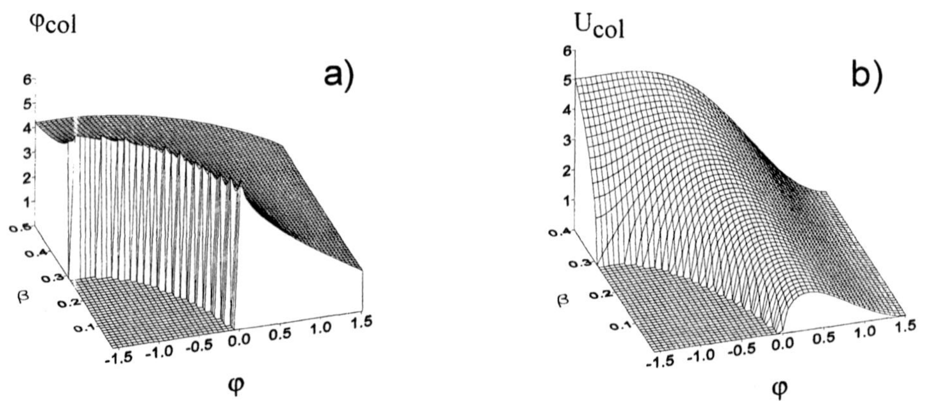

Figure 1. The phase φ_{col} (a) and the energy U_{col} (b) of electron return collisions with parent atoms versus the phase of electron production φ and the rate β of the field amplitude increase.

This can happen for example if the field passes the range of strengths that are critical for the atom decay in the growing phase but close to the pulse center when the amplitude reaches the maximum. Then a freed electron bunch is mainly generated for $\varphi < 0$, while the value β is close to 0. It is interesting to note that if the non-return condition is fulfilled near the axis of the 3D pulse cross-section, at the periphery, vice versa, the excitation of short wavelength bursts does occur, since electrons are released for lower field envelopes near the field maximum $\varphi = 0$. This may cause hollow distributions in the spatial domain where subfemtosecond pulses are generated.

PROPAGATION EFFECTS

From the above simple consideration we conclude that the high energy photon burst emission may be controlled by shaping the driving laser pulse. So, the idea to produce ultrashort bursts needs exploration in the frames of more realistic models. A principal question that arises is how collective processes in the ionized gas affect the characteristics of the burst. In other words, is there an optimal propagation length for producing ultrashort XUV radiation? How is its conversion efficiency affected by the temporal electric field distribution? To investigate this problem quantitatively we have employed a 1+1 simulation model developed in our previous work[8] to consider the self-consistent dynamics of linearly polarized electromagnetic wave propagating in a gas of single-type atoms. It should be noted that propagation model was also developed by Rae, Burnett and Cooper[9] and applied to the high-harmonic emission with long femtosecond laser pulses. The set of equations are the Maxwell equations for a field $E(z,t)$ in the medium written in the simplest form of a one-dimensional scalar wave equation taken in a reduced form and the Schroedinger equation with the intraatomic potential $V(x)$ (x-direction along electric field) for the single electron Ψ-function.

$$\frac{\partial^2 E}{\partial z \, \partial t} = -\frac{1}{2}(E + R),$$

(1)

$$i\frac{\partial \Psi}{\partial t} = -\frac{1}{2}\frac{\partial^2 \Psi}{\partial x^2} - V(x)\Psi + xE(z,t)\Psi.$$

(2)

Here $R(t) = \int |\Psi|^2 \frac{\partial V}{\partial x} dx$ is the source of harmonic emission in the reduced wave equation, z is the dimensionless coordinate along the laser pulse propagation direction, measured in $me^4 c / \omega_p \hbar^3$ units, $t \to t - z/c$ is the local time, and $\omega_p^2 = 4\pi e^2 N / m$ is proportional to the gas density N. The latter scaling ratio demonstrates that the result of the consideration will depend on the product of the gas density and the propagation path, but not on these factors separately. We note that this propagation model (reduced form of the wave equation) neglects the reflection light effect assuming that plasma density appearing during ionization process is much less than the critical one. Another variables of time, intraatomic coordinate, intraatomic potential and electric field in a pulse are normalized by the corresponding atomic values $t_a = \hbar^3 / me^4$, $x_a = \hbar^2 / me^2$, $V_a = me^4 / \hbar^2$, $E_a = m^2 e^5 / \hbar^4$. We have chosen the potential in the form of $V(x) = (1 + x^2)^{-0.5}$ with a binding energy of 0.67 a.u. and considered laser-gas interaction for the few-optical-cycle pulses.

Figure 2 shows the time dependence of the driving field that has the Gaussian envelope with carrier frequency of 0.2 a.u. and the response function $R(t)$, for the field amplitude of 2.1 a.u. Since the R function is responsible for the high-energy photon emission due to the electron-core interaction, we see that it is possible to ionize atom almost completely for one half period. During this time interval the electromagnetic radiation emitted by atoms mainly consists of comparatively low frequency components (as follows from the curve for the response function) and is produced at electron tunneling[10] through the barrier formed by the atomic potential and the driving field.

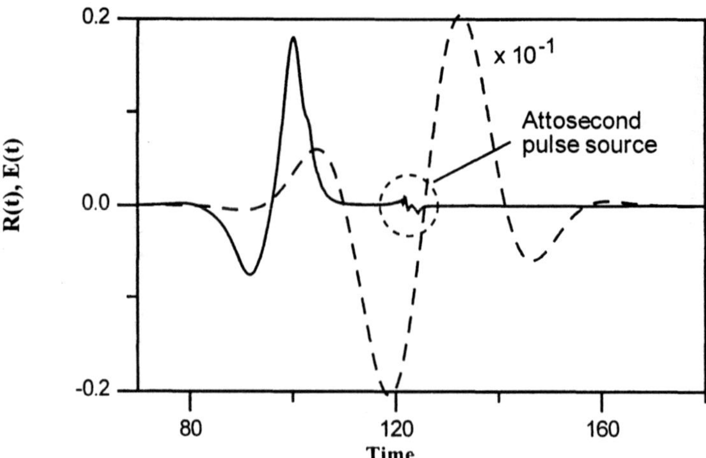

Figure 2. The time dependence of the driving field (dotted line)
and the response function R caused by electron-core interaction (solid line).

For the subsequent half-period, the electron wave packet is free $(R \sim 0)$, and then, returning back and colliding with parent ion (t $\sim$ 125), it generates radiation with higher frequencies. As it is seen in figure 3, where the frequency spectrum of electric field is shown at various distances of pulse propagation, the plateau region is extended over the harmonic order up to $n \sim 80$-90. We also see that spectral intensities in several frequency domains have different dependencies on the propagation distance. In figure 4, we present the value of energy contained in definite intervals of spectrum versus the propagation path. We see that, at the beginning, energies of the generated radiation are increased with the interaction distance, then saturated at a level depending on harmonic frequencies and further they have oscillation character along propagation path.

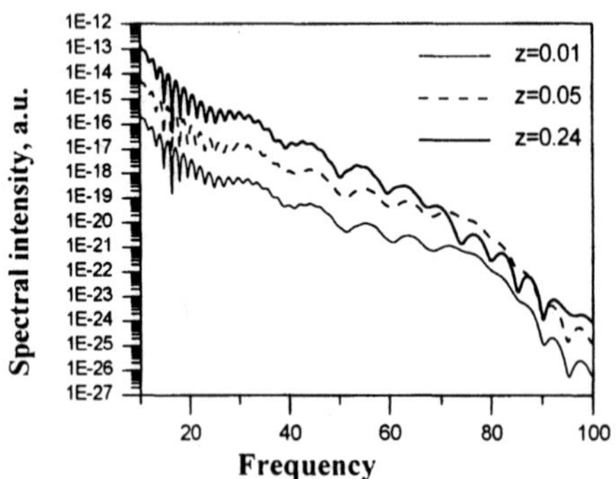

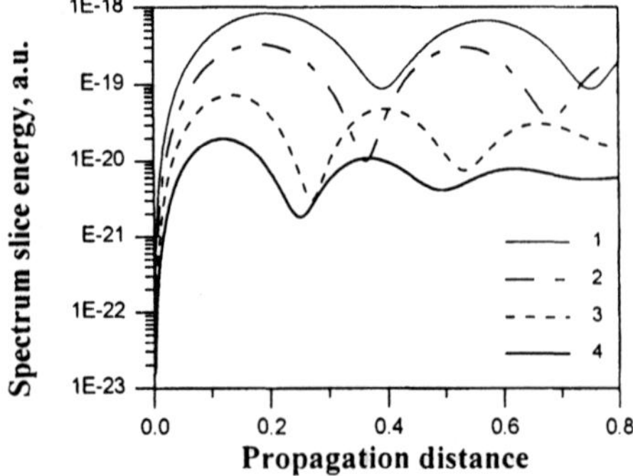

Figure 3. Field spectrum at the various propagation distances.

Figure 4. Energy of high-frequency components integrated over definite spectral intervals versus the propagation distance:
1 – from 55 ω_o to 60 ω_o; 2 – from 60 ω_o to 65 ω_o;
3 – from 65 ω_o to 70 ω_o; 4 – from 75 ω_o to 80 ω_o.

Here we only note that the saturation effect in the energy amplification is mainly caused by wave dispersion in the emerging plasma and it is more essential at higher frequencies (see Fig. 3 and 4).

DYNAMICS OF ATTOSECOND PULSES

In this section we consider attosecond pulse production and propagation effects in our simulations, for the same set of conditions as used previously. As you have seen in figure 2, the source of high-energy photon emission is localized in a very narrow time interval in optical cycle. This means that the pulse duration of high-energy photon bursts generated by sub-10 fs pump lasers must be much shorter than the laser period, i.e. it may lie in few hundred attosecond range. As we are specifically interested in the harmonic emission on a time scale less than an optical cycle, it is necessary to use a time-frequency or wavelet analysis to analyze the time profile of high-energy photon bursts.

Such analysis has been recently used to study the temporal behavior of the harmonic emission process in a single atom response.[11] The form of the analyzing wavelet, which we used here, is the Morlet's wavelet:[12] $W(t,t_0) = \exp(it)\exp\left[-(t-t_0)^2/\sigma^2\right]$ where σ is a fixed parameter and must be choosen as long as $\sigma > 2\pi$.

As an example, we focus on the spectral region expanding from $55\,\omega_o$ to $60\,\omega_o$ (ω_o is the fundamental frequency). Figure 5 shows the nonlinear dynamics of electric field envelope combined from this spectral region along the propagation distance. The pulse duration of the high energy photon bursts is more than 20 times less than pump laser period, i.e. we can expect that by using sub-10 fs, 800 nm laser pulses, single XUV pulses with duration of about 100 attosecond will be directly generated when a gas medium interacts with such ultrafast pump laser.

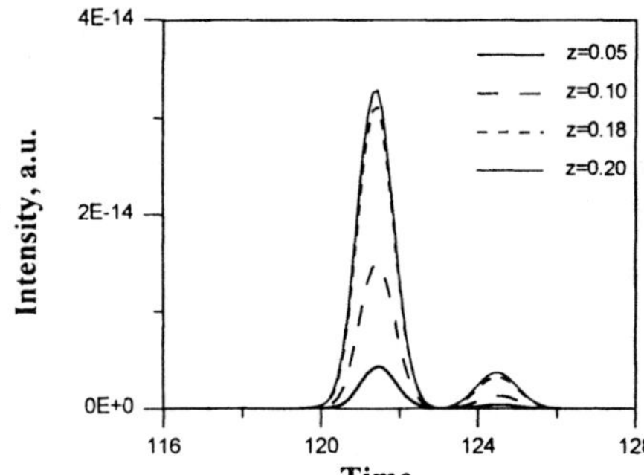

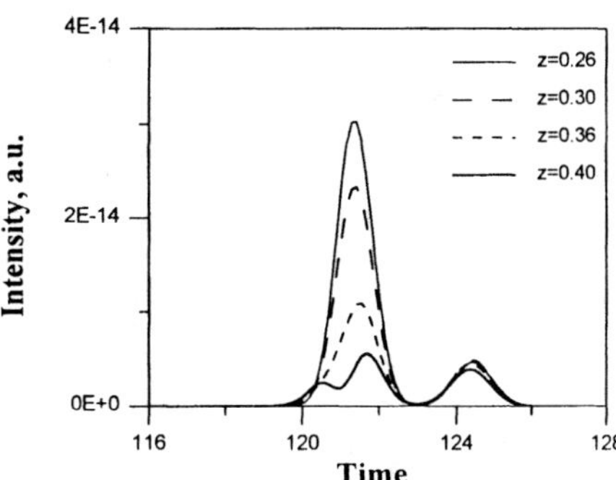

Figure 5. Electric field envelope of the attosecond pulse under the same conditions as Fig. 4 at the various propagation distances. The pulse is combined by spectral interval from $55\omega_o$ to $60\omega_o$.

The main feature of the attosecond pulse generation is its oscillation dynamics (nonmonotonic dependence) along propagation path, as it is also clearly seen in figure 4. So, there is an optimal nonlinear interaction length for efficient attosecond pulse production in a given spectral interval. For higher photon energies this optimal length becomes shorter. Oscillating character of attosecond pulse evolution along pump laser propagation points to the crucial role of the phase mismatching effects. As shown in previous works,[2,9] for the case of long femtosecond pump pulses the main role in efficient high-order harmonics generation plays phase mismatching due to blueshifting of the fundamental frequency which arises self-consistently in the problem of interest. Ionization process is responsible for both plasma-induced blueshifting and high harmonic emission effects. In the "single-cycle" regime, the attosecond pulse source is localized in a narrow temporal interval within one optical cycle (Fig. 2) and, therefore, its position is very sensitive to the temporal distribution of the laser field. As follows from our simulations, plasma dispersion gives the main contribution to the modification of the electric field profile and, thus, defines the efficiency of attosecond pulse production. So, the physical mechanizms of phase mismatching in these two cases, for long femtosecond and few-optical-cycle pulses are quantitatively different.

CONCLUSION

We have shown that attosecond pulses in the XUV range may be generated from gas interaction with strong few-optical-cycle laser pulses. There is an optimal nonlinear interaction length for efficient generation of attosecond pulses which depends on the harmonic frequency. In the "single-cycle" regime, the phase mismatching caused by laser electric field modification due to plasma dispersion restricts the efficiency of energy output of the attosecond pulses.

ACKNOWLEDGMENT

This research was supported in part by the Russian Basic Research Foundation Grants.

REFERENCES

1. J. Zhou, J. Peatross, M.M. Murname et al., Enchanced high-harmonic generation using 25 fs laser pulses, *Phys.Rev.Lett.* 76:752 (1996).
2. A.M. Sergeev, A.V. Kim, and E.V. Vanin, Few-optical-cycle pulse interactions with matter: models and nonlinear effects, *Proc. SPIE* 2701:416 (1996).
3. I.P. Christov, M.N. Murnane, and H.C. Kapteyn, High garmonic generation of attosecond pulses in the "single-cycle" regime, *Phys.Rev.Lett.* 78:1251 (1997).
4. P.B. Corkum, N.H. Burnett, and M.V. Ivanov, Subfemtosecond pulses, *Opt.Lett.*, 19:1870 (1994).
5. P. Antoine, A. L'Huillier, and M. Lewenstein, Attosecond pulse trains using high-order harmonics, *Phys.Rev.Lett.*, 77:1234 (1996).
6. K.J. Schafer and K.C. Kulander, High garmonic generation from ultrafast pump lasers, *Phys.Rev.Lett.* 78:638 (1997).
7. I.P. Christov, J. Zhou, J. Peatross et al., Nonadiabatic effects in high-harmonic generation with ultrafast pulses, *Phys.Rev.Lett.*, 77:1743 (1996).
8. E.V. Vanin, A.V. Kim, M.C. Downer, and A.M. Sergeev, Excitation of ultrashort bursts of harmonics of the radiation during ionization of a gas by an intense light pulse, *JETP Lett.* 58:900 (1993).
9. S.C. Rae, K. Burnett, and J. Cooper, Generation and propagation og high-order harmonics in a rapidly ionizing medium, *Phys.Rev.A* 50:3438 (1994).
10. M.I. Dyakonov and I.V. Gornyi, Electromagnetic radiation by a tunneling charge, *Phys.Rev. Lett.* 76:3542 (1996).
11. P. Antoine, B. Piraux, and A. Maquet, Time profile of harmonics generated by a single atom in a strong electromagnetic field, *Phys.Rev.A* 51:R1750 (1995).
12. J.M. Combes, A. Grossman, and Ph. Tchamitchian, in Wavelets, Springer-Verlag, Berlin (1989).

X-RAY EMISSION FROM RARE GAS CLUSTERS IN INTENSE LASER FIELDS

M. Lezius,[1] S. Dobosz,[1] P. d'Olivera,[1] P. Meynadier,[1] J-P. Rozet,[2] D. Vernhet,[2] N. Normand[1] and M. Schmidt[1]

[1]CEA-Saclay, DSM/DRECAM/SPAM, 91191 Gif-sur-Yvette Cedex, France
[2]Université Paris VI, 2 Place Jussieu, 75251 Paris Cedex 5, France

INTRODUCTION

Recent experimental observations have demonstrated, that the interaction of intense laser light with rare gas clusters leads to extremely energetic states of matter. For example, Xe(M) and Kr(L) shell transition photons with energies of up to 5 keV have been reported by McPherson et al.[1]. Furthermore, highly charged (>30+) xenon ions with startup energies reaching 1 MeV have been observed by Ditmire et al.[2], using irradiation intensities of 10^{16} W/cm^2. Earlier, Snyder et al.[3] were able to resolve xenon charge states as high as q=20 using a reflectron type time of flight mass spectrometer and laser intensities of about 10^{15} W/cm^2. The very effective energy transfer of the laser light into highly ionized atomic states embedded in clusters is currently interpreted as a result of small scale collective effects like coherent electron motion[45], ionization ignition[6] or collisional absorption into cluster-sized nano-plasmas[2,7]. To shed more light onto the ongoing discussion concerning the primary heating and ionization mechanism, e.g. the absorption of the laser light and the production of highly charged ion states, We present here a quantitative study on the x-ray yield as well as a function of the laser intensity (up to 10^{17} W/cm^2), the target material (argon, krypton, xenon) and the cluster size (up to $2{\times}10^6$). Our results contain also relevant data for possible future applications of clusters in strong fields as a renewable and very small x-ray source.

EXPERIMENTAL

The experimental set-up has been described in detail recently[7,8]. We use a 130 fs Ti-Sapphire Laser delivering up to 100 mJ peak power that is focused with a f=170 mm off axis parabolic mirror into the center of our cluster beam. The irradiation intensities of up to $5{\times}10^{17}$ W/cm^2 have been calibrated by comparison with the optical field ionization of a low density neon target using the barrier suppression ionization model[9]. In order to characterize our target, the clusters are generated within a pulsed adiabatic expansion using a standard type conical nozzle with d=150 µm diameter and a half opening angle of α=5°. According to Hagena and Oberst[10] the use of such a nozzle (at temperature T_0=20°) enables us to derive the mean cluster size N from the stagnant pressure P_0:

$$N \approx A_0 \cdot \left[k \frac{P_0}{T_0^{2.29}} \left(\frac{d}{\tan \alpha} \right)^{0.85} \right]^{1.95} \tag{1}$$

The dimensionless constants k and A_0 are taken from Hagena et al.[10] and Farges et al.[11]. It is well established, [12] that in a perfect adiabatic expansion the overall atom density ρ in the center of the molecular beam will be proportional to the backing pressure. We have controlled the overall atom density by a direct measurement of a small voltage signal induced on two charged condensor plates (distance 60 mm, diameter 200 mm, U=2000 V). During the laser pulse (130 fs, 790nm) a small plasma with a dielectric constant ε is created in the geometrical center between the plates, and the resulting variation of U can be directly detected using fast electronics. The total charge induced on the condenser is proportional to the number of charges created in the focus. We have calibrated and tested our technique using background gas densities between 10^{-6} and 10^{-5} mbar and a standard vacuum gauge. We are capable to measure beam densities down to 10^{-6} mbar within a time window defined by the laser pulse length at an accuracy of 20%. Fig. 1 demonstrates that the total atom density in the cluster beam is proportional to the backing pressure P_0. It can be seen that the experimental value remains within a constant offset of 1×10^{14} below the theoretical curve. We suggest, that the theory should indicate an upper limit due to losses by non-perfect valve opening functions and perturbations caused by the skimmer. Nevertheless, the linearity of our signal ensures perfect adiabatic expansion conditions over the pressure range employed.

The x-ray production from the interaction zone is monitored using two Si(Li) semiconductor detectors with well-known characteristics[13]. The detectors have a 10 mm

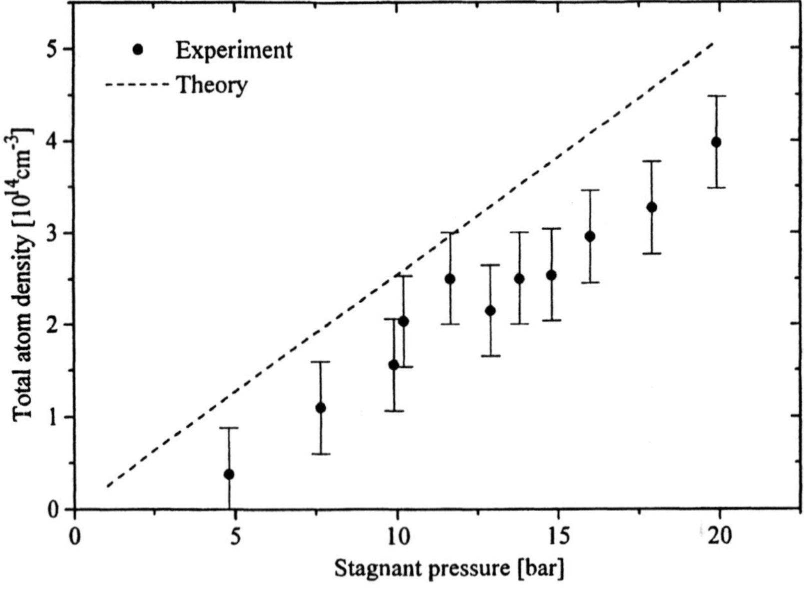

Figure 1. Total atom density in the cluster beam obtained from the dielectric constant variation (see text). The experimental result is compared to the theoretical upper limit of the beam density using the adiabatic flux calculated after Hagena[12].

active diameter which is sealed with a 20 µm beryllium foil. All electrons and ions, as well as photons with energies below 1 keV, are efficiently shielded and do not contribute to the signal. The detectors can be positioned at different angles (0°, 100°, 145°) with respect to the laser beam.. Such different arrangements allow us to study the emission pattern of the observed radiation, and to ensure that all detected photons originate from the interaction zone. The solid angle of detection can be reduced using copper diaphragms. Thus, the counting rates can be limited to less than 0.05 photons per laser pulse and in such a case the detector signal can be assumed to be directly proportional to the photon energy. The energy resolution at 2 keV is approximately 160 eV (full width at half maximum) and the energy scale has been calibrated using low intensity standard x-ray sources. Omitting the copper diaphragms usually leads to the detection of several photons per laser shot, i.e. pileup occurs. In this case the signal is proportional to both, the photon energy and the number of photons. Such spectra correspond well to a Poissonian statistic and the total count of photons per laser shot having a mean energy hv is given by

$$n_p = n_e \frac{E}{h\nu}.$$

(2)

Here n_e corresponds to the total number of events divided by the total number of laser shots and E denotes the mean energy measured by the detector. We employ such pileup spectra for an accurate determination of the photon emission from the target region over more than three orders of magnitude. By varying the detector positions for both horizontal and vertical polarization of the laser beam We have clearly observed an isotropic emission pattern. Thus, respecting the overall detector efficiency, the total photon emission was calculated by integration over the solid angle 4π. A more detailed description of the application of Si(Li) detectors within the present experiment was recently published[8].

RESULTS AND DISCUSSION

Fig. 2 shows typical x-ray spectra of argon, krypton and xenon clusters that have been recorded at reduced counting rates as low as 0.05 counts per laser shot. The respective mean cluster sizes derived from equation 1 are indicated. The highest observable photon energies of 4.1 keV are obtained from xenon, corresponding to $3l \rightarrow 2l'$ transitions of highly charged ions. Moreover, the total power conversion η of the incoming laser power W_L into n x-ray photons of energy $h\nu$ can be directly determined

$$\eta = \frac{nh\nu}{W_L}$$

(3)

For xenon a highest total emission observed was 9.1×10^5 photons, corresponding to a power conversion of $\eta_{Xe} = 7.7 \times 10^{-10}$. Regarding the shape of the spectrum, We currently attribute the broad underlying feature between 1000 eV and 6000 eV to a mixture of a bremstrahlung continuum resulting from scattered highly energetic electrons and, additionally, $4l \rightarrow 3l'$ transitions. In comparison, krypton targets also show a clear $3l \rightarrow 2l'$ transition line, peaking at 1.74 keV. The shape of this line corresponds well to the envelope of previous high-resolution x-ray data by Rhodes and coworkers,[1] that has been assigned to inner shell transitions in Kr^{q+} (q=24-27). In the krypton case We determine a maximum power conversion of $\eta_{Kr} = 2.1 \times 10^{-8}$. Thus, more than two orders of magnitude more efficient than for xenon. The substantially higher conversion corresponds well to krypton's comparably low L-shell ionization potential, that apparently favors transitions around 2 keV at the electron temperatures obtained in our experiment. This interpretation is confirmed when comparing the xenon and krypton results with the power conversion coefficient for argon targets. We have determined a maximum of $\eta_{Ar} = 2.2 \times 10^{-11}$ for the $2l \rightarrow 1l'$ transition at

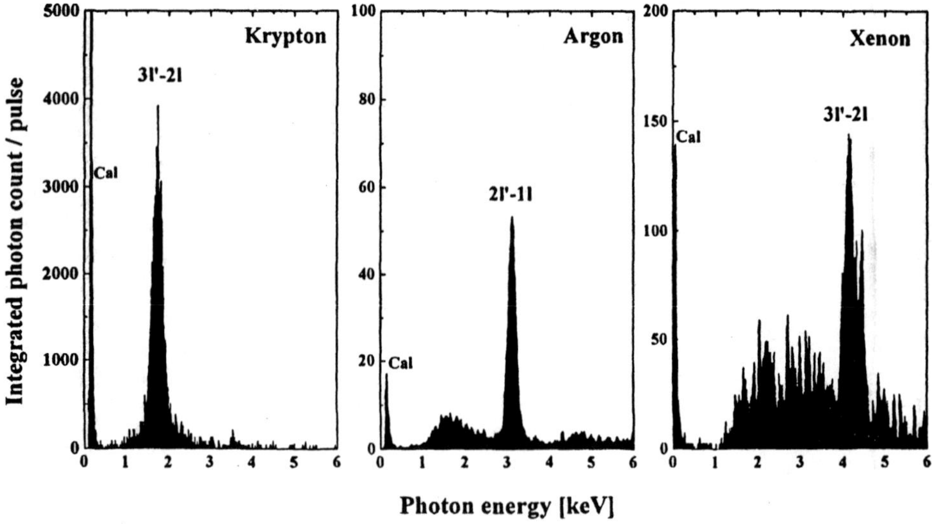

Figure 2. Low intensity x-ray spectra obtained from argon, krypton and xenon clusters. The photon energies correspond to 3l→2l' for Kr (24+..27+) and 2l→1l' for argon (13+..15+). For xenon the photon energy of 4.1 keV can be attributed to 3l→2l' transitions but accurate determination of the average charge state remains difficult. All spectra are calibrated with an electronic signal near 200 eV (Cal).

around 3.1 keV. The shift of the argon x-ray line towards higher energies corresponds to an average atomic charge between 13 and 15. Compared to the photon energies of xenon, the total x-ray emission of argon is rather weak. However, under the present experimental conditions the number of xenon atoms per cluster is about ten times larger compared to argon.

Additionally, it is interesting to determine the dependence of the x-ray emission on the cluster size. In our set-up this can be easily achieved by variation of the stagnant pressure. A typical experimental result is demonstrated in Fig. 3 for the krypton case. It demonstrates, that after an offset of P_c the total x-ray emission of the target region rises proportional to the cube of the stagnant pressure $(P_0-P_c)^3$. This result can be understood by the following arguments. Supposed, that the x-rays are produced by electron-ion collisions within the cluster, the probability of such collisions should scale with the square of the number of cluster constituents, i.e. P^4 when using equation 1. On the other hand, the cluster density in the target region is proportional to P^{-1}. Thus a total x-ray emission following P^3 can be easily understood and is in fact in good aggreement with a collisional heating model. In comparison, ionization ignition should rather be independent on the cluster size, as in that case the ionization occurs during the very first optical cycles when atomic distances are still at minimum. We suggest that the shift P_c of the curve towards approximately 3 bar denotes the minimal cluster size necessary for efficient x-ray production from clusters.

Furthermore, it is interesting to scale the typical x-ray emission of a cluster target to the incoming laser intensity I. A typical intensity dependence obtained from a krypton target is shown in Fig. 4. The data points can be directly compared to a fit function proportional to $I^{3/2}$. Note, that such a dependence appears very similar to results obtained from optical field ionization for low density rare gas targets[9]. From such experiments it is well known that, as soon as the ion yield of a given charge state enters a saturation regime, the signal will increase further with the growth of the focal volume, which is proportional to $I^{3/2}$. We conclude that for the cluster sizes employed the efficiency of the process appears to be saturated already at intensities around 10^{16} W/cm^2. Thus, x-ray generation should be observable at even lower intensities when using larger focal volumes.

106

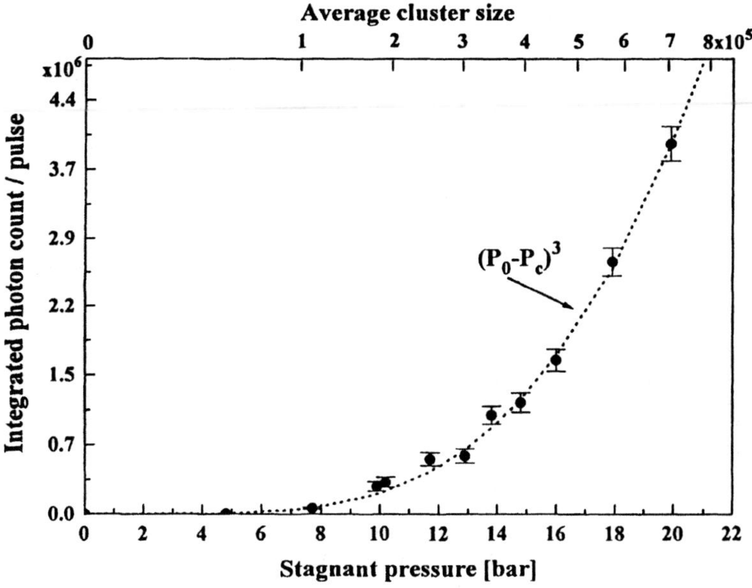

Figure 3. Total number of emitted photons versus stagnant pressure (lower scale) and cluster size (upper scale) for a krypton cluster beam. The emission is compared to a fit using $I=(P_0-P_c)^3$, with $P_c=3$ bar.

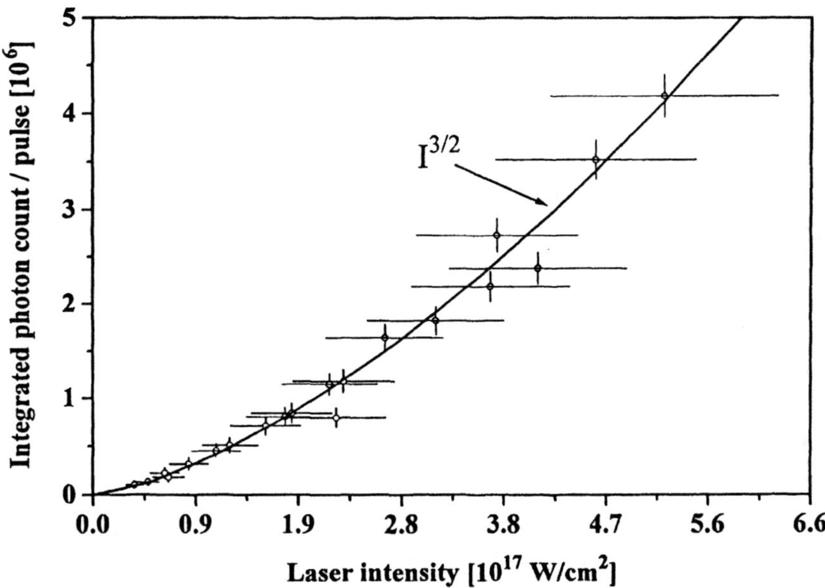

Figure 4. The total number of emitted photons versus the laser intensity I for a krypton cluster beam. The data can be fit with $I^{3/2}$, thus indicating a saturated emission only proportional to the increase of the focal volume. The horizontal error bars are due to the calibration using optical field ionization of a low density neon target. The vertical error bars are based on the Poisson statistics of the x-ray detectors.

Our observations are clearly in variance with scaling laws, recently proposed by Rhodes and coworkers[1], which have suggested threshold intensities exceeding 10^{17} W/cm^2 for the cluster sizes used in our experiment. Moreover, it is possible to compare our data to the theoretical predictions of the ionization ignition model[6]. From the numerical simulations by Rose-Pentruck et al.[6] one should expect a resonant behavior for the x-ray emission at a laser intensity of around 10^{16} W/cm^2. Our experimentally determined intensity dependence suggests that for our data coherent electron motion and ionization ignition models are not applicable in their present form. It is worth mentioning, that our x-ray data correspond well to the recently reported[3] production of Kr^{18+} from laser irradiated cluster targets at approximately 10^{15} W/cm^2 and our observation[7] of Ar^{9+} already at around 10^{14} W/cm^2.

The conversion efficiencies obtained here are several orders of magnitude lower compared to those which are usually observed from solid targets[14,15]. However, this is mainly due to the low density of clusters in the focus. From comparison with our density measurement in Fig. 1 we estimate a very high conversion per atom eventually reaching up to several percent. We currently can not exclude that at certain cluster sizes the average power conversion per atom could be comparable or even larger than in bulk targets. Finally, it is worth mentioning that in contrast to current experiments of Kondo et al.[16] We are unable to observe a notable change of the x-ray yield when switching to the second harmonic (λ=395 nm) while keeping all other experimental conditions constant. We conclude that enhanced electronic coupling does not occur when using shorter wavelengths down to 395 nm. Instead our results suggest that energy conversion into x-rays will be very sensitive to the laser pulse length, since for long laser pulses heating by inverse bremstrahlung occurs over many more optical cycles.

ACKNOWLEDGMENT

The authors acknowledge enlightening discussions with P. Monot and T. Auguste and important technical assistance by E. Caprin and M. Bougard. This work was partially supported by the European union (Grant. No ERBF MBICT 950421).

REFERENCES

[1] A. McPherson, *et al.*, *Phys. Rev. Lett.* **72**:1810 (1994)

[2] T. Ditmire, *et al.*, *Nature* **386**:54 (1997) and *Phys.Rev.Lett.* **78**:2732 (1997)

[3] E. M. Snyder, *et. al.*, *Phys.Rev.Lett.* **77**:3347 (1996)

[4] B.D. Thompson *et al.*, *J. Phys. B* **27** :4391 (1994)

[5] A.B. Borisov *et al.*, *J. Phys. B.* **28** :2143 (1995)

[6] Rose-Petruck et al., *Phys. Rev. A* **55** (2) : 1182 (1997)

[7] M. Lezius, *et al.*, *J. Phys. B* **30** : L251 (1997)

[8] S. Dobosz, *et. al.*, *Phys.Rev.A.:Rapid Comm.* **56**(4) (1997)

[9] T. Auguste, *et al.*, *J. Phys. B* **25**: 4181 (1992)

[10] O. F. Hagena, *Z. Phys. D* **84**:291 (1987)

[11] J. Farges, *et al.*, *J. Chem. Phys.* **84**: 3491 (1986)

[12] O. F. Hagena, *Surf. Sci.* **106**: 101 (1981)

[13] P. Lechner and L. Strüder, *Nucl. Instrum. Methods Phys. Res. A* **354**: 464 (1995)

[14] D. G. Stearns, *et al.*, *Phys. Rev. A* **37**: 1684 (1988)

[15] S. P. Gordon, *et al.*, *Opt. Lett.* **19**: 484 (1994)

[16] K. Kondo, *et al.*, *J. Phys. B* **30**: 2707 (1997)

HARD X-RAY EMISSION FROM FEMTOSECOND LASER INTERACTION IN OVERDENSE PLASMAS

A.A. Andreev,[1] V.N. Novikov,[1] K.Yu. Platonov,[1] and J.-C. Gauthier[2]

[1] ILPh, SC "S.I.Vavilov State Optical Institute"
   12, Birzhevaya line, St. Petersburg, Russia
[2] LULI, Ecole Polytechnique,
   91128 Palaiseau Cedex, France

INTRODUCTION

The recent development of ultra-short pulse lasers has made possible the investigation of laser matter interaction at ultra-high intensities. For sub-picosecond pulses, a hot and overdense plasma is produced very rapidly during the rise of the pulse and further laser interaction occurs with this plasma. One of the results of the interaction is the generation of fast electrons and of intense hard X-ray emission. The X-ray pulse duration is determined by the mean free path of the fast electrons in the target material. It can be very short (< 1 ps) and its intensity sufficient to be detected by the usual methods. With high laser pulse repetition rates, it has been demonstrated[1] that one can obtain an instantaneous signature of fast X-ray dense-matter interaction processes. The high energy of the X-ray photons (up to ~ 1 MeV) makes it possible to study small size objects and even to excite nuclear levels.

Numerous papers[2-7] have been devoted to the study of femtosecond laser pulse interaction with plasmas. In this paper, we calculate the electron energy distribution function in the presence of the laser field, the absorption coefficient, and the parameters of the fast electron flux in the plasma. Our absorption results are in agreement with previously published[3-5] results. A new feature of our calculations is the determination of the energy and spectrum characteristics of the hard X-ray pulse produced by the interaction of an intense laser with a solid-state target.

BASIC SET OF EQUATIONS AND NUMERICAL SOLUTION

We consider a plane linearly polarized electromagnetic wave with oblique incidence on a semi-infinite plasma. The plasma temperature is T_e and the electron density n_e is higher than the critical density n_c. The wave with amplitude E_0 and frequency ω_0 is chosen in such a way that the electron quiver velocity v_e is larger than the thermal velocity v_T. The scale length of the electron density gradient is very much smaller than the depth of the skin layer l_s. This is justified by the fact that during the laser pulse (< 100 fs) the movement of the ions is negligible and the plasma edge preserves its sharpness. The dynamics of the plasma electron component is described by a self consistent set of equations involving the collisionless kinetic Boltzmann equation for $0 < x < l_s$, the Maxwell equations for electromagnetic fields (in covariant form), and the kinetic Fokker-Plank equation for $x > l_s$. Oblique incidence is boosted into normal incidence by using the method of Gibbon.[5] The ambipolar field E_a is determined from the constraint of zero current along the x laser axis.

We use the requirement of mirror-like reflection of the electrons by the edge of the plasma. This model was implemented in our code KINET1D2V. In Figure 1a, we show the distribution function f_e calculated at time 180 $\omega_0 t$ and position x=3l_s. The calculation was carried out with a laser intensity of $I = 10^{18}$ W/cm^2, a plasma temperature of $T_e = 20$ keV, and a density of $n_e = 15 n_c$. One can see that f_e is significantly distorted in comparison with the starting Maxwellian distribution. This indicates that the laser ponderomotive force expels the electrons from the interaction region into the plasma. So, twice during the period of laser oscillation, a flux of fast electrons is produced which propagates inside the plasma and generates intense x-rays.

In Figure 1a, the fast electron flux reveals itself as a tail in the distribution function. The average hot electron energy is ~ 0.5 MeV. The characteristic lifetime of the fast electron $t_h = 1/(n_i\, v_h\, \sigma) \sim 0.1$ ps (here σ is the cross-section for inelastic scattering of the fast electron with velocity v_h in the matter with ion concentration n_i) can exceed the duration of the laser pulse. These fast electrons can leave the ionized region and penetrate into the solid target.

However, the ion density is almost the same in the plasma and the solid matter, so that bremsstrahlung and X-ray line emission have the same behavior in the plasma and the solid target. Analysis of the calculated distribution function at oblique incidence angle θ shows that the constant magnetic field which builds up in the normal incidence frame

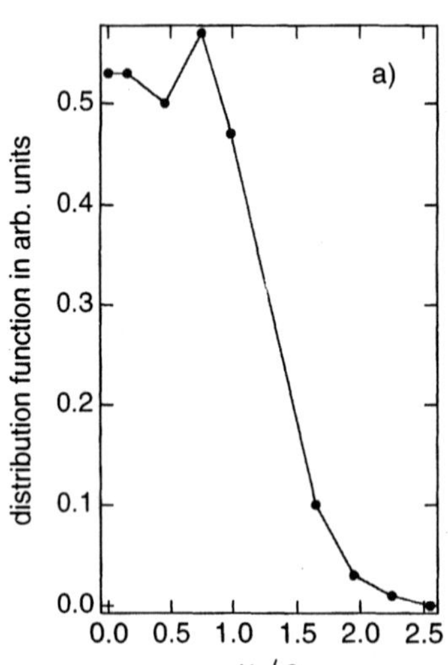

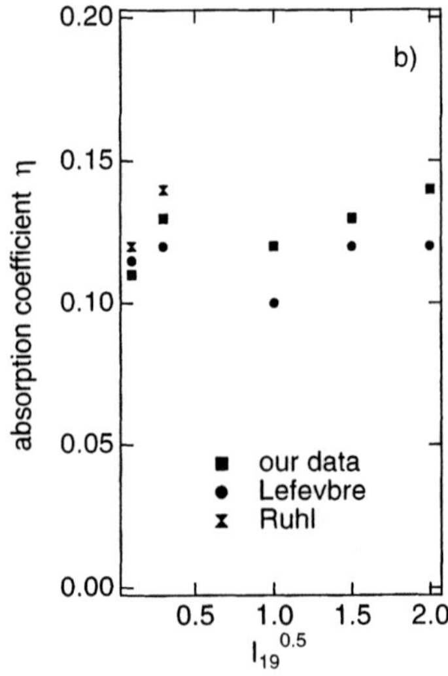

Figure 1. a) Distribution function for normal incidence. x / l_s=3, $\omega_0 t$ = 180. b) Absorption coefficient η as a function of the square-root of laser intensity in units of 10^{19} W/cm^2.

(boosted frame) results in an acceleration of the fast electrons in the longitudinal x direction during one half of the laser oscillation period and in a deceleration during the other half of the period. In Figure 1b, we show the calculated absorption coefficient η as a function of laser intensity. We note that the change in the absorption coefficient is small with laser intensity as in previous works.[4-6]

Noteworthy, for low intensities the absorption depends on the incidence angle with a characteristic maximum for $\theta \sim \pi/2$; for higher intensities ($> 10^{18}$ W/cm^2) such a dependence with the incidence angle is absent. In this case, the absorption coefficient reaches the same value as for normal incidence, similarly to what was found by Ruhl.[4] The reason is that high harmonics generation and strong magnetic field production perturb the plasma modes. The mode frequencies becoming very different from the laser frequency ω_0, this decreases the efficiency of conversion in the plasma waves — the source of the resonant character of the absorption.

EVALUATION OF THE X-RAY PULSE PARAMETERS

Spectral intensity of the hard bremsstrahlung x-ray emission from a unit volume of the hot plasma is determined in our work for an optically thin plasma. This requirement is fulfilled for high energy photons (>10 keV), the case that we consider in this paper. If the intensity of laser radiation is high ($>10^{18}$ W/cm^2) the velocity v_h of the hot electrons approaches the speed of light c. In this case the dependence of the bremsstrahlung emission on the electron velocity and, hence, on the specific distribution function, is rather weak (logarithmic). The x-ray pulse shape of photons with a frequency of $\hbar\omega_x = 30T_e$ is shown in Figure 2a as a function of time for a thick target.

In our numerical calculations, we have used the distribution function f_e that was found before. Pulse duration (~ 0.1 ps) is determined by the hot electrons lifetime t_h. The intensity of the X-ray emission from the plasma I_x can be evaluated from the parameters of the incident wave:

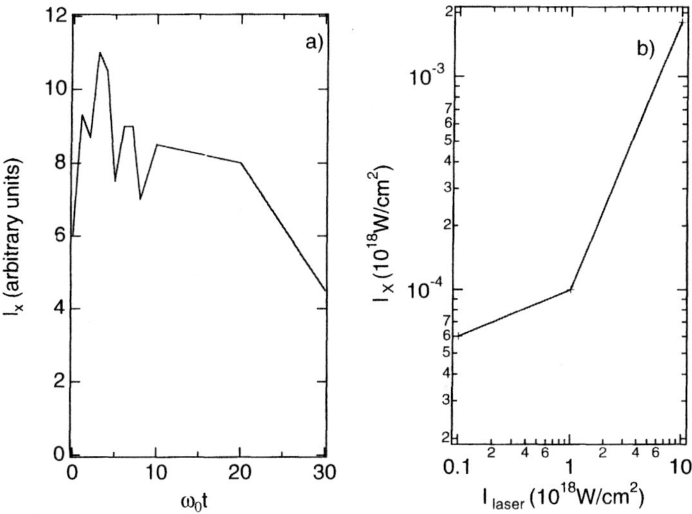

Figure 2. a) Spatially averaged [0, $10l_s$] intensity of X-ray radiation as a function of time $\omega_0 t$. b) Intensity of X-ray radiation as a function of laser intensity in units of 10^{18} W/cm^2.

$$I_x \approx (Ze^7 n_e^2 E_0 l_x)/(m^2 c^3 \hbar \omega_{x0})$$

where Z is the plasma ion charge, ω_0 the x-ray frequency and l_x the photon mean free path length. Spectral intensity depends, to a low extent, on the photon frequencies for the whole spectral range of generated photons from $T_e = 20$ keV to $e^2 E_0^2 / m \omega_0^2$ (~ 500 keV). Laser radiation to X-ray emission conversion efficiency can be as high as 10^{-4}. In Figure 2b, we show the hard X-ray conversion efficiency as a function of laser intensity. We can see an increase of the conversion efficiency, roughly as $I_0^{1.25}$. This dependence is in good agreement with the experimental results of Kmetec.[7] Approximately 0.1 ps (lifetime of hot electrons corresponding to their mean free path) after switching off the laser radiation, X-ray emission does not disappear completely, but is replaced by the equilibrium, comparatively long (~ 1 ps), thermal X-ray emission of the residual plasma.

Acknowledgments

This work was partly supported by the INTAS grant 94-934 and by the Centre National de la Recherche Scientifique.

REFERENCES

1. F. Raksi, K.R. Wilson Z. Jiang, A. Ikhlef, C.Y. Cote, and J.C. Kieffer, Jl. Chem. Phys. 104:6066 (1996).
2. A. Rousse, P. Audebert, J. P. Geindre, F. Fallies, J.-C. Gauthier, A. Mysyrowicz, G. Grillon, and A. Antonetti, Phys. Rev.E 50:2200 (1994).
3. S.C. Wilks, Phys. Fluids 5:2603 (1993).
4. H. Ruhl, Proceedings of Conference LIRPP, (Osaka, Japan, 1995), p.65.
5. P. Gibbon, Phys. Rev. Lett. 68:1535, (1992); *ibid* 73:664 (1994).
6. E. Lefevbre, Ph. D. Thesis, Paris (1996).
7. J.D. Kmetec, Phys. Rev. Lett. 68:343 (1992).

PLASMA WAVEGUIDE: DENSITY DEVELOPMENT AND HIGH INTENSITY GUIDING

Thomas R. Clark Jr., Sergei P. Nikitin, Yuelin Li and Howard M. Milchberg

Institute for Physical Science and Technology
University of Maryland
College Park, MD 20742

ABSTRACT

Recent experiments characterizing the formation, development, and guiding properties of a plasma channel suitable for the guiding of high intensity sub-100 fs laser pulses from a chirped pulse Ti:Sapphire laser system are presented. Time and space resolved interferometry measurements of the plasma density evolution have allowed for the precise determination of the channel size, depth and local uniformity of the guide and have provided the experimental details necessary for the determination of the physical mechanisms involved in its evolution.

THE PLASMA CHANNEL

The basic understanding of the plasma channel formation, obtained from previous guiding experiments and numerical modeling[1], is that it is the result of the expansion of a laser breakdown spark formed in a background gas. The early portion of the laser pulse produces an initial plasma density through multiphoton ionization. These plasma electrons are then heated through inverse bremstrahlung absorption, driving further ionization through collisions to a level which is determined predominantly by the initial gas density and laser intensity. These heated electrons drive the expansion of the ion density into the background gas resulting in the formation of a shock wave due to ion-atom and ion-ion collisions at the plasma-gas boundary and an on-axis ion density minimum. Since the plasma remains essentially charge neutral during this process, an on-axis electron density minimum follows the ion density providing the desired index profile for guiding[1].

Plasma Density Measurements

The electron density profiles in these experiments were measured using a folded wave front interferometer in a pump-probe configuration[2] as shown in Figure 1(a). A Nd:YAG based regenerative amplifier/power amplifier seeded with pulses from a mode-locked oscillator provided both the pump pulse (1.064 μm, 100 ps, 500 mJ) and the frequency doubled probe pulse (0.532 μm, 70 ps, 100 μJ) which was split off earlier in the amplifier chain. The doubled probe having the benefit, compared to the fundamental, reduced refraction due to the λ^{-2} dependence of the critical density. The pump beam produced the plasma in the ~1cm line focus of a 35° base angle axicon with intensities of ~5x10^{13} W/cm^2. The probe pulse of diameter ~1 cm was passed through a corner cube optical delay line providing delay relative to the pump of -1 to 11 ns and then was passed transversely through the cylindrically symmetric plasma. A matched two lens system[3] served to relay the image of the plasma through the interferometer, which was a BK7 glass wedge plate producing two vertically separated complementary interferograms. One of these was imaged onto a charge coupled device (CCD) camera and digitized by a

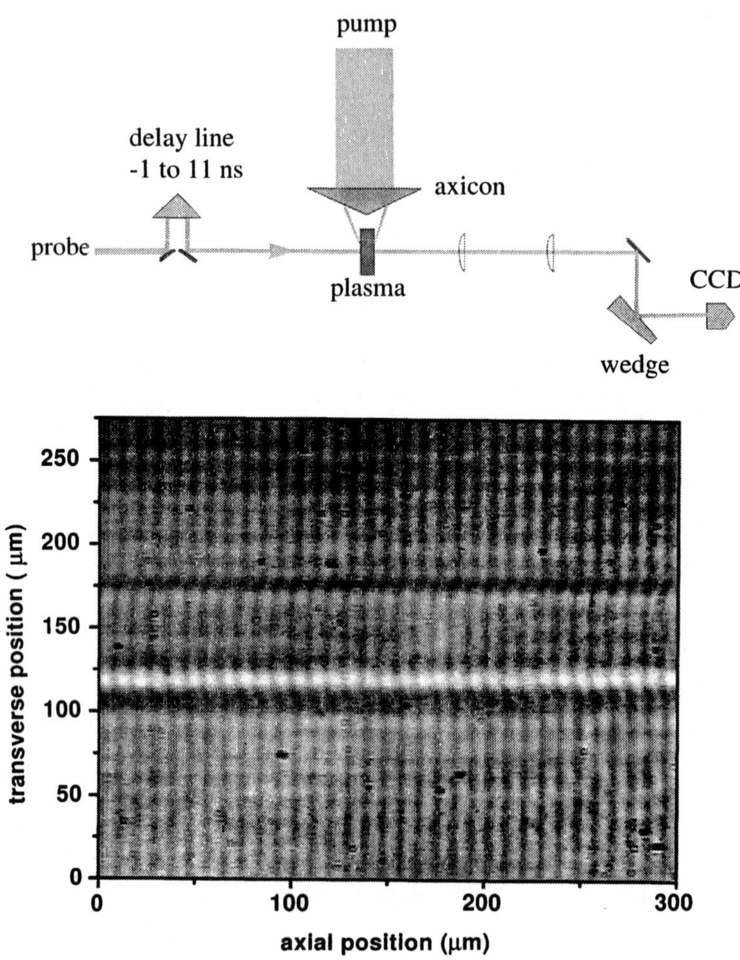

Figure 1: (a) Experimental setup for interferometry measurements. (b) Interferogram of central 300 μm of a ~ 1 cm waveguide formed in a 150 torr Kr gas probed at a delay of 1.1 ns.

frame grabber. The system spatial resolution of ~1.5μm/pixel was set by the magnification of ~25X and the temporal resolution was set by the probe pulse duration of 70 ps.

A sample interferogram is shown in Figure 1(b), where the top and bottom portions of the image are the unperturbed regions where the probe passes by the plasma and the central region shows the plasma induced phase shift as the bending of the vertically oriented fringes. The high degree of axial and radial symmetry is clearly evident from the image and plays an important role in the data analysis as will be seen below. Also apparent in the image is the rapid phase gradient at the plasma-gas boundary where a shock front has formed. The phase was extracted from images like this one using FFT methods[4] which allowed the separation of the interesting plasma induced phase shift which rides on the fringe modulation due to the wedge from the low frequency information due to the background intensity nonuniformity. While some refraction is expected due to the phase gradient in the shock region, a ray trace analysis and a refractive fringe diagnostic[5] employed have been used to determine a refracted transverse distance of $L_T < \lambda/10$ indicating that refraction is a negligible effect for the conditions studied here.

Extraction of the electron density was then reduced to the well known Abel inversion problem for a cylindrically symmetric object from which chordally integrated information is known.[6] Averaging of the raw phase data is used in regions of axial uniformity, such as in Figure 1(b), and this greatly reduces uncertainty from random fluctuations in the phase. Figure 2 shows electron density plots obtained from the Abel

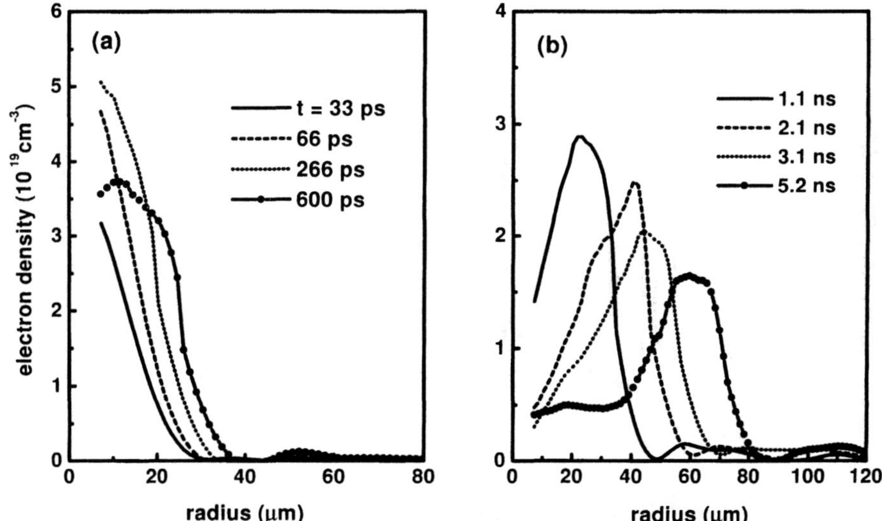

Figure 2: Electron density profiles obtained from Abel inversion of raw phase data for 150 torr Kr gas backfill.

inversion of raw phase data averaged over ~200 lineouts for a gas backfill of 150 torr of Krypton. Experimental uncertainty in the electron density from calibration of the imaging system was measured to be less than 5%.

Density Development Dynamics

The general features of channel formation discussed above are evident in Figure 2 and are consistent with those previously predicted by our calculations and suggested by our guiding experiments. Figure 2(a) shows the electron density evolution up to 600 ps after the plasma forming pulse. The initial electron density is created in the line focus of the axicon and is seen to rapidly grow to a maximum on axis of 5 x 10^{19} cm^{-3}. This growth occurs during and immediately after the laser pulse and continues for ~300 ps past the peak, indicating that at this density, the contribution of heating and subsequent

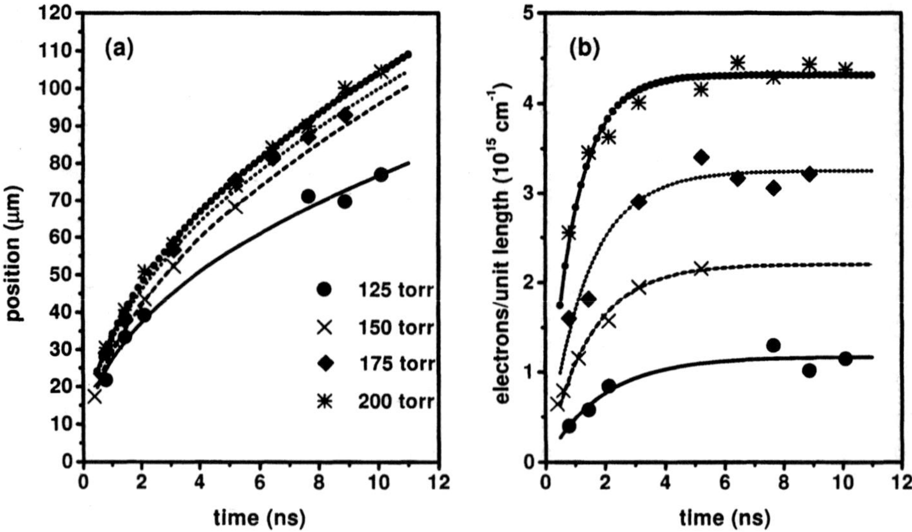

Figure 3: (a) shock position vs. Time for gas backfills of 125, 150, 175, and 200 torr Kr showing cylindrical blast wave dynamics. (b) Number of electrons per unit length calculated as the radial integral of the electron density profiles vs. Time.

collisional ionization is important. The maximum on-axis electron density at early times (< 300 ps), before expansion and shock development occur, can be used to obtain the peak average ionization level through $Z \sim N_e/N_i$, where N_i corresponds to the original atom density in the gas target, obtained from the measured gas fill pressure. Since recombination occurs on a time scale of many nanoseconds, this average ionization level will be maintained for the full range of delays measured here . By 600 ps, the plasma has begun to expand into the background gas resulting in the onset of shock development and an on-axis electron density minimum. Note that the density minimum develops well within 1 ns, consistent with earlier experiments showing the onset of leaky guiding around this delay, where the shock wall width and depth is still insufficient to confine the lowest order guided mode[7]. A channel suitable for guiding is seen to persist to beyond 10 ns after plasma creation, with the on axis section flattening and the expansion slowing considerably as will be discussed below.

The waveguide expansion for three different pressures (under the same peak pump intensity of 5×10^{13} W/cm^2) is shown in the shock position vs. time plots of Fig. 3(a). The data points are shown least squares fitted to the expression $R_s = \alpha t^m$, where the fits give m$\cong$0.5 to within ~10% for all cases, and the fit to α is a function of pressure. This is in good agreement with the expression for self-similar expansion of a cylindrical blast wave[8], $R(t) = \xi_0 (E_{th}/\rho_0)^{1/4} t^{1/2}$, where E_{th} is the thermal energy per unit length initially driving the expansion, ρ_0 is the initial mass density, and ξ_0 is a dimensionless parameter which is a function of the specific heat ratio and is of order unity. This expression applies to the phase of the expansion where a hot 'fluid' enclosed by the shock is cooling through expansion work. Using the fitted values of α for each pressure, and $\xi_0 \approx 1$, gives E_{th} = 2.92, 4.8, 8.78, and 11.8 mJ/cm for the 125, 150, 175, and 200 torr cases. Our measured pump absorption fractions are typically ~5-15% in this range of pressure, indicating that rapid cooling has already taken place at earlier times, within a few hundred picoseconds of the pump pulse. Shock speeds are calculated as the derivative of the curve fits and show a rapid decrease in the expansion rate within the first 2 nanoseconds from values of ~2 x 10^6 cm/sec to ~ 1 x 10^6 cm/sec for all cases. Since the pressure of the surrounding gas is negligible compared to the plasma pressure,

116

the shock speed is related to the temperature T_e by $c_s \approx (ZkT_e/m_i)^{1/2}$. Using this expression and measuring $Z \sim 8$ by the method described above gives a profile average temperature drop of $T_e \sim 40$ eV to ~ 10 eV for the first 2 ns. At longer times up to the maximum delay of 11 ns, the shock speed approaches 5×10^5 cm/sec, corresponding to a temperature of ~ 3 eV. The curve fits of Fig. 3(a) suggest that the temperature drop over the range of delay shown is attributable to expansion work, with other channels contributing weakly if at all.

Additional ionization at the plasma edge would make conductive losses more important. Figure 3(b) shows the number of electrons per unit length of channel, calculated as $Q = \int_0^{r_m} 2\pi r N_e(r)dr$, where r_m is a radial position outside of the plasma. The data is shown fitted to the expression $Q=\Delta Q(1-exp(-t/\tau))$, where τ is a time constant for ionization increase. The fits give $\tau = 1.89$ ns, 1.47 ns, 1.37 ns, and 0.96 ns for the 125, 150, 175, and 200 torr Kr cases respectively. Inspection of the density profiles shows that beyond a few hundred picoseconds any continuing growth in ionization originates near the plasma edge, where collisional ionization of the periphery neutrals takes place. Little additional ionization is seen to occur during the later times where the temperatures of all cases are much less than that required for ionization of a neutral atom (I_p=14 eV for Kr). The decrease in the time constant τ with pressure can be readily explained by considering the scaling with radius and peak electron density of the rate of charge increase, $dQ/dt \approx 2\pi R_s(c_s N_{es} + \Delta R_s(\partial N_{es}/\partial t))$, where N_{es} is the peak density at the shock and ΔR_s is the approximate shock width.

HIGH INTENSITY GUIDING

The above described plasma channel has been shown to be effective in accomplishing a large intensity-interaction length product at modest intensities[1,7]. Here we investigate its application the next generation of high intensity experiments.

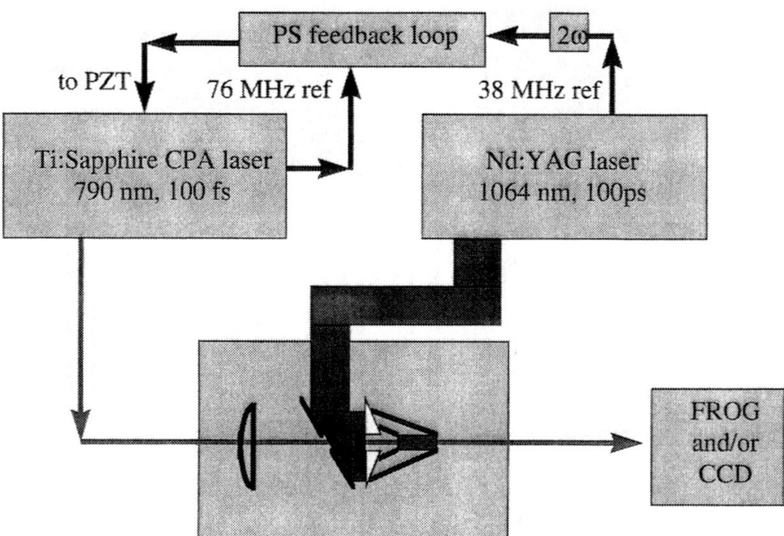

Figure 4: Schematic of guiding experiment. Shown at the top is the phase sensitive (PS) feedback loop with the two 76 MHz reference inputs and a driving voltage for a piezo-electric element mounted on one of the oscillator end mirrors.

Guiding Experiments

Shown in Figure 4 is a schematic of the giding experiments. As seen in the above discussion of the plasma channel, the relevant time scales for channel formation and evolution are in the range of 0.1 to 1 ns, where the fastest time of 0.1 ns is the growth time of the shock wave which forms the channel boundary and the longer time of ~1 ns is the scale of overall radial expansion. From this we would expect that synchronization of a few hundred picoseconds between the plasma forming pulse and pulse to be guided would be sufficient. This level of synchronization was achieved between the plasma forming laser (100 ps Nd:YAG laser system) and a sub-100 fs Ti:Sapphire chirped pulse amplification system using a feedback loop to tune the cavity length of the Ti:Sapphire oscillator. A piezo-driven mirror was adjusted to match the phase of the Ti:Sapphire pulse train with the phase of the reference radio frequency of the Nd:YAG oscillator acousto-optic modelocker. This synchronization was verified by producing a sum frequency mixing signal with the output of both oscillators in a 1mm KDP crystal. The full width at half maximum of the cross-correlation signal was found to be 109 ps, indicating a synchronization of less than one Nd:YAG pulsewidth.[9] The output of the Ti:Sapphire laser system (0.78 µm, 90 fs, currently up to 20 mJ) was focused at f/18 onto the entrance of the preformed plasma channel. In order to limit nonlinear phase distortions due to self-phase modulation in the chamber windows and focusing optics, the pulse energy was kept below 10 mJ. The delay between the 10 Hz repetition rate amplified Nd:YAG pulses and 10 Hz amplified Ti:Sapphire pulses was adjusted by passing the modelocker reference signal through a variable transmission line delay. A lens recollimated the channeled output and a beam splitter allowed simultaneous FROG and exit mode image measurements. The image at the channel exit was compared to the vacuum focus image to determine the coupling efficiency. Interferometry measurements were also made of the coupling and guiding process using the folded wavefront interferometer.

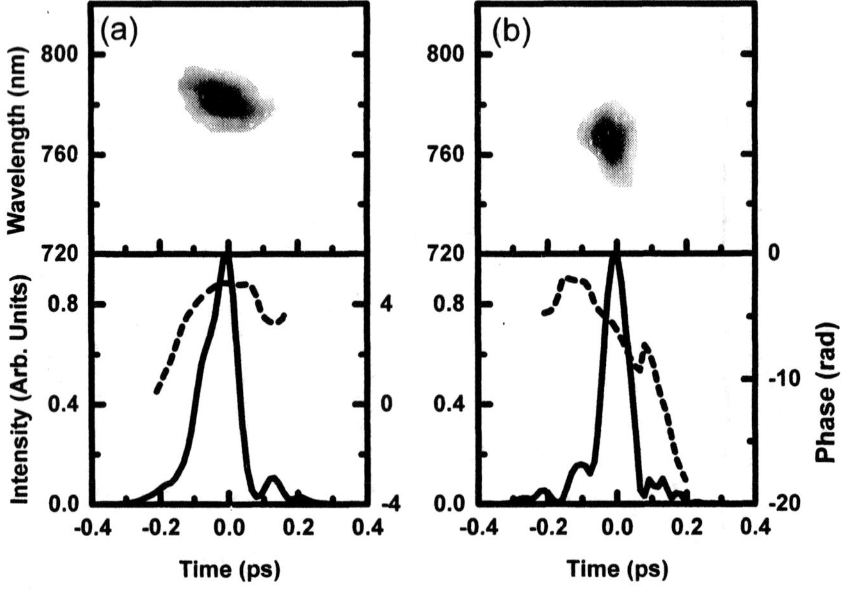

Figure 5: (a) FROG image and retrieved pulse amplitude and phase for an evacuated chamber (b) FROG image and retreived pulse amplitude and phase for a guided pusle intensity of ~5 x 10^{15} W/cm^2.

Pulse Guiding Dynamics

Plasma channels were formed in a gas mixture of 300/50 torr He/N₂O for these guiding experiments. Helium was used to limit the amount of ionization due to the Ti:Sapphire pulse as it approached the focus prior to reaching the channel entrance. However, the injection of the femtosecond pulse was still accompanied by a rapid field ionization of neutral gas in advance of the entrance region, as evidenced by the lengthening of the spark at the channel entrance. The coupling efficiency into the lowest order mode, of ~25 μm FWHM spot size, was measured to be 30% resulting in a guided intensity of 5×10^{15} W/cm². For small channels (delays of <2-3 ns), the channel exit plane image typically consisted of a central guided mode surrounded by blue shifted light outside the channel, indicating time-dependent ionization-induced defocusing at the channel entrance. At delays longer than ~5 ns, the blue shifted component appeared in the spectrum of the now spatially multimode guided beam, suggesting guiding of the blue shifted scattered light. Spatially dependent blue shifting attributed to laser-induced ionization and defocusing has been measured previously in short pulse-gas interaction experiments.[10]

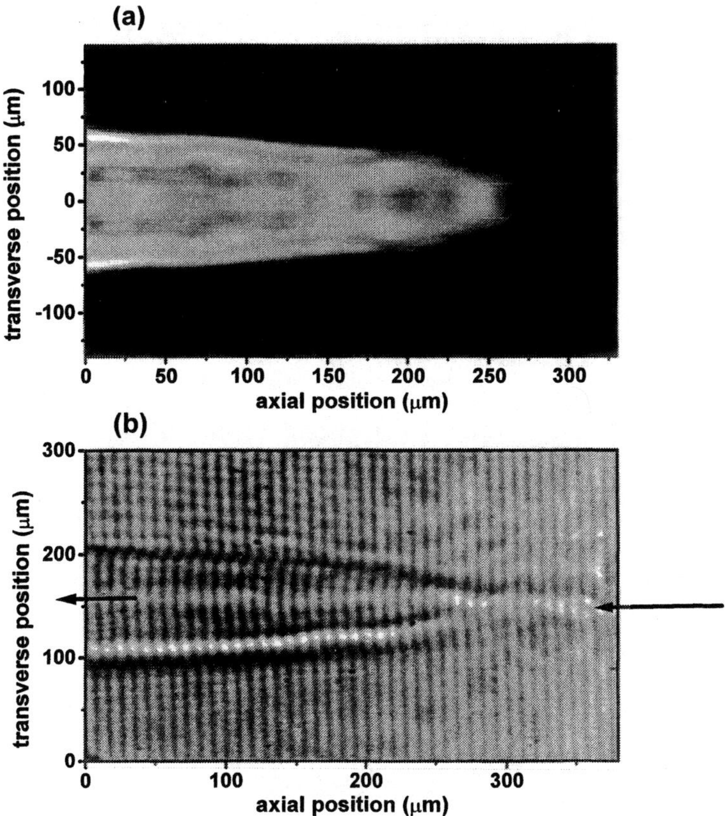

Figure 6: (a) Grayscale image of the electron density at the end of a ~1 cm waveguide formed in a 380/20 torr Ar/N₂O gas backfill and probed at a delay of 4 ns. The whitest portions indicate $N_e > 3 \times 10^{19}$ cm⁻³, the central gray regions are typically ~1 × 10¹⁹ cm⁻³ and the black region indicates $N_e < 10^{18}$ cm⁻³. (b)Interferometric image showing the effect on the electron density of the coupling of a 100 fs, 10 mJ Ti:Sapphire pulse focused at f/12 (vacuum intensity ~10¹⁷ W/cm²) into the end of the channel of (a). The delay between the interferometry probe and the Ti:Sapphire pulse is 700 ps. The Ti:Sapphire pulse path can be followed from right to left and is indicated by the two arrows.

The effect of injection and guiding on the envelope and phase of the intense 90 fs FWHM pulse was studied using the polarization gated frequency resolved optical gating (PG-FROG) technique.[11] Shown in Figure 5 is the FROG trace and retrieved pulse phase and amplitude for the case of the evacuated chamber (~channel input) and the output of the channel for a 2 ns delay. The input pulse of Fig. 5(a) shows a small amount of chirp due to self-phase modulation in the chamber windows and focusing optics and shows a pulse duration of ~100 fs. The output pulse of Fig. 5(b) shows a shorter pulse duration of ~70 fs and a negative going phase shift indicative of ionization occuring during the pulse. Further studies of the propagation of an intense 100 fs pulse in a gas jet and backfill of He indicate that the pulse shortening is due to refraction of the tail of the pulse by a plasma created by field ionization as the beam is focusing down to the channel entrance.[9]

The coupling of the intense femtosecond pulse into the channel was studied using the interferometric techniques discussed above. A grayscale image of the electron density at the entrance of a channel formed in a gas backfill (380/20 Ar/N_2O) and probed ~ 4 ns after plasma generation with the guided beam blocked is shown in Figure 6(a), which has been reflected through the axis to show the channel for the convienience of the reader. Here the whitest portions at the top and bottom edges correspond to electron densities > 3 x 10^{19} cm^{-3} and the gray regions within the channel are typically ~1 x 10^{19} cm^{-3}. An on-axis minimum in electron density begins within the first 50-100 μm of the experimentally measurable end of the channel, indicating there would be little refraction of the input pulse away from the channel due to the tapered end. Figure 6(b) shows the coupling into and guiding of the pulse by the plasma channel. The Ti:Sapphire pulse enters the picture from the right (indicated by an arrow) and the resulting ionization due to the guiding of the pulse is clearly evident as the brighter track running from right to left. Here the channel end has been probed 700 ps after the guided pulse has passed. The higher atomic fill pressure was chosen to allow the better visualization of the ionization track left by the guided pulse.

CONCLUSION

We have experimentally characterized the formation, development, and guiding properties of a plasma channel and shown it to be suitable for the guiding of the sub-100 fs laser pulses. The fastest time scales relevant to the plasma channel, for shock development and channel formation, have been measured to be a few hundred picoseconds and we have demonstrated the ability to synchronize two laser systems to well within that time. The precise determination of the channel size, depth and local uniformity of the guide provides a new level of control for determining the desired initial conditions for many applications.

REFERENCES

1. C. G. Durfee III and H. M. Milchberg, Phys. Rev. Lett. **71**, 2409 (1993).
2. T. R. Clark and H. M. Milchberg, Phys. Rev. Lett. **78**, 2373 (1997).
3. J. A. Stamper, S. H. Gold, S. P. Obenschain, E. A. McLean, and L. Sica, J. Appl. Phys. **52**, 6562 (1981).
4. M. Takeda, H. Ina, and S. Kobayashi, J. Opt. Soc. Am. B **72**, 156 (1981).
5. M.M. Michaelis and O. Willi, Opt. Comm. **36**, 153 (1981).

6. I. H. Huchinson, *Principles of Plasma Diagnostics* (Cambridge University Press, Cambridge, 1987).
7. C. G. Durfee III, J. Lynch, and H. M. Milchberg, Opt. Lett. **19**, 1937 (1994).
8. Y. A. Zeldovich, and Y. P. Raizer, *Physics of Shock Waves and High-Temperature Hydrodynamic Phenomena*, (Academic Press, New York 1966).
9. S. P. Nikitin, T. M. Antonsen, T. R. Clark, Yuelin Li, and H. M. Milchberg, Opt Lett. **22**, 1787 (1997).
10. S. P. LeBlanc and R. Sauerbrey, J. Opt. Soc. Am. B **13**, 72 (1996); W. M. Wood, C. W. Siders, and M. C. Downer, Phys. Rev. Lett. **69**, 772 (1992).
11. D. J. Kane and R. Trebino, Opt. Lett. **18**, 823 (1993).

EXPLOSION OF ATOMIC CLUSTERS HEATED BY HIGH INTENSITY, FEMTOSECOND LASER PULSES

T. Ditmire*, J. W. G. Tisch, E. Springate, M. B. Mason, N. Hay, J. P. Marangos, and M. H. R. Hutchinson

Blackett Laboratory, Prince Consort Road
Imperial College of Science, Technology, and Medicine
London, SW7 2BZ
United Kingdom

*Current address: Laser Program, Lawrence Livermore National Laboratory, L-440, Livermore, CA 94550

INTRODUCTION

Atomic clusters have long been studied by chemists and physicists because of the unique position that clusters hold as an intermediate state between molecules and solids [1]. Many studies have traced the properties of materials from their monatomic characteristics to their bulk state characteristics through an examination of the material as it forms larger and larger clusters. Recently, there has been much activity in extending these studies to very high intensity, ultrashort laser pulses with peak laser intensities $>10^{15}$ Wcm^{-2} and pulse widths of 0.1 to 10 ps [2-11]. There has also been some preliminary theoretical work in this area as well [6,12]. In this parameter regime the physics governing the laser cluster interaction is fundamentally different than in previous studies. At these intensities the laser interaction is non-perturbative and very high order multiphoton ionization and strong electric field tunnel ionization are possible. Consequently, highly charged ions can be produced [2,5,8,10]. Furthermore, the short pulses used are comparable to or shorter than the disassembly times of a cluster in the laser field [6] and, so, the entire laser pulse interacts with an inertially confined body of atoms.

In this paper we present an investigation of the physics of intense short pulse interactions with noble gas clusters over a range of cluster sizes, species and laser wavelengths. Direct measurements of the electron and ion energies resulting from the interactions of laser pulses with isolated clusters have been obtained. Both electrons and ions are ejected from the heated cluster with substantial kinetic energy. We find that electrons with energies up to 3 keV and ions with energies of up to 1 MeV are produced. These experimental observations are well explained by a theoretical model of the cluster as a small plasma sphere that explodes following rapid electron collisional heating by the intense laser pulse.

Applications of High Field and Short Wavelength Sources
Edited by DiMauro *et al.*, Plenum Press, New York, 1998

DESCRIPTION OF THE EXPERIMENT

In our experiment a beam of atomic clusters, produced in the expansion of a high-pressure gas into vacuum, is irradiated by a focused, high-intensity, femtosecond laser beam. The electrons and ions expelled from the clusters with velocities perpendicular to both the cluster beam and the laser beam propagate along a time-of-flight (TOF) tube and are detected by a micro-channel plate detector (MCP). The ion energies are then determined by measurement of their flight time; the electron energies are found by measuring the decrease in the MCP signal as a retarding voltage is applied to a grid placed between the focus and the MCP.

The laser used was a high-power Ti:sapphire laser, based on the principle of chirped pulse amplification, which delivers 150 fs pulses at a wavelength of 780 nm. The laser is focused to a peak focused intensity of ~2 x 10^{16} Wcm^{-2} with 20 mJ of laser intensity. A solenoid pulsed gas jet valve produced the noble gas clusters in our experiment. The clusters produced by the gas jet were collimated by a skimmer into a low density cluster beam which intercepted the laser beam at the focus.

The electrons emitted in the interaction were detected by a two-stage micro-channel plate (MCP) placed at the end of a 17 cm flight tube, oriented perpendicular to both the cluster beam and the laser beam. The area subtended by the MCP detector limited the detection angle to 3.5°. Two grids (spaced 3mm apart) were placed immediately behind the entrance to the flight tube. The first grid was charged to a voltage Φ and the second was grounded, introducing a potential barrier to electrons with energy less than $e\Phi$. The front plate of the MCP was grounded and the back held at +2 kV. The electron signal was measured as a function of retarding voltage and differentiated to give the electron energy spectrum. Each data point is the average of 50 shots, taken within a ± 10% laser energy bin. The ion energies were determined by measuring their flight time in the field free drift tube, which was extended to 38 cm or 80 cm for these measurements. The front plate of the MCP was held at -2 kV and the back plate was at earth. A grounded metal grid placed ~2 mm before the MCP ensured that the flight tube was field free.

EXPERIMENTAL RESULTS

The measured energy spectrum of electrons emitted along the direction of the laser polarization during the irradiation of clusters of ~2100 xenon atoms (4.5 bar backing pressure) with an intensity of 1.5 x 10^{16} Wcm^{-2} is shown in figure 1. There are two distinct features in the electron energy spectrum. The first, broad peak consists of what we shall call 'warm electrons' with energies ranging from 0.1 to 1 keV. A second, sharper peak (referred to as the 'hot electrons' throughout this paper) appears at 2.5 keV. Both peaks are only present when the laser interacts with Xe clusters. When the interaction region contains only monatomic xenon (in a static fill) we detect no electrons with energy above 100 eV.

The most remarkable aspect of this energy distribution are the high electron kinetic energies, with a large fraction of the electrons having energies between 2 and 3 keV. Previous measurements of ATI spectra from single atoms at this intensity and pulse duration have indicated that the vast majority of electrons produced have energies below 100 eV [13]. Only a very small fraction of electrons (typically 10^{-3} to 10^{-4}) have higher energy, with no detectable electrons having an energy of above 1 keV [13]. The spectrum observed from Xe clusters clearly indicates a much greater coupling of laser energy to electrons than is present during the irradiation of single atoms. Furthermore, this spectrum indicates that the laser - cluster interaction produces even hotter electron temperatures than a laser - solid interaction at this intensity, where electron temperatures of 100 - 500 eV are typical [14].

The presence of two distinct peaks in the electron energy spectrum suggests that these two groups may be produced under different conditions at different times in the cluster expansion. This assumption is supported by examining the angular dependence of the electron emission with respect to the laser polarization (Figure 2). The angle between the direction of polarization and the detector was varied with a $\lambda/2$ plate placed before the

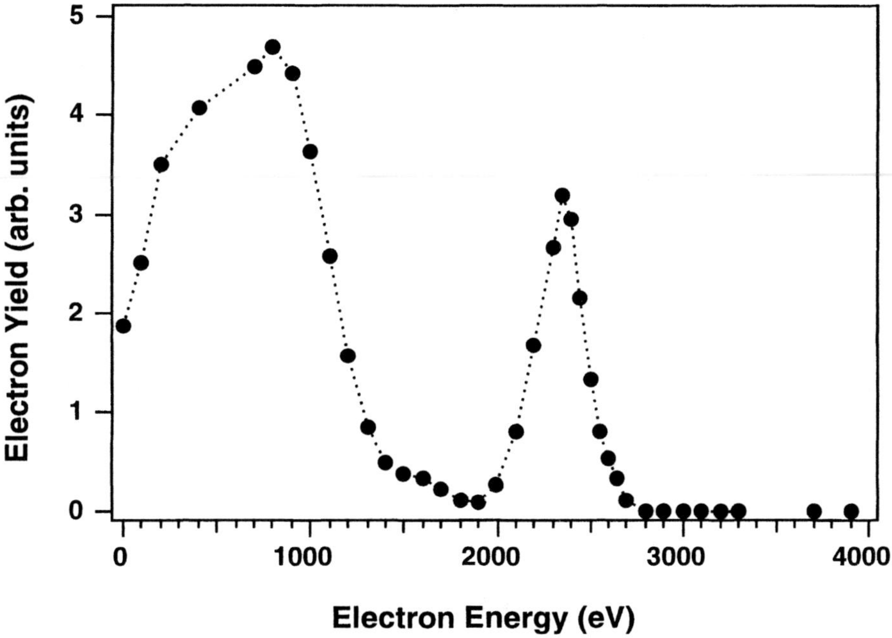

Figure 1. Measured electron energy distribution from xenon clusters with a peak intensity of 1.5 x 10^{16} Wcm^{-2}. The gas-jet backing pressure was 4.5 bar, corresponding to an average cluster size of 50 Å.

entrance to the vacuum chamber. We exploited the different flight times of the two sets of electrons to discriminate them and then integrated the MCP signal over an appropriate time-gate. The measured angular distribution of warm electrons is shown in Figure 2a, and that of the hot electrons in Figure 2b.

The angular distributions of the two groups of electrons are markedly different. The hot electron emission is completely isotropic, having no variation with respect to the direction of the laser polarization, while warm electron emission is peaked along the laser polarization. The warm electron peak has a full-width at half-maximum of about 60°. Both these distributions are significantly different from the angular distributions associated with single atoms. The electrons from high-order ATI are expected to have a much narrower angular distribution. In high field tunnelling ionization, the narrow angular distribution stems from the much higher tunnelling rate in the direction of the laser field. The electrons observed in our experiment cannot, therefore, be interpreted as simply resulting from the tunnel ionization of individual atoms. Some rescattering of the electrons by ions in the cluster is necessary to explain the broadening observed in the warm electron distribution. The isotropic distribution of the hot electrons indicates that these electrons have undergone many electron-ion collisions in the laser field, completely randomizing their velocity distribution.

The remarkably high energies of the electrons produced in the intense laser-cluster interaction suggests that highly charged ions with large kinetic energies may also be ejected from the cluster. Charge separation of these hot electrons will inevitably accelerate the cluster ions to high velocities. The energy spectrum of ions resulting from the interaction of ~2500-atom Xe clusters with a laser pulse of intensity ~2 x 10^{16} Wcm^{-2} is shown in figure 3. The most remarkable aspect of this energy distribution is the presence of ions with energies up to 1 MeV. This energy is four orders of magnitude higher than has previously been observed in the Coulomb explosion of molecules [15] and about 1000 times higher than the energy of Ar ions ejected in the disintegration of small clusters of up to six argon atoms irradiated at 10^{15} Wcm^{-2} reported by Purnell *et al* [4]. The average ion energy of this distribution is 45 ± 5 keV. Thus the average laser energy deposited per ion is also substantial.

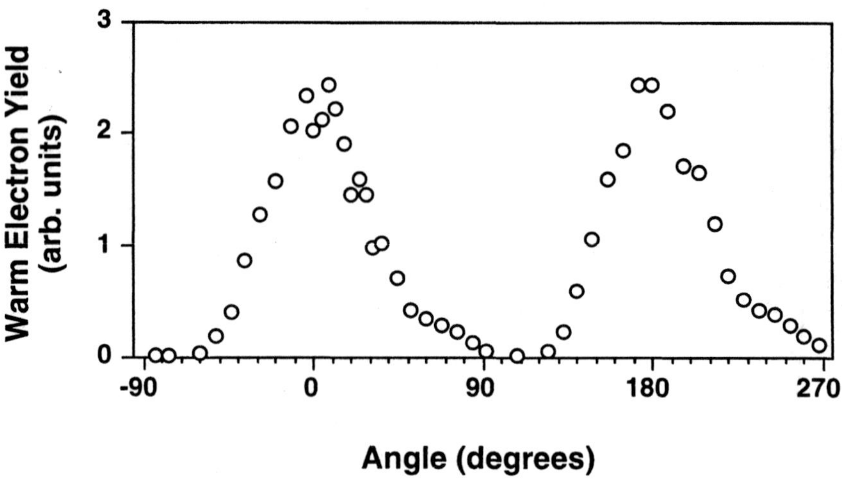

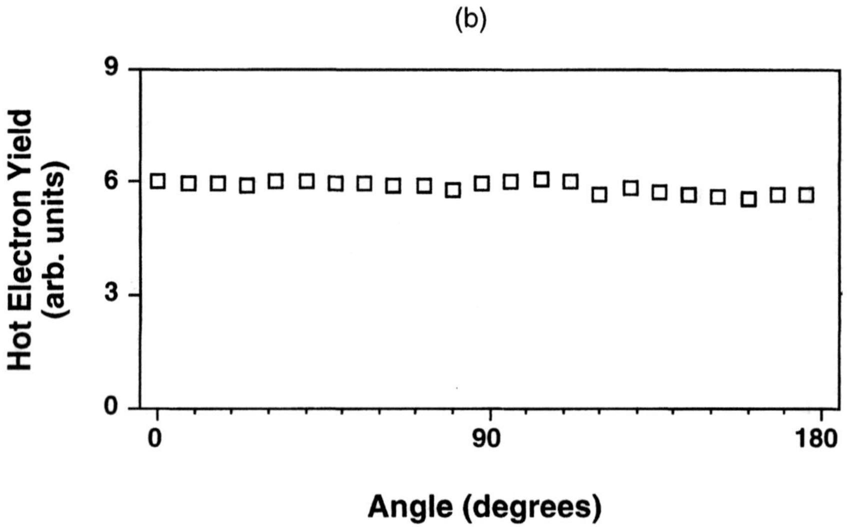

Figure 2. (a) Angular distribution of 'warm electrons' with energies ranging from 0.3 to 1 keV. Emission along the direction of laser electric field polarization is defined as being at 0° and 180°.(b) Angular distribution of 'hot electrons' with energies ranging from 2 to 3 keV. The gas-jet backing pressure was 4 bar, corresponding to an average cluster size of 1600 atoms.

Our ability to control the cluster size by changing the backing pressure of the gas-jet enables us to examine the scaling of the ion energy distribution with cluster size (Figure 4). We find that both the maximum energy, E_{max}, (defined as the energy at which the signal drops to 10^{-5} of its maximum) and average energy, $\overline{E}$, of the ion distribution increase slowly with increasing cluster size. At a pressure of 5 bar, corresponding to 2500 atoms/cluster, E_{max} is 1 MeV and $\overline{E}$ is 41 keV, while at 2 bar (400 atoms/cluster), E_{max} is 200 keV and $\overline{E}$ is 29 keV. Though there is a slight drop in the energies of the ions produced, there is no dramatic shift in the shape of the ion distribution as the cluster size is increased, This suggest that the mechanism driving the cluster ion explosion does not dramatically change as the cluster size is varied from a few hundred to a few thousand atoms per cluster.

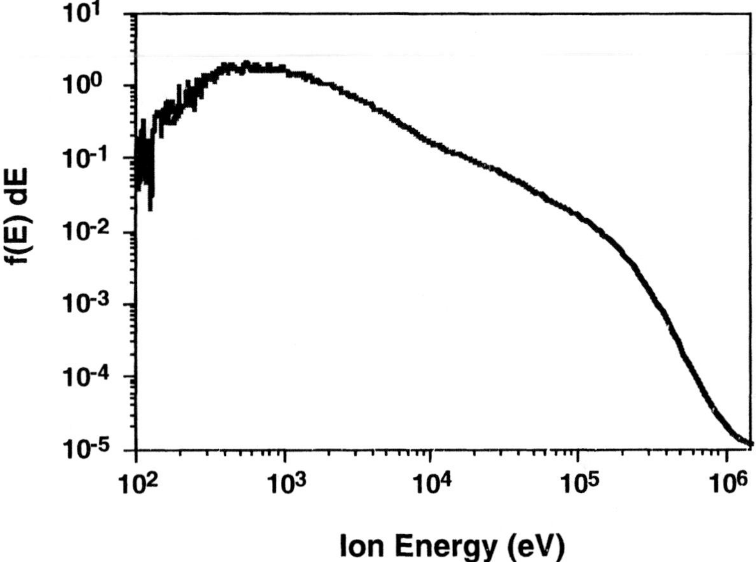

Figure 3. Ion energy spectrum from clusters of 2500 Xe atoms (backing pressure 5 bar) irradiated by a peak intensity of 2 x 10^{16} Wcm^{-2}

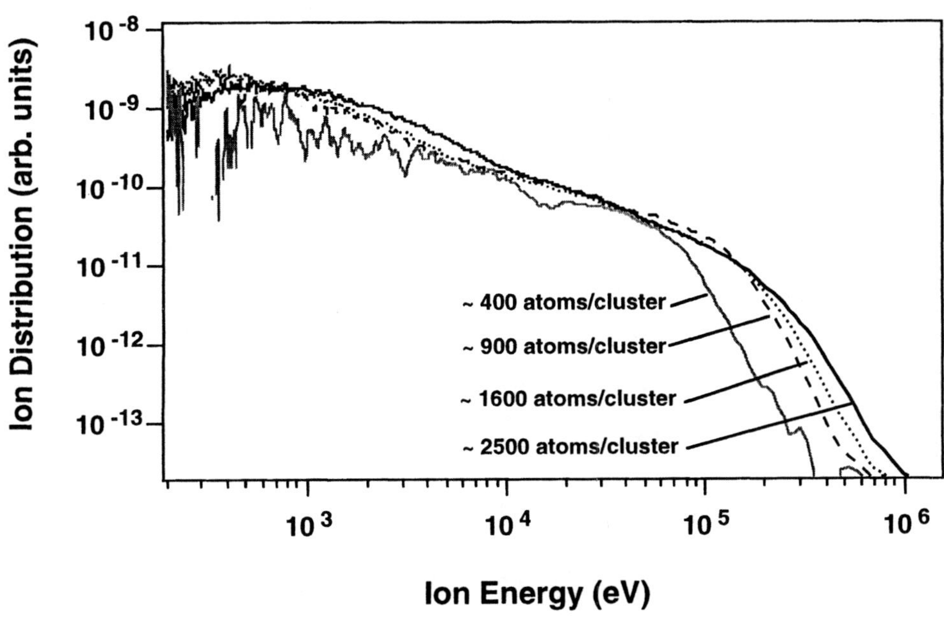

Figure 4. Ion energy spectra from clusters of 400, 900, 1600 and 2500 xenon atoms (backing pressure 2, 3, 4 and 5 bar respectively) irradiated by a peak intensity of 2 x 10^{16} Wcm^{-2}. The average ion energies are 29 keV for 400 atoms/cluster, 38 keV for 900 atoms/cluster, 39 keV for 1600 atoms/cluster and 41 keV for 2500 atoms/cluster.

Though all of the data presented to this point have been taken with near infrared radiation at a wavelength of 780 nm, we have also conducted a preliminary study of the nature of the cluster ion explosion in ultraviolet light. For these experiments, the laser was frequency doubled to a wavelength of 390 nm using a 3mm thick KDP crystal. Up to 5 mJ of light was focused into the TOF chamber with a lens of 20 cm focal length. Figure 5 shows a comparison of the ion energies obtained when clusters of ~2500 xenon atoms were irradiated with 780 nm and 390 nm light at intensity of 2 x 10^{16} Wcm^{-2} . In general we find that the shape of the ion TOF trace produced with UV light is very similar to that produced with the IR pulse. The ions from the UV irradiation appear to be slightly hotter.

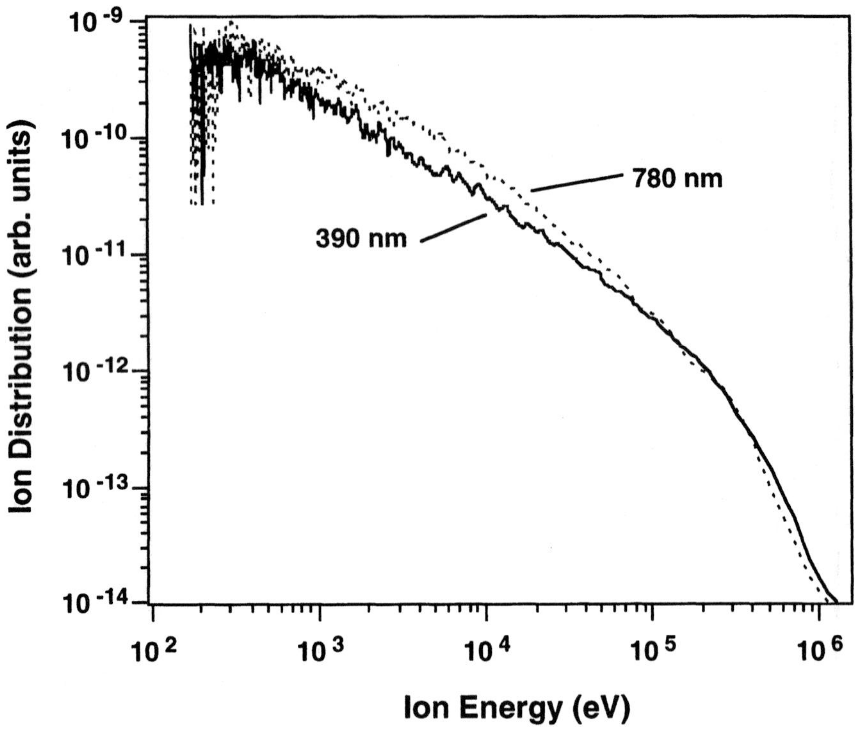

Figure 6. Ion energy spectrum for clusters of 2500 Xe atoms irradiated with a peak intensity of 2 x 10^{16} Wcm^{-2} at 780 nm, and at 390 nm.

DISCUSSION OF RESULTS

The very high energy particles observed experimentally are dramatically different to those typically produced in strong field laser interactions with molecules. Such energies are typical of particles produced in the interaction of a high intensity laser with solid density plasmas. This suggests that the appropriate way to explain the exploding cluster behavior is to treat it as a small microplasma. This picture of the cluster implies a number of interesting consequences. Because of the high electron and ion densities within the cluster, electron collisional processes will be very important. In particular, the electrons will undergo rapid heating by the laser field due to electron-ion collisions (inverse bremsstrahlung). This process converts the coherent oscillation energy of the electron cloud to random thermal energy. Electron collisional ionization will also be important, stripping the constituent atoms to very high charge states.

The production of hot electrons through inverse bremsstrahlung seen in the data can drive a very energetic explosion of the cluster. This explosion is manifested in the very high energy ions observed. Charge separation of the hot electrons will inevitably drive a rapid expansion of the cluster. The explosion of the cluster can be driven by two forces. The first is the Coulomb repulsion between the highly charged ions in the cluster. If all the free electrons are retained in the cluster, the cluster is quasineutral and this force is negligible. However, the free streaming of electrons from the cluster will cause a charge build-up of the plasma sphere, and a Coulomb "pressure" will develop. The second force important in driving the cluster explosion is the hydrodynamic pressure of the free electrons in the cluster. This force is present even if the cluster plasma remains neutral. The hot electrons in a plasma will set up a radial ambipolar potential which then accelerates the cluster ions. This hydrodynamic pressure is the same force that drives the expansion of a solid target plasma into vacuum. Numerical modeling indicates that hydrodynamic forces dominate the dynamics for large clusters because of the large space charge forces associated with large clusters which confine the electrons to the cluster [6].

The reasons for the very high energy ions observed can be very easily explained by a simple model. In the hydrodynamic expansion of the cluster both electrons and ions ultimately reach a velocity given roughly by the sound speed of the cluster plasma, $c_s = (ZkT_e/m_i)^{1/2}$ (where kT_e is the electron thermal energy and m_i is the ion mass). Most of the resulting kinetic energy is, however, contained in the ions due to their much greater mass. On the basis of this statement, we expect that the average ion energy will be of the order of $^1/_2 m_i c_s^2 \sim ZkT_e$. This implies, for example, that the average Xe ion energy will be ~ 50 keV if we assume that the electron temperature is given by the high energy electron feature in figure 1, $i.e.$ $kT_e \sim 2.5$ keV and the average charge state is $Z \sim 20^+$ (the charge state observed in these experiments [8]). This is in good agreement with our observed average Xe ion energies.

CONCLUSION

In conclusion, we have studied the photoionization of noble gas clusters by a high intensity, femtosecond laser pulse. We have examined the energies of the electrons and ions produced in the explosion of the clusters and found that their kinetic energies are remarkably high. The electron energy distribution from the exploding clusters contains electrons with energies as high as 3 keV, which is several orders of magnitude higher than the energies observed in ATI from single atoms or small molecules. The ions produced in the explosion have mean energies of tens of keV and the maximum ion energy observed was 1 MeV.

The high ion energies observed in the explosion of clusters of a few hundred to a few thousand atoms in an intense laser field are very much like those of the ions observed in the expansion of a laser-heated solid-density plasma into vacuum. They contrast dramatically with the low-energy, low charge-state ions produced in the Coulomb explosion of small molecules and clusters of only a few atoms in strong laser fields. Clusters of more than a few hundred atoms, therefore, represent an important transition in the dynamics of intense laser-matter interactions from molecules to solids.

Acknowledgements

This work was supported by the EPSRC and the Ministry of Defence.

REFERENCES

1.   A.W. Castleman and R.G. Keesee, "Gas-Phase Clusters: Spanning the States of Matter," Science **241**, 36 (1988).

2.   A. McPherson, B.D. Thompson, A.B. Borisov, K. Boyer, and C.K. Rhodes, "Multiphoton-induced X-Ray Emission at 4-5 keV from Xe Atoms with Multiple Core Vacancies," Nature (London) **370**, 631 (1994).

3.   A. McPherson, T.S. Luk, B.D. Thompson, A.B. Borisov, O.B. Shiryaev, X. Chen, K. Boyer, and C.K. Rhodes, "Multiphoton Induced X-Ray Emission from Kr Clusters on M-Shell and L-Shell Transitions," Phys. Rev. Lett. **72**, 1810 (1994).
4.   J. Purnell, E.M. Snyder, S. Wei, and A.W.C. Jr., "Ultrafast Laser-Induced Coulomb Explosion of Clusters woth High Charge States," Chem. Phys. Lett. **229**, 333 (1994).
5.   T. Ditmire, T. Donnelly, R.W. Falcone, and M.D. Perry, "Strong X-ray Emission from High Temperature Plasmas Produced by Intense Irradiation of Clusters," Phys. Rev. Lett. **75**, 3122 (1995).
6.   T. Ditmire, T. Donnelly, A.M. Rubenchik, R.W. Falcone, and M.D. Perry, "Interaction of Intense Laser Pulses with Atomic Clusters," Phys. Rev. A **53**, 3379 (1996).
7.   Y.L. Shao, T. Ditmire, J.W.G. Tisch, E. Springate, J.P. Marangos, and M.H.R. Hutchinson, "Multi-keV Electron Generation in the Interaction of Intense Laser Pulses with Xe Clusters," Phys. Rev. Lett. **77**, 3343 (1996).
8.   T. Ditmire, J.W.G. Tisch, E. Springate, M.B. Mason, N. Hay, R.A. Smith, J. Marangos, and M.H.R. Hutchinson, "High-Energy Ions Produced in Explosions of Superheated Atomic Clusters," Nature (London) **386**, 54 (1997).
9.   T. Ditmire, J.W.G. Tisch, E. Springate, M.B. Mason, H. Hay, J.P. Marangos, and M.H.R. Hutchinson, "High Energy Ion Explosion of Super-Heated Clusters: Transition from Molecular to Plasma Behavior," Phys. Rev. Lett. **78**, 2732 (1997).
10.   E.M. Snyder, S.A. Buzza, and A.W. Castleman, "Intense Field-Matter Interactions: Multiple Ionization of Clusters," Phys. Rev. Lett. **77**, 3347 (1996).
11.   M. Lezius, S. Dobosz, D. Normand and M. Schmidt, "Hot Nanoplasmas from Intense Laser Irradiation of Argon Clusters," J. Phys. B: At, Mol. Opt. Phys. **30**, L251 (1997).
12.   C. Rose-Petruck, K.J. Schafer, K.R. Wilson, and C.P.J. Barty, "Ultrafast Electron Dynamics and Inner-Shell Ionization in Laser Driven Clusters," Phys. Rev. A **55**, 1182 (1997).
13.   B. Walker, B. Sheehy, L.F. DiMauro, P. Agostini, K.J. Schafer, and K.C. Kulander, "Precision Measurement of Strong Field Double Ionization of Helium," Phys. Rev. Lett. **73**, 1227 (1994).
14.   R. Shepherd, D. Price, W. White, S. Gordan, A. Osterheld, R. Walling, W. Goldstein, and R. Stewart, "Characterization of Short Pulse Laser-Produced Plasmas," J. Quant. Spec. Rad. Trans. **51**, 357 (1994).
15.   C. Cornaggia, M. Schmidt, and D. Normand, "Coulomb Explosion of CO_2 in an Intense Femtosecond Laser Field," J. Phys. B: At. Mol. Opt. Phys. **27**, L123 (1994).

SOLID TO PLASMA TRANSITION IN FS-LASER-IRRADIATED FE: COLLAPSE OF THE SPIN-ORBIT GAP

M. K. Grimes, Y. -S. Lee, M. C. Downer

The University of Texas at Austin
Department of Physics, Austin, Texas 78712, USA

Quantitative measurements of the optical conductivity of iron under earth core conditions are important in modelling geomagnetism[1]. We approximate such conditions transiently[2] by exciting an Fe, and a control Al, surface in a vacuum or helium environment with 620 nm, 120 fs FWHM laser pulses with 10^5 peak-background contrast ratio at .6 ps focussed to peak intensities $10^{11} < I < 10^{15}$ W/cm^2 on target. Figs. 1 and 2 present p- and s- polarized self-reflectivity $R_{p,s}(\theta, I)$ for constant incident angle θ and peak intensity I, respectively. Geometric correction for the dependence of spot shape on θ has been made in plotting the data. Using very linear pulse energy reference monitors, reproducibility to within $\Delta R/R \leq 0.1\%$ was achieved. This allowed measurement of very slight reflectivity changes, as shown in the inset of Fig. 1. This figure also shows that the reflectivities of Fe and Al in the solid to plasma transition region $(10^{13} - 10^{15}$ W/cm$^2)$ approach each other as I increases, consistent with their similar total conduction electron densities $(n_e \simeq 1.8 \times 10^{23}$ cm$^{-3})$ when the Fe d-electrons are included. This suggests qualitatively that progressive unbinding of the d-electrons dominates the changes in optical properties of Fe in this regime. Fig. 2 provides evidence that a density gradient develops during the laser pulse. It is well known[3] that θ_{min}, the angle at which the minumum p-polarized reflectivity occurs, shifts toward smaller angles as the density gradient scale length grows. Although the largest angle of incidence attainable in the experiment was smaller than θ_{min} under these conditions, the data clearly indicate a shift toward smaller values, seen for Fe in Fig. 2 as a crossing of the p-polarized data sets.

Quantitative analysis of the data requires that the perturbation of Fresnel reflectivity by small scale length $(L/\lambda \leq 0.1)$ surface density gradients, if present during the pulse, be identified unambiguously and distinguished from ΔR caused by simultaneous intensity-dependent changes in dielectric properties of the plasma. We achieve this separation by calculating the reflectivity of such gradients under a wide range of conditions, and identifying those values of L/λ and the dielectric constant $\epsilon = \epsilon_r + i\epsilon_i$ that best reproduce the measured results. This reflectivity calculation numerically solves the Helmholtz wave equations[4] assuming a Riemann solution[5] of the form $\rho(z) = \rho_0(1 - z/L)^3$ to describe the density profile, and assuming that $\epsilon - 1$ varies linearly with density throughout the gradient. The qualitative features of the fit result are however largely reproduced when both assumptions are relaxed to require only that ϵ drop monotonically from the bulk toward the vacuum value of 1. Fig. 3 shows the resulting values of ϵ_r and ϵ_i vs. laser intensity for Fe. The corresponding results for the

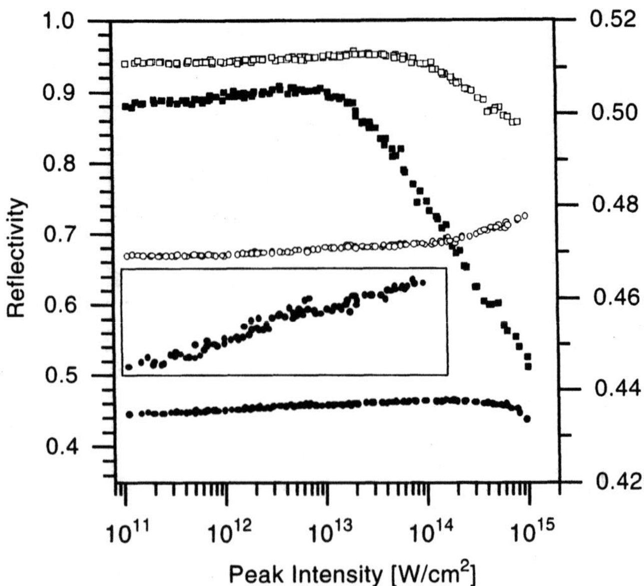

FIG. 1. Reflectivity of 120 fs pulses from Al and Fe surfaces vs. peak pulse intensity, at a 45° angle of incidence. Al data are shown with squares, Fe with circles. Measurements with polarization parallel to the plane of incidence (p-pol.) are shown with filled points; open points are used for polarization perpendicular to the plane of incidence (s-pol.). A section of the Fe, p-polarized data is shown in the inset, magnified vertically as indicated by the right-side axis.

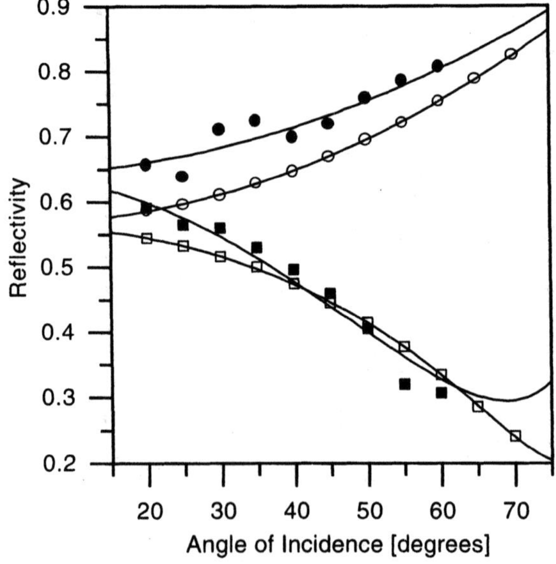

FIG. 2. Reflectivity of Fe vs. angle of incidence at $I = 10^{11}$ W/cm^2 (open points) and 2×10^{14} W/cm^2 (filled points) for s-polarization (circles) and p-polarization (squares). The reflectivity of a density gradient calculated from the corresponding best fit results are shown with lines.

132

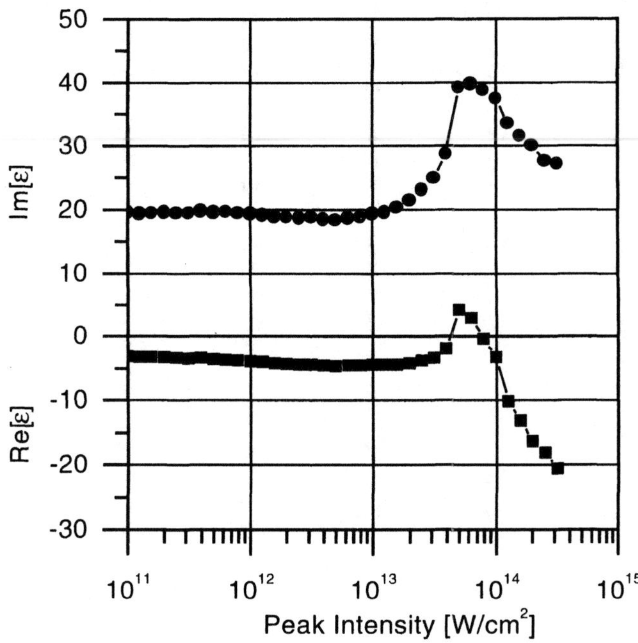

FIG. 3. Best fit results for the complex dielectric constant ϵ of Fe vs. peak laser intensity.

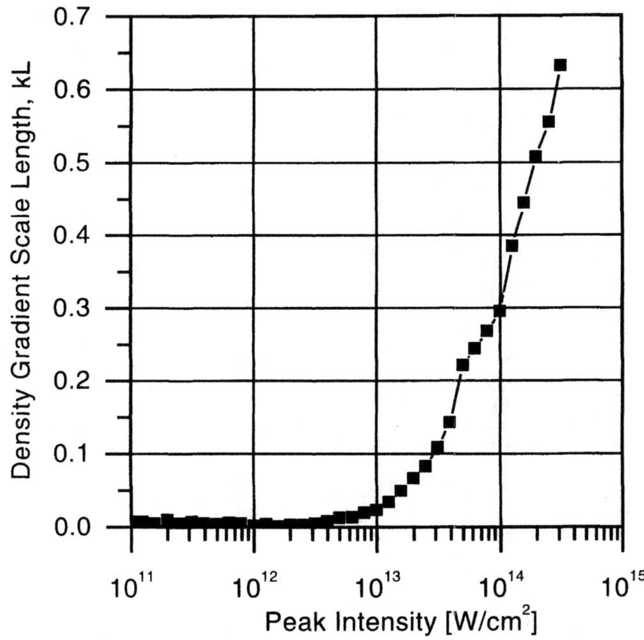

FIG. 4. Best fit results for the dimensionless density gradient scale length $kL = 2\pi L/\lambda$ of Fe vs. peak laser intensity.

density gradient scale length are shown in Fig. 4. Note that at the highest measured intensity the gradient scale length is still small compared to the laser wavelength, with $L/\lambda \lesssim .1$.

The dielectric function of Fe at optical frequencies is well described at room temperature by a Drude-Lorentz model of the following form[6]:

$$\epsilon(\omega) = 1 + \epsilon_b - \frac{\omega_p^2}{\omega(\omega + i\nu)} + \frac{a^3}{\omega\omega_0}\frac{\omega + i\Gamma}{\omega_0^2 - (\omega + i\Gamma)^2}, \tag{1}$$

where $\omega_p = 4\pi n_e e^2/m^*$ is the plasma frequency with effective electron mass m^*, ω is the laser frequency, ν is the Drude collision frequency, and a describes the strength of the Lorentz oscillator resonance centered at ω_0 of spectral width Γ. The term ϵ_b describes a frequency-independent background contribution to the polarizability. At the laser frequency $\hbar\omega = 2.0$ eV there is a convenient separation for Fe — the real part of ϵ is dominated by the free-electron, Drude term, and the imaginary part is dominated by the intraband, Lorentz term. This simplifies the interpretation of the changes in ϵ for low intensities. The nature of the interband, Lorentz resonance term is illuminated by the band-structure calculations of M. Singh et al.[7], which incorporate the effects of spin-orbit coupling critical to the description of the optical properties of transition metals. A collection of majority-spin states just below the Fermi energy and a collection of minority-spin states just above it lead to a joint density of states peaked at 2.4 eV.

The changing dielectric constant of Fe (Fig. 3) passes through three distinct stages as the intensity increases, each attributable to its own causes: first an initial decrease in both ϵ_r and ϵ_i, then a resonance feature, followed finally by a steep drop in ϵ_r and a milder decrease in ϵ_i. The initial drop in ϵ_r, the term dominated by the Drude free-electron term, is a consequence of a shrinking effective mass. At room temperature $m^* \simeq 8m_e$, but as the electron temperature rises with laser intensity the predominately d-orbital character of the conduction electrons in Fe will become more s-like. The greater mobility of these s-band states leads to a reduction in the effective mass and a corresponding increase in the collision frequency and $|\epsilon_r|$. The anticipated increase in collision frequency ν with temperature will offset this change somewhat, allowing the determination that at $I_{abs} = 10^{13}$ W/cm^2 the effective mass has dropped to $m^* \lesssim 5m_e$. In the second stage ($10^{13} < I < 10^{14}$ W/cm^2) the interband resonance term becomes dominant as the resonant frequency ω_0 drops from its initial value at 2.4 eV into the laser frequency at $I = 6 \times 10^{13}$ W/cm^2. This resonance is similar to those observed by Glezer, et al.[8] in fs-laser-induced bandgap collapse in GaAs. Finally, for $I > 10^{14}$ W/cm^2 ϵ_r becomes strongly negative again, indicating both collapse of the spin-orbit band gap and transition to a fully ionized plasma described entirely by the Drude model.

[1] J. A. Jacobs, *The Earth's Core*, Academic Press, London, 1987.

[2] M. C. Downer, H. Ahn, D. H. Reitze, D. M. Riffe, X. Y. Wang, in *Laser Interactions with Atoms, Solids, and Plasmas*, edited by R. M. More, Plenum Press, New York, 1994.

[3] O.L. Landen, D.G. Stearns, E.M. Campbell, Phys. Rev. Lett. **63**, 1475 (1989); O.L. Landen, B.T. Vu, D.G. Stearns, W.E. Alley, in SPIE Conference Proceedings Vol. 1413, *Short-Pulse High-Intensity Lasers and Applications* (1991).

[4] H. M. Milchberg, R. R. Freeman, J. Opt. Soc. Am. B **6**, 1351 (1989).

[5] Ya.B. Zeldovich and Yu.P. Raizer, *Physics of Shock Waves and High-Temperature Hydrodynamic Phenomena*, Ch. 1, Academic Press, New York, 1966.

[6] N.W. Ashcroft and K. Sturm, Phys. Rev. B, **3**, 1898 (1971).

[7] M. Singh, C.S. Wang, and J. Callaway, Phys. Rev. B **11**, 287 (1975).

[8] E.N. Glezer, Y. Siegal, L. Huang, and E. Mazur, Phys. Rev. B **51**, 6959 (1995).

PRESSURE IONIZATION AND DENSITY DIAGNOSTICS
IN SUBPICOSECOND LASER-PRODUCED PLASMAS

M. Nantel, G. Ma, S. Gu, C.Y. Côté, J. Itatani, and D. Umstadter

Center for Ultrafast Optical Science University of Michigan
1006 IST Building, 2200 Bonisteel Blvd
Ann Arbor, MI 48109-2099

INTRODUCTION AND MOTIVATION

The atomic physics of high-density plasmas is studied extensively for its relevance to astrophysics[1], inertial confinement fusion,[2,3] x-ray lasers,[4] and to the interaction of ultrashort lasers with solids.[5-7] Of utmost importance is the knowledge of the plasma parameters of electron density, N_e, and temperature, T_e, as they govern the atomic physics in the plasma, from its ionization balance to its emission and absorption. The structure and behavior of atoms and ions, for example, can be radically affected by the presence of strong fields in high-density plasmas[1], leading to such effects as extreme line broadening and pressure ionization.[1,2,9] Pressure ionization and line-merging have been used in laboratory plasmas as a density diagnostic of spatially- and/or temporally-integrated spectra.[2,10-13] But in laser-produced plasmas, conditions often vary rapidly over time and space, so it is important to resolve both these dimensions for accurate diagnostics. Furthermore, several models are available to quickly extract densities from spectroscopic data but are very different and need to be carefully benchmarked in order to identify which apply for any given set of plasma parameters. Precise data for model validation is rare and usually comes from plasmas limited in density and temperature range.[13] Here, we compare four models under a wide range of densities and temperatures in plasmas created with ultrafast laser pulses. These 100-fs laser pulses have the advantage over nanosecond pulses of depositing the energy of the laser impulsively, in a small target layer. Thus, the spectroscopic measurements are conducted after the laser pulse, in a freely expanding plasma, without the added complication of further energy deposition during the plasma evolution.

MODELS OF CONTINUUM LOWERING AND LINE-MERGING

Pressure ionization, also known as "continuum lowering" or "ionization potential depression", is a fundamental concept of atomic physics in plasmas, defining their existing energy states, as well as their emission and absorption properties. In laser-produced plasmas, continuum lowering has been observed in solid-density compressed plasmas through shifts in absorption edges [14-16] and in high-density ablation plasmas as a merging of the emission lines with the continuum.[2,10,12] Since it has an impact on the ionization state and radiation transfer in the plasma, continuum lowering is included in most comprehensive plasma atomic physics simulations, and several models are used to account for it.[1,9,17-20] As illustrated schematically in figure 1, the simpler continuum lowering

Applications of High Field and Short Wavelength Sources
Edited by DiMauro *et al.*, Plenum Press, New York, 1998

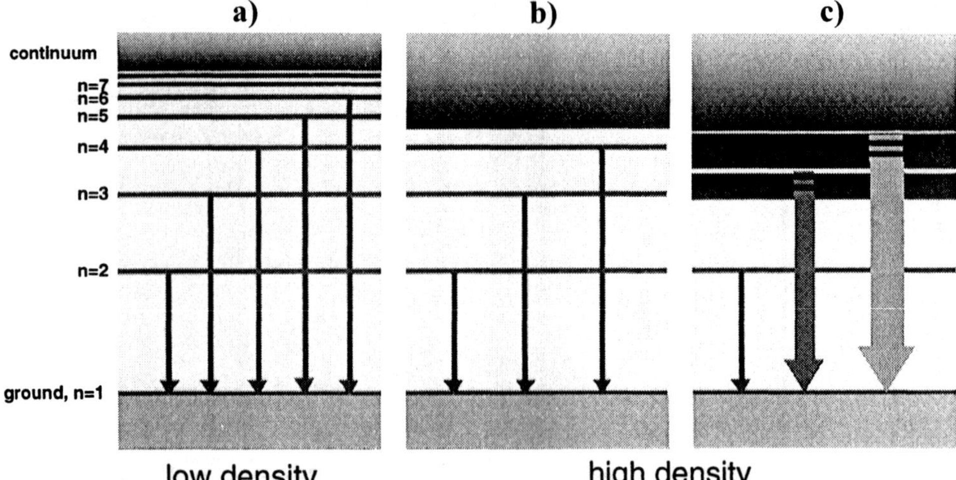

Figure 1. Schematic representation of the effect of pressure ionization in He-like ions. a) In vacuum, the continuum is at the vacuum ionization potential of the ion, and all the excited states exist. b) When the ion is immersed in a plasma, the average fields of the free electrons and of the other ions bring down the energy necessary to further ionize the ion and the continuum lowers, eliminating the higher excited states. c) When the fluctuating microfields in the plasma are also taken into account, there is a further lowering of the apparent continuum due to merging of the broadened energy states near the continuum. The elimination and/or broadening of the excited states will be reflected in the line emission of the ions (downward arrows).

models (ion-cell, Debye-Huckel, Stewart-Pyatt)[1,17,18] usually define a sharp cut-off for the excited state levels n from the average plasma fields without considering the plasma fluctuation perturbations (Fig. 1b). The Inglis-Teller limit[21] can be used to calculate the detectable position of the continuum in the plasma emission spectrum, by taking into account—albeit approximately—the plasma microfield fluctuations. The fluctuations lead to broadened levels that merge into one another to further lower the apparent ionization potential (Fig.1c). Essentially, the Inglis-Teller limit predicts the density at which the Stark broadening of two adjacent levels is more than the energy separating them, effectively when they hydridize to cover more than one ion. More sophisticated models (generalized ion-cell, Dharma-wardana-Perrot)[19,20] take into account these microfields in a more self-consistent way, but require much computation; they will not be considered in this paper.

EXPERIMENTAL RESULTS AND ANALYSIS

The experiments were conducted with the 10-Hz, 100-fs Ti:sapphire laser at the Center for Ultrafast Optical Science. Laser pulses of 50-mJ in energy at 780 nm were focused with a MgF_2 lens on solid targets to an intensity of 10^{17} W/cm^2. The emission spectra from the plasma were recorded with a grazing-incidence flat-field imaging XUV spectrometer coupled to an x-ray streak camera with CCD readout. With this system, we had a spatial resolution of 80 μm or better in the direction normal to the target plane and a spectral resolution of 0.15 Å. The laser contrast was improved by a factor of 100 from its original value of 10^5 through high-energy seeding of the regenerative amplifier.[22] This higher contrast of 10^7 ensures the quality of the solid-target interaction. The targets were solid disks of carbon on a rotating stage. To obtain the best possible temporal resolution and signal-to-noise figure on our spectra, we used a subpicosecond x-ray streak camera[23,24] which we coupled to a jitter-free accumulation sweep system.[25] This new camera (recently made available commercially through Axis Photonique Inc.) has a single-shot temporal resolution of 800 femtoseconds, and has been operated with the jitter-free system to resolutions as good as 1 ps over 5000-shot accumulations with a kHz laser. In our experiment, the temporal resolution was limited to 4 ps over 600 accumulated shots, mostly due to laser pulse energy variations on the expectedly less stable 10-Hz laser.

To concentrate on the plasma region exhibiting the most continuum lowering, we present data from the 80 μm closest to the target plane. Figure 2 shows a composite graph including three carbon spectra taken at 0 ps, 50 ps and 225 ps. These lineouts are integrated over 4 ps, and $t = 0$ ps is arbitrarily defined as the time of the start of the emission. The He-like and H-like np-$1s$ transition series are identified, starting with the $1s2p$-$1s^2$ at 40.268 Å and $2p$-$1s$ at 33.737 Å, respectively. The vacuum ionization potential (IP) of the He-like ion is marked at 31.62 Å (392.09 eV).[26] One of the most striking differences between the three spectra is in the position of the detectable recombination continuum edge, which has been measured to be at 35.3 Å (351.3 eV) for 0 ps, at 33.0 Å (375.2 eV) for 50 ps, and at 32.3 Å (383.6 eV) at 225 ps. The shaded area between the vacuum IP and the edge highlights for each spectrum the region of continuum lowering/hybridization, which can be as high as 40.8 eV, or >10% of the IP for the spectrum at 0 ps. To the best of our knowledge, this is the highest percentage recorded to date in a laboratory plasma. Also, one can identify the presence of He-like emission line from $1s2p$-$1s^2$ to $1s5p$-$1s^2$ at least for the spectrum at 225 ps, while only the $1s2p$-$1s^2$ line is unambiguously identified at 0 ps, which gives lowering values in agreement with the continuum edge position measurement. For the spectrum at 50 ps, one can readily see the $1s3p$-$1s^2$ line; the $1s4p$-$1s^2$ may be in the low-wavelength shoulder of the H-like $2p$-$1s$ line.

To test the continuum lowering and line-merging models as density diagnostics, we must first independently extract the density and temperature from the experimental spectra We establish the electron temperature by the slope of the He-like continuum. We then generate artificial spectra using the FLY atomic physics code[27] and find a match for the

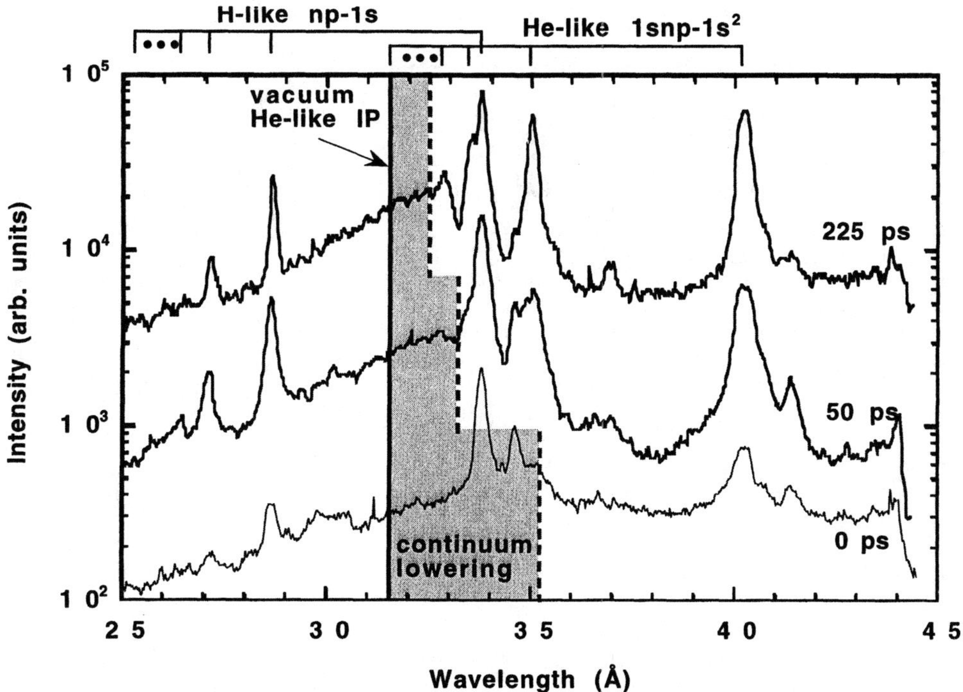

Figure 2. XUV spectra in H-like and He-like carbon from a plasma created by focusing the laser on solid target to an intensity of 10^{17} W/cm^2. The three spectra have been offset vertically to improve visibility, and correspond to the emission in the first 80 μm from the target plane, for 4-ps integrations starting at times 0 ps, 50 ps, and 225 ps. Time 0 ps is arbitrary and corresponds to the start of the emission. The position of the vacuum ionization potential (IP) of He-like carbon is marked by the solid vertical line at 31.62 Å, and the position of the detectable edge is indicated for each spectrum by a dashed vertical line. The continuum lowering/hybridization is the difference of the two, in the shaded area.

137

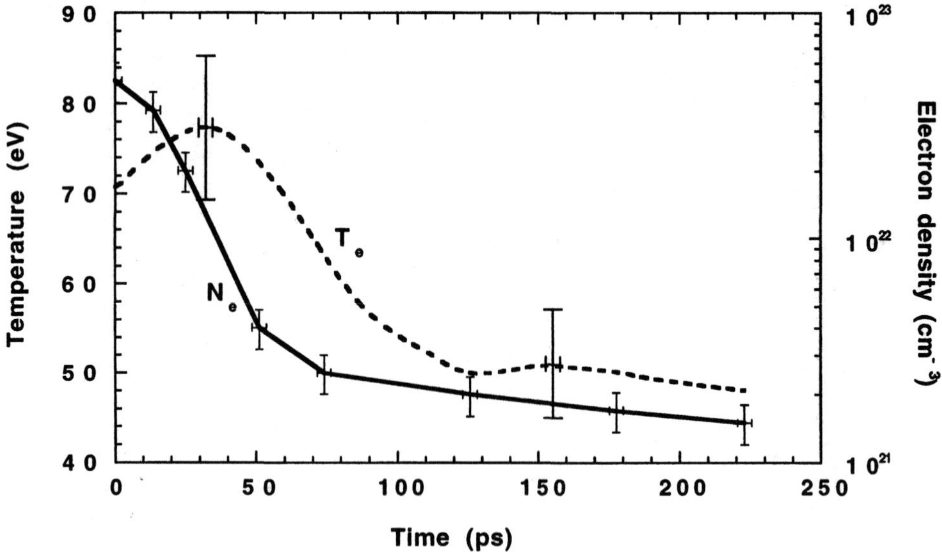

Figure 3. Electron temperature (dashed line) and density (solid line) as a function of time for a carbon plasma created with a 100-fs laser pulse at an intensity of 10^{17} W/cm^2.

experimental spectra by varying the density in the simulation. The artificial spectra include instrumental broadening, continuum lowering, Stark broadening and opacity. The temperature and density obtained are plotted in figure 3, for the first 225 ps of the emission from the slice near the target. A highest density of 5 x 10^{22} cm^{-3} is inferred, corresponding to one-tenth of the solid density for a carbon plasma of average ionization $z = 4.5$.

Figure 4 shows a comparison of the density diagnostic obtained using FLY to those from the simple continuum lowering models (Fig. 1b: ion-cell, Debye-Huckel, Stewart-Pyatt) and from the Inglis-Teller limit line-merging model (Fig. 1c). For the three microfield-free models, we measured the continuum lowering with the position of the He-like free-bound recombination edge; for the Inglis-Teller diagnostic, we identified on the experimental spectra the last detectable and first undetectable $1snp$-$1s^2$ He-like transitions, which give the higher and lower bounds for the density, respectively. It becomes readily obvious from Fig. 4 that the choice of continuum lowering model will dramatically influence the diagnostic. Choosing Stewart-Pyatt, which is meant to apply to any density regime, one overestimates the electron density by a factor 2, while the ion-cell model, which should be the model of choice for high-density plasmas, gives densities up to 10 times higher than the full FLY atomic physics diagnostics. Despite being designed to handle only low densities, the Debye-Huckel model appears to show the best agreement of all three simple models used. This is rather surprising, given that in our density/temperature regime one can calculate that there are on average less than two particles in the Debye sphere, rendering the very concept of Debye shielding inapplicable.[28] The Debye-Huckel model thus *coincidentally* gives the right density but is phenomenologically wrong for our plasma regime. The agreement of the Debye-Huckel diagnostics partly arises from the fact that, of the three simple models, it is the one giving the lowest density for a given value of the continuum lowering. There is good agreement for the diagnostics using the Inglis-Teller limit—particularly at high density early in the plasma evolution—but with a relatively large uncertainty (shade area) dictated by the separation between the excited levels. At high densities, Stark broadening of the levels near the continuum will be such that they will start to overlap—and their emission lines merge—moving the detectable edge to longer wavelength than with the ion-cell or Stewart-Pyatt models (see Figs. 1b and 1c). This is why the Inglis-Teller model agrees best earlier in time: the actual value of the true continuum lowering is hidden by the line-merging and hybridization The importance of the microfield effects are directly related to the low

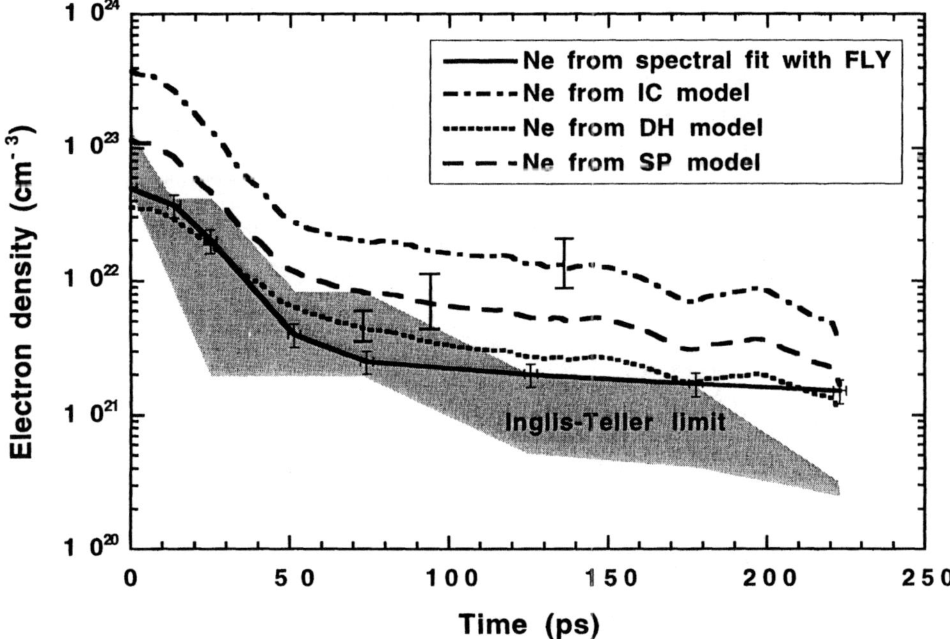

Figure 4. Comparison of the electron density as a function of time obtained by a fit of the experimental spectra to FLY atomic physics simulations (solid line) or using four models of continuum lowering: Stewart-Pyatt (SP, long-dashed line), Debye-Huckel (DH, short-dashed line), ion cell (IC, dot-dashed line), and Inglis-Teller (shaded area). The fits to FLY spectra were obtained for eight times only (identified by error bars on that curve), as was the diagnostic using the Inglis-Teller limit. The other models were applied to experimental spectra taken at each 4 ps, and we show one representative error bar for each smoothed curve taking into account the scatter in the data and the error on the diagnostic itself.

number of particles in the Debye sphere, which causes the emitting ion to be more susceptible to rapid plasma field fluctuations.[28]

CONCLUSION

We presented time- and space-resolved XUV spectra of carbon plasmas created by focusing 100-fs laser pulses on a solid target to an intensity of 10^{17} W/cm^2. The recorded spectra exhibit severe continuum lowering and line-merging (a maximum of 40 eV in He-like carbon or 10% of the ionization potential) from the high-density/low-temperature conditions in the plasma. We compare the density diagnostics from four different continuum lowering/hybridization models with that obtained with full FLY atomic physics calculations, and identify that while the Debye-Huckel model coincidentally gives a good diagnostics, it is the Inglis-Teller limit that should be used for quick diagnostics in dense plasmas, as it properly takes into account the rapidly fluctuating fields (i.e. the Stark effect) in such plasmas.

ACKNOWLEDGMENTS

This work is supported by the Centre for Ultrafast Optical Science under NSF Grant No. STC PHY 8920108, and by DOE contract DE-FG02-96ER14685. M. Nantel is supported in parts by the FCAR fund. We would like to thank J. Faure and M. Finlayson for help with the experiment, and Spectrogon for the laser compressor gratings.

REFERENCES

[1] S. Brush and B.H. Armstrong, *Proc. Workshop on Lowering of the Ionization Potential*, JILA report **79**, (Univ. of Colorado, Boulder, CO 1965).

[2] C.M. Lee and A. Hauer, Appl. Phys. Lett. **33**, 692 (1978).

[3] B.A. Hammel, C.J. Keane, M.D. Cauble, D.R. Kania, J.D. Kilkenny, R.W. Lee, and R. Pasha, Phys. Rev. Lett. **70**, 1263 (1993).

[4] *Atomic Processes in Plasma*, AIP Conf. Proc. **257**, E.S. Marmar, J.L. Terry, eds. (American Institute of Physics, New York, NY, 1992), and references therein.

[5] A. Rousse, P. Audebert, J.P. Geindre, F. Falliès, J.C. Gauthier, A. Mysyrowicz, G. Grillon, and A. Antonetti, Phys. Rev. E **50**, 2200 (1994).

[6] Z. Jiang, J.C. Kieffer, J.P. Matte, M. Chaker, O. Peyrusse, D. Gilles, G. Korn, A. Maksimchuk, S. Coe, and G. Mourou, Phys. Plasma **2**, 1702 (1995).

[7] J. Workman, A. Maksimchuk, X. Liu, U. Ellenberger, J.S. Coe, C.Y. Chien, and D. Umstadter, Phys. Rev. Lett. **75**, 2324 (1995).

[8] *Spectral Line Broadening by Plasmas*, H.R. Griem (Academic Press, New York, NY, 1974).

[9] R.M. More, J. Quant. Spectrosc. Radiat. Transfer **27**, 345 (1982).

[10] G.A. Kyrala, R.D. Fulton, E.K. Wahlin, L.A. Jones, G.T. Shappert, J.A. Cobble, and A.J. Taylor, Appl. Phys. Lett. **60**, 2195 (1992).

[11] P.G. Burkhalter, G. Mehlman, D.A. Newman, M. Krishnana, and R.R. Prasad, Rev. Sci. Instrum. **63**, 5052 (1992).

[12] D. Riley, L.A. Gizzi, F.Y. Khattak, A.J. Mackinnon, S.M. Viana, and O. Willi, Phys. Rev. Lett. **69**, 3739 (1992).

[13] D.J. Heading, G.R. Bennett, J.S. Wark, and R.W. Lee, Phys. Rev. Lett. **74**, 3616 (1995).

[14] D.K. Bradley, J. Kilkenny, S. Rose, and J.D. Hares, Phys. Rev. Lett. **59**, 2995 (1987).

[15] L. DaSilva, A. Ng, B.K. Godwal, G. Chiu, and F. Cottet, Phys. Rev. Lett. **62**, 1623 (1989).

[16] J. Workman, M. Nantel, A. Maksimchuk, and D. Umstadter, Appl. Phys. Lett. **70**, 312 (1997).

[17] J.C. Stewart and K.D. Pyatt, Jr., Astrophys. J. **144**, 1203 (1966).

[18] J.A. Kunc and W.H. Soon, Astrophys. J. **396**, 364 (1992).

[19] B.J.B. Crowley, Phys. Rev. A **41**, 2179 (1990).

[20] M.W.C. Dharma-wardana and F. Perrot, Phys. Rev. A **45**, 5883 (1992).

[21] D.R. Inglis and E. Teller, Astrophys. J. 90, 439 (1939).

[22] J. Itatani, J. Faure, M. Nantel, G. Mourou, and S. Watanabe, submitted to Opt. Commun. (1997).

[23] C.Y. Côté, J.C. Kieffer, P. Gallant, J.C. Rebuffie, C. Goulmy, A. Maksimchuk, G. Mourou, D. Kaplan, and M. Bouvier, SPIE Proc. **2869**, 956 (1997).

[24] P. Gallant, Z. Jiang, J. Fuchs, J.C. Kieffer, H. Pépin, D. Gontier, A. Mens, N. Blanchot, J.L. Miquel, J.F. Pelletier, and M. Sutton, to be published in SPIE Proc. **3157**.

[25] A. Maksimchuk, M. Kim, J. Workman, G. Korn, J. Squier, D. Du, D. Umstadter, G. Mourou, and M. Bouvier, Rev. Sci. Instrum. **67**, 697 (1996).

[26] R.L. Kelly, J. Phys. Chem. Ref. Data **16**, suppl. 1, 1 (1987).

[27] R.W. Lee, B.L. Whitten, and R.E. Stout, II, J. Quant. Spectrosc. Radiat. Transfer **32**, 91 (1984).

[28] M. Nantel, G. Ma, S. Gu, C.Y. Côté, J. Itatani, and D. Umstadter, submitted to Phys. Rev. Lett. (1997).

TIME-DEPENDENT SCHRODINGER EQUATION FOR THE INTERACTION BETWEEN A LASER PULSE AND A METAL

P. Martin and G. Petite
Commissariat à l'Energie Atomique, DSM/DRECAM/SRSIM, CEN Saclay,
91191 Gif sur Yvette, France
e.mail : martin@drecam.cea.fr

As well known, solving the time-dependent Schrödinger equation allows an accurate non pertubative description for the interaction between an intense optical field and an electron in a potential [1].

In this work, we apply this method to the case of a linear chain of atoms which mimics the periodic potential « seen » by a test electron in a quasi-free electron metal. This « crystal » potential is built as a series of screened coulombic potentials. Outside the lattice, the potential is constant and adjusted in order to reproduce the work function of the metal considered. After diagonalization we obtain well known band structures with characteristic gaps at the end of each one dimensional Brillouin zones (figure 1). For eigenstates below the vacuum level, we obtain typical Bloch functions extending in the whole « crystal ». Above the vacuum level, we find, as normal, that, the electron has a larger probability to be outside the lattice.

The laser is supposed to impinge on the chain at grazing incidence and the electric field is parallel to the chain. The electric field is, in the case of metallic reflection, space dependent : constant in vacuum and damped on a length taken equal to the skin depth inside the chain. In this framework, the notion of « surface » appears naturally because the laser irradiates only one side of the chain. This has some important consequences

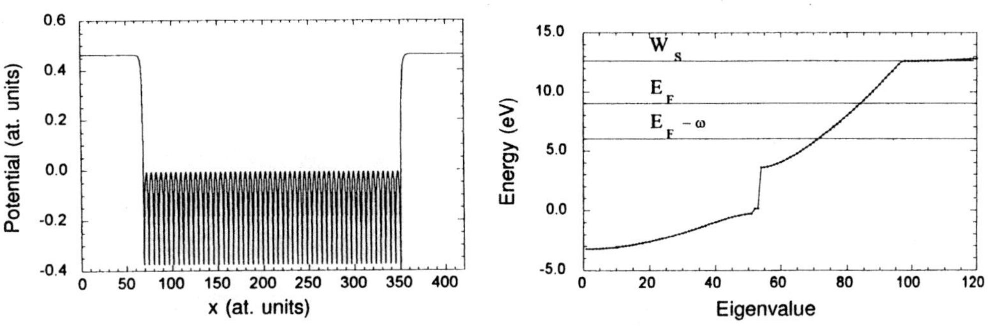

Figure 1. The potential and its corresponding eigenvalues. E_F: Fermi level location, Ws : vacuum level, ω : photon energy

Applications of High Field and Short Wavelength Sources
Edited by DiMauro *et al.*, Plenum Press, New York, 1998

especially on the harmonic spectra because, despite of fact that we are dealing with a fully centro-symetric static potential, the electric field introduces a non symmetric component in the (time-dependent) wave-function, allowing for the observation of odd and even harmonics. As well known, this is not the case in atoms.

In order to simulate the interaction properly, we must take into account the fact that only the eigenstates between the Fermi level and the Fermi level minus the energy of the photon must be propagated. In that case, only transitions between unoccupied states take place and this accounts for the Pauli exclusion principle. On the other hand, because we adopt a one-electron potential model, we must sum the contributions from each level for electron as well as harmonics spectra.

We present a few examples of calculations which have been performed for 1.55 eV photons and 24 optical cycles. The width of the (trapezoidal) laser pulse is in that case of the order of 50 fs. The calculation is performed for a chain of 200 unit potentials which represents an improvement compared to the calculation of référence [2] (50 atoms and 16 optical cycles). Below the vacuum level, the spacing between two successive levels is inversely proportional to the number of atoms included in the calculation. Around the Fermi level, this spacing is of the order of 50 meV. This is, of course, considerably larger than for a real metal, but, considering the energy broadening of the laser pulse, this is enough to consider that each transition presents a quasi-resonant character. Under these conditions, we propagate up to 26 eigenstates. The skin depth is for Aluminum and for 1,55 eV photons, of the order of 100 atomic units.

We present in figure 2 (a,b), electron spectra for two laser intensities. For $I=1\times10^{10}$ W/cm^2, we note that the width of the spectrum is less than the photon energy showing that only levels requiring a minimum number of photons (3 in our conditions) to overcome the 4.2 eV corresponding to the work function contribute to the photo-current. For $I=5\times10^{11}$ W/cm^2, we note that the spectrum is « hotter » because it involves a small contribution from deeper states (4 photons needed to reach the vacuum for levels around E_F-ω) and a high energy contribution from Fermi electrons having absorbed 4 photons.
We observe well resolved harmonic spectra through the whole range of intensities explored. In particular, for $I=1\times10^{13}$ W/cm^2, we note a « double plateau » structure which was already present in the 50 atoms calculation where the harmonic structure was barely visible.

Figure 4 shows the dependence of the total current on the laser intensity. Below, $I=1\times10^{12}$ W/cm^2, the log-log plot shows a lowest order perturbative behaviour, the slope being equal to 3. However, we observe a saturation of the current for higher intensities. This is due to two important and intricate points: the first one is that for such intensities,

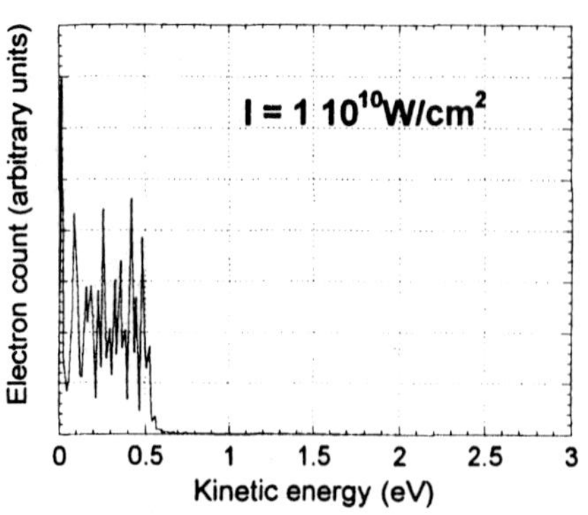

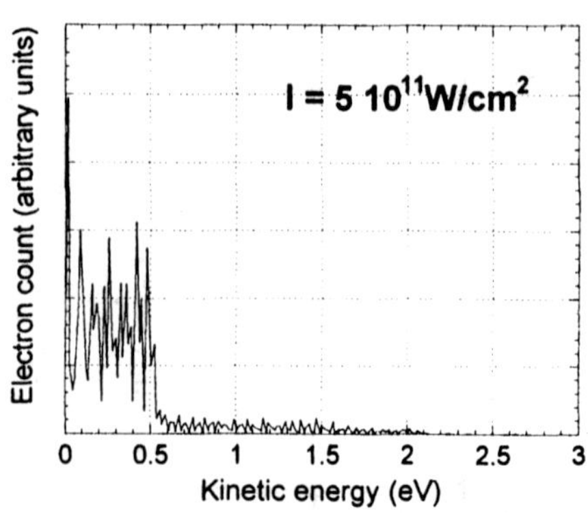

Figure 2. electron spectra for For $I=1\times10^{10}$ W/cm^2 (a) and For $I=5\times10^{11}$ W/cm^2 (b)

the initial population is depleted and, the second is that we have reached the so-called Keldysh [3] regime. The Keldysh parameter (ratio of the time necessary for an electron to tunnel through the barrier over the half optical period) is equal to 0.2. This regime appears earlier than in an atom for two main reasons : the electron velocity around the Fermi energy (12 eV) is high and the work function (i.e. the analogue of the ionisation potential) is low (4.2 eV).

We present also, in figure 3 (a,b,c), the harmonic spectra obtained for three intensities.

The model presented above has the advantage of simplicity. However, many improvements are needed if we want to perform more accurate simulations of the interaction of a laser beam with a metal. Two of them can be added : in a metal, the life-time of an electron hole pair close to the Fermi level is in general very short compared to the laser pulse duration. That means that during all the interaction, the propagated levels are never depopulated. This is not the case here, especially in high intensity limit. This can

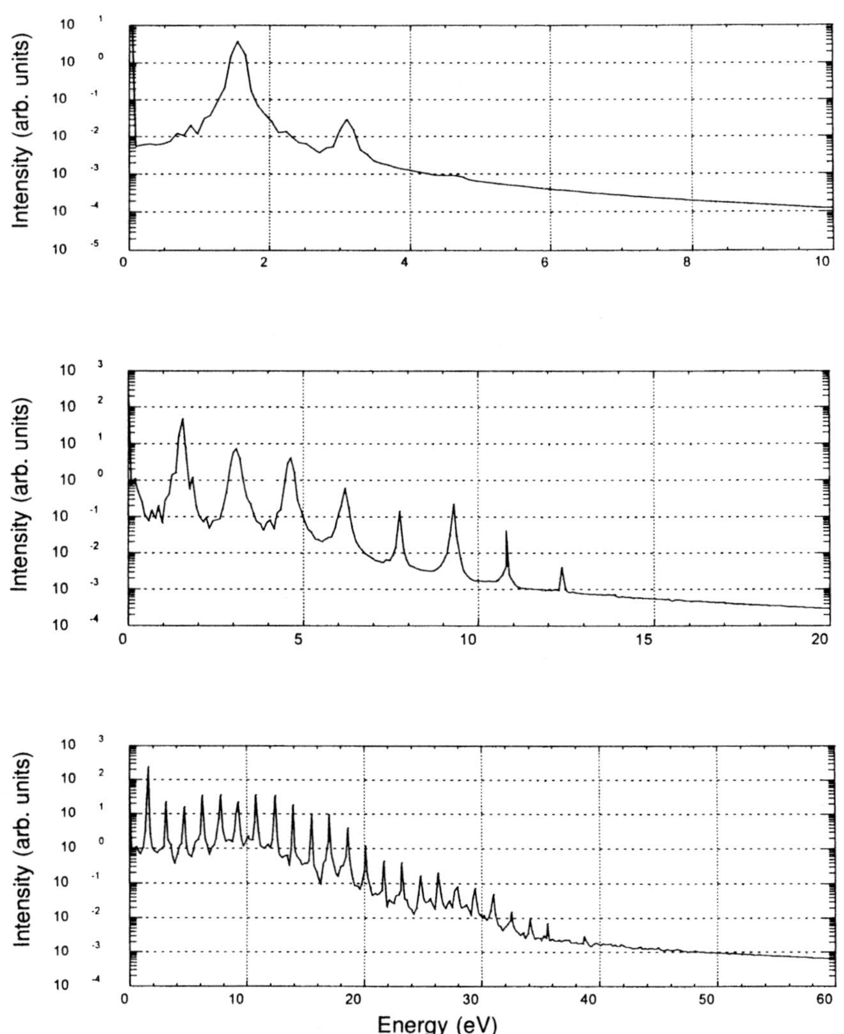

Figure 3. harmonics spectra for For I=1x10^{10} W/cm^2 (a) , 5x10^{11} W/cm^2 (b) and 1x10^{13} W/cm^2 (c)

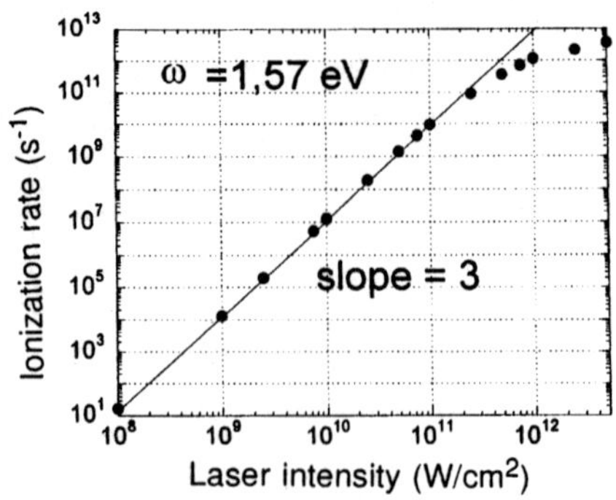

Figure 4. total ionisation rate as a function of the laser intensity. The slope of the straight line is equal to 3.

strongly influence the conclusions on the «Keldysh regime». More importantly, this calculation, neglects the role of the collisions during the laser pulse. These collisions are of fundamental importance because they introduce incoherence in the interaction process. In the visible or near infra-red part of the spectra, where the optical cycle is of the order of 2 fs and the pulse duration of the order of 40 fs, it is reasonable to neglect collisions processes because, typically, the mean collision time for an electron in a metal like aluminum is, at room temperature of the order of 10 fs [4]. This is of course not the case if an infra-red (10 µm) laser is used because the optical cycle becomes longer than the collision frequency. In such a case, the collisions associated with the electron-phonon coupling become very important because, as in indirect semi-conductors, the associated phonons can supply momentum to the electron during transitions and thus increase the total photocurrent and destroy the coherent process needed to observe harmonics.

RÉFÉRENCES

1. Kulander K., Schafer K.and Krause L., *Atoms in Intense Laser Fields*, (Gavrila M. ed) Academic press, San Diego, 1992
2. P. Martin, J.Phys.B : At. Mol. Opt. Phys, **29**, L635, 1996
3. L.V. Keldysh,, Sov. Phys. JETP **20**, 1307 (1965), F.V. Bunkin and M.V. Federov, JETP **21**, 896 (1965)
4. N.W. Ashcroft and N.D. Mermin, Solid state physics, internationnal edition, 1976

144

FEMTOSECOND STIMULATION OF ATOMIC AND NUCLEAR PROCESSES IN HIGH INTENSITY LASER PLASMAS

V.S. Rozanov[1], M.C. Richardson[2], N. Demchenko[1], S. Gus'kov[1] & D. Salzmann[3]

[1] P.N. Lebedev Physical Institute, Moscow, Russia

[2] Laser Plasma Laboratory, CREOL, University of Central Florida
4000 Central Florida Blvd., P.O. Box 162700, Orlando, FL 32816-2700

[3] Soreq Nuclear Research Center, Yavne, Israel

Tel: 407-823-6819, Fax: 407-823-6819, Email: mcr@creol.ucf.edu

ABSTRACT

The dynamics of the interaction of intense femtosecond laser pulses with dense plasmas is examined. In particular we consider the effect of the high fields generated in plasmas created from heavy isotope of hydrogen on the ion kinetics and neutron generation.

INTRODUCTION

The interaction of intense femtosecond duration laser pulses with dense plasmas is replete with new plasma phenomena and exciting experimental opportunities. The intense fields within laser radiation with the power levels of 10^{19} W/cm^2 range interacting with dense plasmas leads to the generation of enormous bursts of energetic electrons (j ~ 10^{15} A/cm^2 that then give rise to the production of intense hard x-ray emission and unprecedented magnetic fields that could be in the GigaGauss range [1,2]. The role these phenomena play on the particle dynamics in these plasmas poses many interesting questions. As the development of femtosecond laser systems reaches these conditions [3,4], the examination of these phenomena becomes possible.

In this paper we examine the interaction of intense 100 fs laser light with targets of the heavy isotopes of hydrogen. The interaction is modeled with a hydrodynamic code (RAPID-SP)[5,6] that includes approximations for the radiation coupling physics and the subsequent particle transport and kinetics. In particular we examine the dynamics of the ions within the plasma, and the generation of neutrons in these plasmas.

When high intensity laser pulse ($q\lambda^2 > 10^{15}$ μm^{2}W/cm^2) interacts with the target, the laser flux is transformed mainly into a flux of superthermal electrons [1,7,8]. The collisional mechanism accounts for a very small fraction of the absorbed flux. The energy of the fast electrons is much greater than the thermal energy. The fast electrons are ejected by the field with an energy range significantly exceeding the thermal level. The process is similar to electron beam propagation in a plasma.

Applications of High Field and Short Wavelength Sources
Edited by DiMauro *et al.*, Plenum Press, New York, 1998

The ponderomotive forces and the related ion nonlinearities play an important role in plasma profile formation for these high flux densities. The oscillating ponderomotive potential modulates the plasma density and velocity profiles. The dissipation of the plasma perturbations leads to an increase of the internal ion and electron energies. The energy dissipation within this type of plasma is mainly determined by the ions.

DIRECT ION HEATING

When the flux density is relatively low, the collisional dissipation of the ion perturbations takes place due to the ion viscosity [9]. The ions are heated in this case. The rate of heating Q per unit plasma mass can be expressed in the form (linearized approach for homogeneous plasma)

$$Q = \frac{\partial \bar{\varepsilon i}}{\partial t} = \frac{2\mu k_0^2 u_0^2 G}{\rho_0 c^2 \rho_0^2 c_s^4} q_0 q_1$$

where

$$G = \left[\left(M^2 - 1^2 \right)^2 + \left(\frac{2\mu k_0 \beta_0 M}{\rho_0 c_s} \right)^2 \right]^{-1}$$

and ε_1 is the specific internal ion energy averaged over the oscillation period; $\mu \approx \rho_0 \nu_{Ti} l_I$, the ion viscosity coefficient; $k_0 = \omega / c$, the wave number; ω, the laser frequency; u_0, the velocity with respect to ponderomotive potential maxima; $c_s = \left[\left(ZT_e + T_i \right) / m_i \right]^{\frac{1}{2}}$, the sound velocity; $M = u_0 / c_s$, the Mach number; ρ_0, the nonperturbed plasma density; ρ_c, the critical density; q_0 and q_1, the incident and reflected fluxes of laser radiation; β_0, the refractive index. This heating can be considered as the work of an effective friction force between the electromagnetic field beatings and the plasma when the velocity of relative motion equals u_0. The friction force F_1 per one ion can be defined be the expression

$$F_1 = \frac{m_1 Q}{u_0}$$

By considering the damping factors of the longitudinal field in the ion wave, assuming negligible inertia in the ions, it is possible to show that in the strong field limit, the relative damping factors of ions to electrons is proportional to $\Delta(A / Z)^{\frac{1}{2}}$. This damping can be described by viscosity coefficients for each one species μ_i where

$$\mu_i = \frac{2 \wp \ell_0}{k^z}$$

This relation permits us to take account of collisionless mechanisms in a hydrodynamic model.

HYDRODYNAMIC SIMULATIONS

The code RAPID-SP simulates the interaction process on the basis of hydrodynamic equations taking into account the ponderomotive force, effective ion viscosity and the energy transport by ions, thermal and superthermal electrons, and on the base of Maxwell equations for oblique incidence of s- and p-polarized waves. The

146

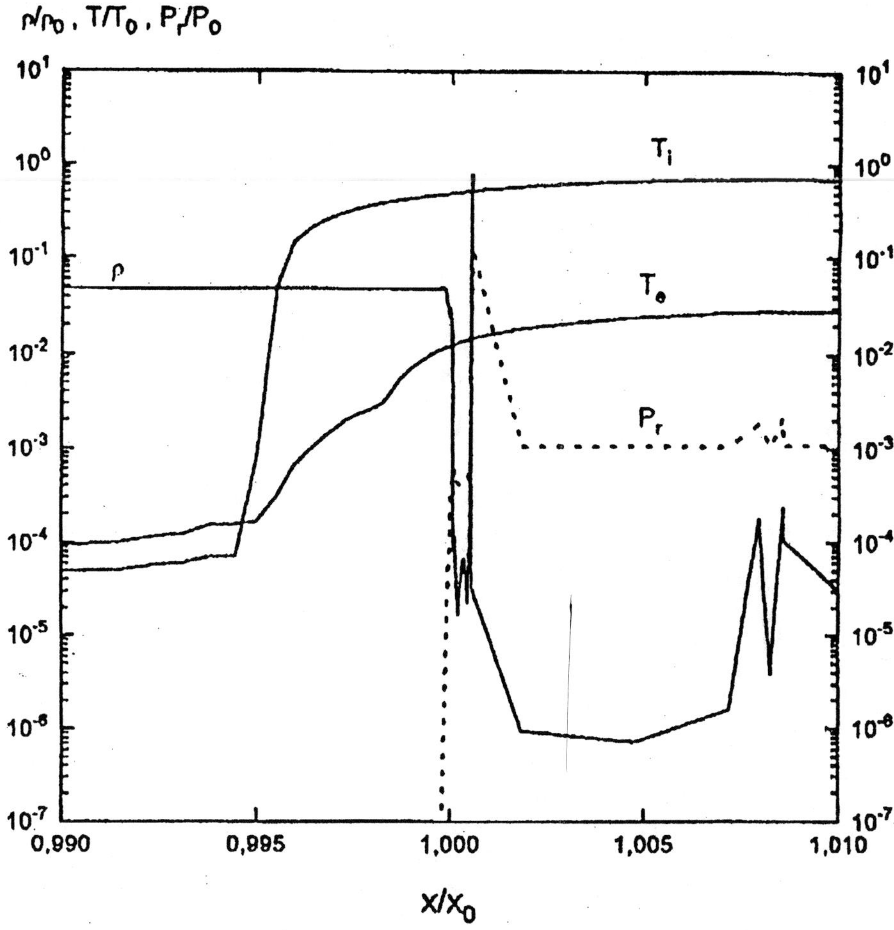

Fig. 1 Hydro code simulations of the electron and ion temperature T_e and Ti, and the plasma density, ρ, and pressure, P.

relativistic increase of the electron mass [10] and density (due to Lorentz frequency is an invariant quantity.

An example of the condition predicted to exist in these plasmas is shown in Fig.1, which shows the spatial dependence of the electron and ion temperatures, the plasma density and pressure midway through the interaction of a 100 fs (FWHM) 850 nm, 1 J laser pulse focused to a 10 μm diameter spot on a C_8D_8 target. These conditions are expected to be generated by the upgraded CrLiSAF laser system currently nearing completion at CREOL [6]. Of particular interest are the extremely high pressures generated in the interaction region, and the very high ion temperatures predicted (~ 100 keV).

The calculated neutron yield is quite high in these simulations, of order ~ 10^8 neutrons. Similar calculations of the interaction of 1ps duration, 20 J pulses with C_8D_8 targets showed reasonable agreement of the estimated neutron yield (~ 10^9) with experiments performed at Rutherford Appleton Laboratories [12].

The extremely high electric and magnetic fields generated in the plasma will also have a major impact on the atomic structure and possible also the near nuclear field. Magnetic fields in the GigaGauss region will tend to create modified states of matter consisting of a distribution of atoms transformed into a pseudo-molecular state in which outer electron orbits of ions and atoms envelope other nuclei in directions normal to the field. The conductivity of this distribution will consequently be strongly dependent on the magnitude and direction of the field.

CONCLUSIONS

These calculations suggest that plasmas produced from heavy hydrogen targets by femtosecond high intensity lasers, could potentially provide a useful bright pecosecond point source of neutrons. Such a source may be useful for short pulse calibration capability, and could also be used as a point projection imaging source.

This work was supported by Russian Foundation of Basic Researches (96-02-16678-a), INTAS (93-2571, DoE and by the State of Florida.

REFERENCES

1. S. Wilks Phys Fluids B 5, 2605, (1993); S.C. Wilks, W.L. Kruer, M. Tabak & A.B. Langdon, Phys Rev Lett 69, 1383 (1992).
2. J. Denavit, Phys Rev Lett 69, 3052 (1992).
3. P.A. Beaud, M. Richardson, & E.J. Miesak, IEEE, J. Quant. Electron, 31.317 (1995).
4. C.P.J. Barty, Laser Focus, p 93, (June 1996).
5. Demechonko N.N. et al Pizika Plasmy 16, 812-817 (1990).
6. Kalashnikov M.P. et al Phys. Rev. Lett. 73, 260-263 (1994).
7. Brunnel F. Phys. Rev. Lett 59, 52-55 (1987).
8. Gibbon P., Bell A.R. Phys. Rev. Lett. 68, 1535-1538 (1992).
9. Demchenko N.N. and Rozanov V.B. JETP 76, 997-1001 (1993).
10. Kaw P. and Dawson, J. Phys. Fluids 13, 472-481 (1970).
11. N.N. Demchenko & V.B. Rozanov, JETP Lett. (1996).
12. Fews A.P. et al Annual Report RAL Report TR-95-025 p.30-32 (1994-95).

TIME RESOLVED OPTICAL PROBING OF SUPERSONIC IONISATION FRONTS IN SHORT PULSE-SOLID TARGET INTERACTIONS

E. T. Gumbrell,[1] R. A. Smith,[1] T. Ditmire,[1,2]
A. Djaoui,[3] S. J. Rose,[3,4] and M. H. R. Hutchinson[1,3]

[1]Blackett Laboratory, Imperial College of Science, Technology and Medicine, London SW7 2BZ, United Kingdom
[2]Present address: Lawrence Livermore National Laboratory, Livermore, California, 94550
[3]Rutherford Appleton Laboratory, Chilton, Didcot, OX11 0QX, United Kingdom
[4]Clarendon Laboratory, University of Oxford, Parks Road, Oxford, OX1 3PU, United Kingdom

INTRODUCTION

With pulse durations ranging from a few picoseconds down to tens of femtoseconds, and substantial free electron quiver energies at the high intensities available, CPA lasers enable the production of transient, small scale length plasma at both high temperature and solid density. The study of energy transport in these interactions is complicated, as there are several competing mechanisms which can rapidly transport absorbed laser energy away from where it is first deposited.

On progressively longer time scales, electron thermal transport,[1] hydrodynamic expansion[2] and shock wave propagation are known means by which plasma energy leaves the interaction region. Supra-thermal electron transport must also be considered, although the contribution from this mechanism varies substantially between different experiments. It is also understood that even when the fraction of energy subsisting in thermal radiation is small, the energy transport can become dominated by energetic photons if the radiation temperature is sufficiently high, and the temperature gradients are sufficiently steep.[3] Since high intensity, short pulse laser generated plasmas are a natural choice for meeting these criteria, they should suit laboratory studies of radiation transport. In fact, recent studies have demonstrated the importance of radiation transport resulting from 2psec laser interactions with solid density targets.[4] In these experiments, ionisation fronts driven by thermal radiation transport from the laser absorption region were inferred, with measured velocities up to 40 times higher than those seen using nsec laser pulses.[5]

Here, we report on further experiments, accompanied by detailed 1-D planar simulations, which more completely address the issue of the energy transport processes responsible for driving ultra-fast ionisation fronts. Specifically, we have conducted picosecond optical probing of high intensity interactions with planar fused silica in order to determine the kinematics of supersonic ionisation fronts driven into the bulk of these targets. These ionisation fronts, launched from high energy density plasmas at the target surface, have been investigated as a function of laser intensity, wavelength and target angle. The laser absorption has been characterised for our simulations with a radiation hydrodynamics and an opacity generating code. Consideration has been given to thermal and supra-thermal electron transport. It is shown that the ionisation front kinematics are best explained with a radiation transport model.

OPTICAL PROBING EXPERIMENTS

The CPA laser system used in our experiments is based on Nd:Glass. The pulse duration is 2psec and the central wavelength 1053.4nm (ω_0). In the new experimental work the heating pulse was frequency doubled and residual ω_0 light rejected to provide a pre-pulse free second harmonic ($2\omega_0$) heating beam. F/10 refractive optics were used to focus the heating beam, which was measured as 1.5 times diffraction in $2\omega_0$. With our high intensity, short pulse it is an appropriate assumption that collisional absorption dominates for normal incidence interactions.[6] The space and time integrated laser absorption f_A was measured using an Uhlbricht sphere as an input for our simulations. We note that there was no significant difference in f_A between normal incidence and 45° incidence for the high contrast p-polarised 526.7nm irradiation, thus confirming that the suprathermal electron fraction was small compared to the collisional fraction.

The experiment is illustrated in Fig. 1. SiO_2 targets were irradiated on a planar face. The plasma evolving inside the target was imaged using a probe pulse generated by exciting a 620nm Raman transition in ethanol with a $2\omega_0$ component of the main laser pulse. Varying the delay between the heating and probe beam enabled a series of shadowgrams to be obtained of the volume evolving plasma above the probe critical density $n_{crit} \approx 3\times10^{21}$ cm^{-3}. The probe pulse allowed discrimination of the extent of this plasma against scattered heating laser radiation and harmonic generation arising within the plasma. The ionisation front extent x_{front} corresponding to n_{crit} was determined as a function of time. The time resolution was $\approx \pm1$psec and the spatial resolution to which we could determine the maximum extent of the heat front from the target face was estimated to have upper bounds $\Delta x \approx \pm 2 \mu$m.

PROBE IMAGE INTERPRETATION AND PLASMA SIMULATIONS

In Fig. 2 we show a typical ionisation front image from $2\omega_0$ irradiation at normal incidence. Our physical interpretation uses the assumption that the SiO_2 plasma is in LTE. We then use the equivalence $T_e^{x_{front}} \approx 1$eV in our simulations, which is derived using Ref. 1. The LTE radiation hydrodynamics[7] and opacity codes[8] we employ can describe electron thermal transport, either by diffusive heat flow[9] or a properly delocalised heat flux,[10] and supra-thermal generation and transport is also explicitly included. Radiation transport is

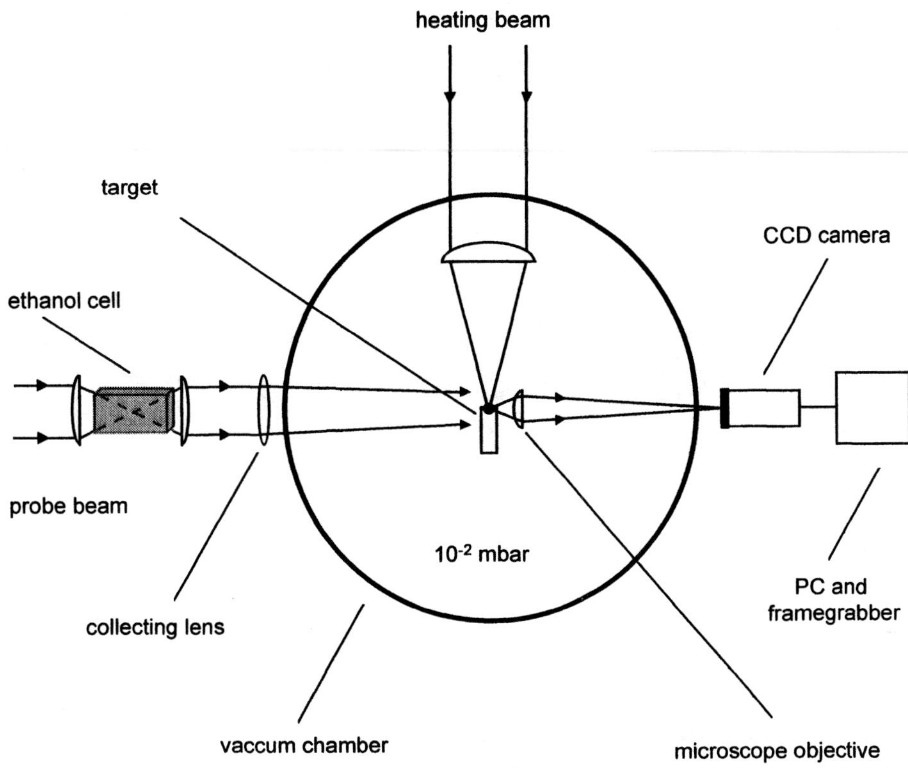

Figure 1. The experimental configuration for the optical probing experiments.

calculated with the assumption of an instantaneous distribution of emission and absorption sources determined by the local plasma conditions at each fluid element, including accounting for ionisation and excitation energies. The frequency resolved Planckian opacities are generated in an average atom model, a locally Planckian source function is assumed and our simulations enable recovery of both the optically thick (diffusive) and optically thin limits to the transport of ionising radiation.

In Fig. 3 the results of our electron transport simulations using Spitzer heat flow[9] are presented against the experimental data for ω_0 irradiation; results for $2\omega_0$ are also shown. Here, allowing for known experimental uncertainty, a nominal peak intensity $I_0^{\omega_0} = 1.5 \times 10^{17}$ Wcm^{-2} is simulated for the 1053.4nm heating beam. We have chosen to use the upper limit on the measured f_A and have set the harmonic flux-limiter for the maximum free streaming limit of unity. Evidently, this classical electron thermal conduction theory cannot adequately explain our results. If instead the nonlocal heat flux prescription is applied, we are unable to reproduce an increase in heat front penetration and velocity, owing to the high particle densities. A radiation-hydrodynamics simulation for the $2\omega_0$ irradiation data is also shown in this figure.

Within the uncertainties inherent in the experiments, the simulations correlate well with the data, although the codes overestimate the heat front speed dx_{front}/dt at early time because the source function in the experiment differs from the assumed Planckian. The ionisation front penetration arising during the interaction is evident with the rapid

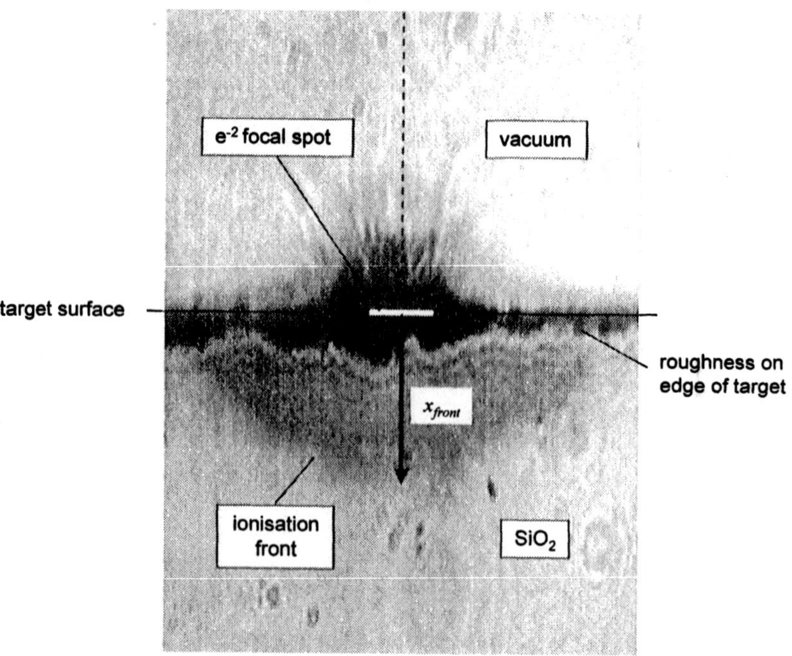

Figure 2. Time and space resolved ionisation front image, indicating the extent of plasma evolving within a planar SiO_2 target consequent upon *linearly* polarised 526.7nm irradiation at normal incidence. This image was captured at late time after the initial visible onset of the burn into the target. The estimated peak intensity was 7×10^{16} Wcm^{-2}. The measured focal spot size (20μm) is shown, and this can be compared to the extent of the ionisation front $x_{front} = 72$ μm.

dx_{front}/dt about $t = 0$, which we estimate as $\leq 10^9$ cms^{-1}, consistent with up to 10μm penetration over 1psec. We have also shown that dx_{front}/dt arising from irradiation of SiO_2 targets coated with a submicron Pb layer is faster still, and this is discussed in a future paper.[11]

Supra-thermal electron transport simulations (allowing for space and time dependencies in the supra-thermal generation) and electron stopping distance data suggest that we would observe prompt heat front penetration if these particles dominated the energy flow.[11] Another indicator of the significance of supra-thermal electron transport comes from comparison of the late time penetrations for $2\omega_0$ and ω_0 irradiation as a function of peak intensity I_0, where the measured scalings were $I_0^{0.6}$ and $I_0^{0.7}$ respectively. Although these are too similar to reconcile with empirical supra-thermal electron scaling relationships,[11] they closely agree with the background temperature scaling for radiation thermal conduction.[3]

The absence of change in $x_{front}^{\omega_0}(t)$ with p-polarised irradiation of 45° angled targets over irradiation at normal incidence,[4] lead us to use the pre-pulse free $2\omega_0$ beam at various oblique incidences in order to modify the hot electron generation - here, with a cleaner pulse, the tunnel distance to heating laser critical surface should be reduced. However, there was still no angularly maximised dependence in $x_{front}^{2\omega_0}(t)$ in evidence.

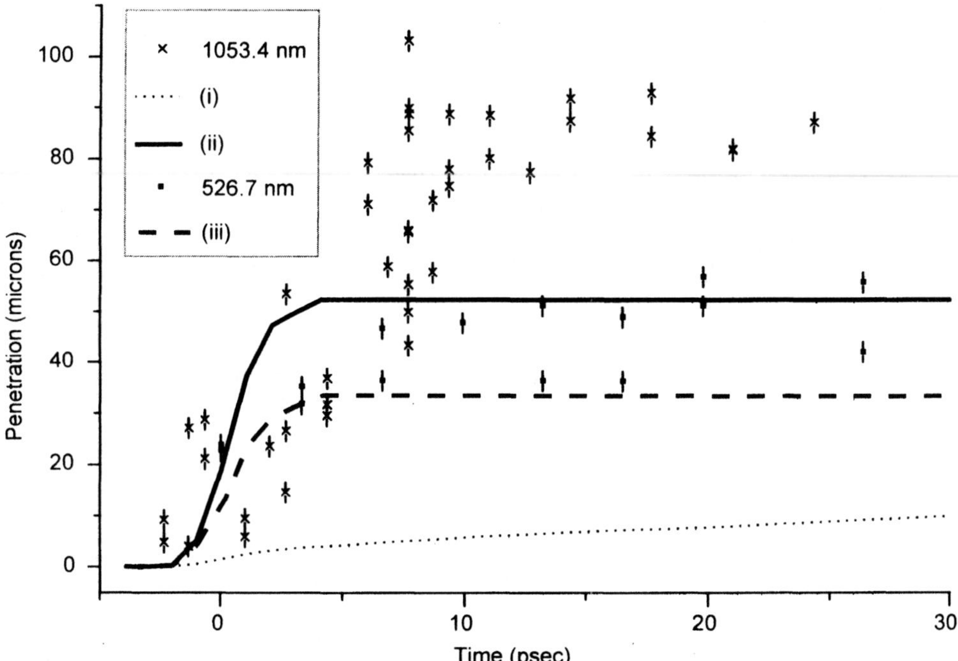

Figure 3. Ionisation front kinematics $x^{\omega_0}_{front}(t)$ and $x^{2\omega_0}_{front}(t)$ for *circularly* polarised 1053.4nm irradiation and *linearly* polarised 526.7nm irradiation at normal incidence. Overlaid are results of MEDX radiation hydrodynamics simulations, including for $f^{\omega_0}_A$ = 30% and the upper limit on the measured intensity, (i) the thermal transport simulation clamped to the free-streaming limit, and (ii) the same case with radiation transport activated. Simulation (iii) is for the $2\omega_0$ interaction, the radiation drive algorithm is also turned-on and the measured $f^{2\omega_0}_A$ = 40%, $I^{2\omega_0}_0$ = 7.0 $\times$ 10^{16} Wcm^{-2} are applied as input.

Therefore, we cannot conclude that suprathermal electron generation from resonance absorption was responsible for the measured ionisation kinematics from angled target shots. Conversely, the absence of angular scaling is consistent with the radiation transport hypothesis.

ACKNOWLEDGEMENTS

We would like to thank P. Ruthven, A. Gregory and S. Hanif for technical support. Financial support from both the UK EPSRC and MOD is gratefully acknowledged.

REFERENCES

1. B.-T. V. Vu, A. Szoke, and O. L. Landen, Phys. Rev. Lett. 72, 3823 (1994).
2. D. D. Meyerhofer, H. Chen, J. A. Delettrez, B. Soom, S. Uchida, and B. Yaakobi, Phys. Fluids 5 2584 (1993).
3. Y. B. Zeldovich and Y. P. Raizer. *Physics of Shock Waves and High Temperature Hydrodynamic Phenomena* (Academic Press, New York and London ,1966).

4. T. Ditmire, E. T. Gumbrell, R. A. Smith, L. Mountford, and M. H. R. Hutchinson, Phys. Rev. Lett. **77**, 498 (1996).

5. T. Afshar-rad, M. Desselberger, M. Dunne, J. Edwards, J. M. Foster, D. Hoarty, M. W. Jones, S. J. Rose, P. A. Rosen, R. Taylor, and O. Willi, Phys. Rev. Lett. 73, 74 (1994).

6. D. F. Price, R. M. More, R. S. Walling, G. Guethlein, R. L. Shepherd, R. E. Stewart, and W. E. White, Phys. Rev. Lett. **75**, 252 (1995).

7. A. Djaoui, J. Quant. Spectrosc. Radiat. Transfer 54 143 (1995).

8. S. J. Rose, J. Phys. B: At. Mol. Opt. Phys. 23 1667 (1992).

9. L. Spitzer and R. Härm, Phys. Rev. 89, 977 (1953).

10. J. F. Luciani, P. Mora, and J. Virmont, Phys. Rev. Lett. 51, 1664 (1983).

11. E. T. Gumbrell, R. A. Smith, T. Ditmire, A. Djaoui, S. J. Rose, and M. H. R. Hutchinson, submitted to Phys. Plas.

ULTRAFAST INCOHERENT X-RAY SOURCE
UTILIZING ULTRA THIN FREELY SUSPENDED FILMS

V.M.Gordienko, M.A.Joukov and A.B.Savel'ev

Physics faculty and International Laser Center of M.V.Lomonosov
Moscow State University
Vorobyevy gory, Moscow, Russia, 119899
e-mail: savelev@femto.ilc.msu.su

INTRODUCTION

For quite some time already, femtosecond laser induced plasma induces steady interest of many research groups world-wide[1-5]. Among many prospective uses of FLP one can name such applications as: microscopy (including biological "water window" microscopy[6], especially if used in conjunction with x-ray resonant mirrors as has been done in[7]), microlithography, new x-ray lasers schemes, x-ray spectra of FLP bear previously unavailable information on this unique state of matter allowing deeper study of this promising phenomena. Ability to produce and to study analytically such plasmas also made possible advances in study of some problems of quantum electrodynamics, nuclear physics and CLTS, astrophysics, and in creation of essentially novel generation of ultra-short VUV, X-ray and Γ-ray sources[1-4].

To optimise for such properties as x-ray yield, mean ionisation degree, maximum electron temperature, one has to study the processes which lead to desired effect, and adjust experimental conditions accordingly. Obviously it can be done through increasing intensity of "ignition" laser pulse: intensities of $\sim 10^{17} W/cm^2$ already provide increase in the mean electron temperature up to $\sim 1 keV$[5,8-11]. However, alternative approach proves to be both easier and cheaper to implement - it is based on applying various modifications to the target itself that result in a drastic increase in aforementioned parameters. As it was shown in[12-15] this enables one to achieve the same values of plasma parameters as when using higher intensity lasers to generate FLP on conventional, bulk targets.

One of the choice here is freely suspended ultra thin films (UTFs). First irradiated with laser beam by authors of[14], UTFs were studied in a detailed way (both experimentally, theoretically and numerically) in a number of works[15-17] and it was shown one can achieve unusually high electron temperatures (of $\sim 1 keV$) by making UTF depth smaller than that of heat wave propagation distance on the timescale of laser pulse duration and, subsequently, to maximise FLP x-ray yield in harder spectral region.

Applications of High Field and Short Wavelength Sources
Edited by DiMauro *et al.*, Plenum Press, New York, 1998

Subject for this particular work is to compare in detail experimental and numerical data obtained for water window x-ray from carbon films, as well to discuss application of high-Z UTF to generate hard x-rays. We show that by using targets of the proper type one can control critical plasma parameters such as electron temperature, ionisation state, radiative capability, non-linear susceptibility, spectral composition.

"OVERHEATING" OF FLP IN FREELY SUSPENDED ULTRA THIN FILMS (UTF) MADE OF LIGHT ELEMENTS

Parameters dynamics of plasma layer being formed as well as their maximum values are seriously affected by presence of unheated bulk. Indeed, laser pulse is absorbed in a layer approximately as thick as skin-layer, whereas depth of layer that is heated by non-linear heat wave on the timescale of the order of laser pulse duration can be much larger[18].

One can estimate skin layer as:

$$l_s = 6.8 \cdot \sqrt{\frac{A \cdot \nu_{e-i}}{Z \cdot \rho \cdot \omega_0}} \ \text{nm}, \qquad (1)$$

where A is the atomic mass (in atomic mass units), ω_0 - frequency of incident light field, Z - mean ion charge, ρ - density (g/cm^3), $\nu_{e-i} = 4\sqrt{27}\pi n_e q_e^4 \Lambda \dfrac{Z}{m_e^{1/2} T_e^{3/2}}$ - electron-ion collision frequency, with Λ being Coulomb logarithm, and T_e - the electron temperature, assessable as (according to[2]):

$$T_e \cong 710 (\tau)^{2/9} (I_{16})^{4/9} \text{eV}, \qquad (2)$$

where I_{16} is in units of 10^{16}W/cm^2, τ - in hundreds of femtoseconds, and ne$_{22}$ is electron density in units of 10^{22}cm^{-1}.

If applied to a carbon target (atomic weight A~12, Z=6, density ρ~2g/cm^3) and laser pulse with intensity ~10^{15}W/cm^2, wavelength ~600nm and duration ~200fs, gives l$_s$~5.4nm. On the other hand, heat wave travel length can be calculated from[18]:

$$l_{th} = \frac{2.8 \cdot 10^3}{Z} (I_{15})^{5/9} \left(\frac{\tau}{n_{e_{22}}} \right)^{7/9} \text{nm}, \qquad (3)$$

(here I_{16} is in units of 10^{16}W/cm^2, and Z_{13} is normalised to 13) to be ~7nm, thus ratio l$_{th}$/l$_s$ being ~1.3. For I=5x10^{15}W/cm^2 this ratio becomes (due to increase in l$_{th}$) ~3.2, and rises to ~4.7 (l$_{th}$~25nm) with intensity reaching 10^{16}W/cm^2. This means that on the timescale of 100 fs, major part of absorbed energy is re-distributed away from laser-target interaction area into the bulk. "Suppression" of this thermal flux into the bulk of the target should lead to increase in specific energy per atom, and electron temperature could surpass 1keV, even when using "moderate" intensities of $10^{15} \div 10^{16}$W/cm^2, with subsequent increase in ionisation degree and harder x-ray yield. To illustrate this, let us neglect plasma ablation, assume that all the energy deposited by the laser beam remains in the area being directly heated and calculate the temperature. Taking reflection coefficient to be ~0.6 (a typical value for conditions under consideration), intensity 10^{15}W/cm^2, duration 200fs, focal spot 5x5μm, one can estimate T_e for carbon to be ~700÷900eV depending on percentage of energy deposited in ionisation process. This number reaches value of several keV for intensity of ~10^{16}W/cm^2. This rough estimate is the upper limit of how high temperature can go with thermal flux limitation technique discussed here. Thus, with UTF targets, our effort is aimed at getting as close as possible to energy deposition regime just described, by limiting heat carry-away from the area being heated.

In 1990 authors of[14] have conducted first experiments with 10nm and 30nm carbon UTFs irradiating them with 100fs 620nm pulses of various energies. They've reported a tenfold increase of x-ray yield in >750eV range and suggested that these x-ray pulses have to be very short due to fast plasma cooling by ablation.

Advantages arising from femtosecond plasma formation in ultra thin films instead of bulk target were discussed by us in[18] based on computer simulation.

In experiments described in[15-17], freely suspended ultra thin carbon films 10-60nm thick served as targets. As a "thick" target 1μm mylar film was used. Authors employed linear polarised radiation l~600nm, τ~200fs, I~10^{15}-10^{16}W/cm^2 incident on the target surface at the angle of $\Theta \sim 45°$. (for detailed laser system description see[15]).

Conversion efficiency η was measured with two x-ray p-i-n diodes, with main part of the emission from carbon plasma being provided by characteristic lines of C^{4+}(4.02 nm, 0.308 keV) and C^{5+}(3.37 nm, 0.368 keV) ions[11]. 750nm Al or 1μm mylar films were used as filters, which made it possible to measure conversion into the following spectral regions: >300eV (Al) and 100-290eV, >700eV (mylar).

Such "scanning" through carbon UTF depths yielded the fact that there's an optimal thickness at around 20÷30nm where x-ray yield reaches maximum, and at the depth of ~10nm there begins a fall in conversion, discussed in more detail below. It should be noticed here that used in[12] 100 nm carbon foil was too thick to get desired effect.

In order to simulate x-ray emission of this carbon FLP, we have used an approximated 6-state transition scheme[19], being part of SOVSEM 1D single-fluid, two-temperature hydrodynamic code with transient ionisation kinetics and limited thermal flux[18,19].

It allows to evaluate emission from transition $1s^2$-1s2p for He-like and 1s-2p for H-like ions. The scheme consists of ground state of Li-like ion, ground and first excited states of He- and H-like ions and also a fully stripped atom. According to[19] taking into account more states gives only an insignificant correction to data obtained from 6-state scheme, i.e. schemes with 1 and 5 excited states of an ion differ by less than 20%.

Fine structure of ion states can be also safely neglected here, since temperature of such plasma greatly exceeds typical separation energy of a state.

When applied to conditions of experiments described in[14] our code yielded results which well coincided with those obtained from real experiments. Thus, ~300÷400eV quanta yield suffered a ~(1÷2)x10^3 times increase when moving up the energy absorbed by the target from 5 J/cm^2 to 20 J/cm^2 according to[14] (for 10nm UTF), while our code predicted it to be ~10^3 times.

On fig.1a there is a dependence of conversion efficiency η on films thickness D for p- and s- polarised light measured experimentally in[15] (dots) and simulated curves produced by our code. Simulated curves of dependencies of η on carbon films depth appear to be good approximation of measured data, once normalised for the "thick" film data and p-polarisation.

Striking difference in dependencies for different polarisations should be attributed to higher absorption taking place for p-polarised light[20]. Latter fact occurs due to a well known effect of resonant absorption of an oblique p-polarised light, which leads to reflection lowering or more effective energy deposition. Thus, reflection R for s- and p- polarised light

($I\sim10^{15}W/cm^2$), falling at the angle of 45°, changes from 78% to 64% respectively, showing greater absorption in case of p-polarisation.

Notable peculiarity of acquired data is the fall of η for the thinnest film of 10nm. It can be explained with plasma's fast ($\sim10^7cm/sec$ according to simulation data) ablation into void. Calculated maximum density temporal dependence for a number of UTF depths is presented on fig.2. It can be seen that for ultra thin films maximum plasma density growth cuts off sharply and is replaced with its rapid fall, becoming under-critical for D~30nm at t~0.5ps, and t=0 for D~10nm (t=0 corresponds here to the maximum of laser pulse). Thus, decrease in conversion η is primarily connected with fast decrease in plasma density.

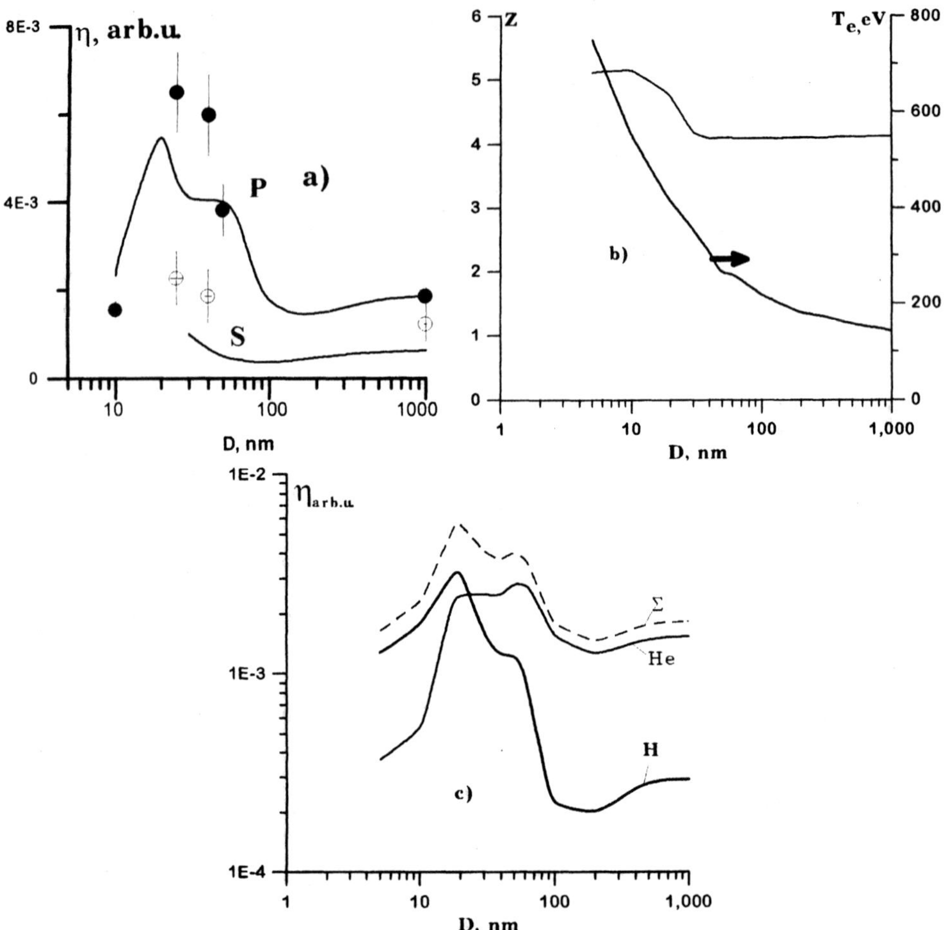

Figure 1. Dependence on carbon UTF thickness D of: a) to x-ray conversion efficiency η, b) peak electron temperature T and ionisation degree Z, c) partial inputs of H- and He- like ions and sum of the both. ($I\sim2x10^{15}W/cm^2$, λ=616nm, $\tau\sim200fs$)

Indeed, when, still very hot (~600eV), FLP is ablating into the void at such velocities, that ionisation state of the plasma "freezes" - leading on one hand to lengthening of x-ray pulses as is shown on fig.3. There, there is a dependence of H-like ion line (λ=33.7A) emission duration on target film thickness. As can be seen, it increases from ~220fs for 100nm film (being of the same order for one 1μm thick) to ~1ps for 10nm one. On the other hand, (since photorecombination probability is directly proportional to plasma density) it leads to the fall of the conversion efficiency η for thinner UTFs (since they get overheated and ablate faster).

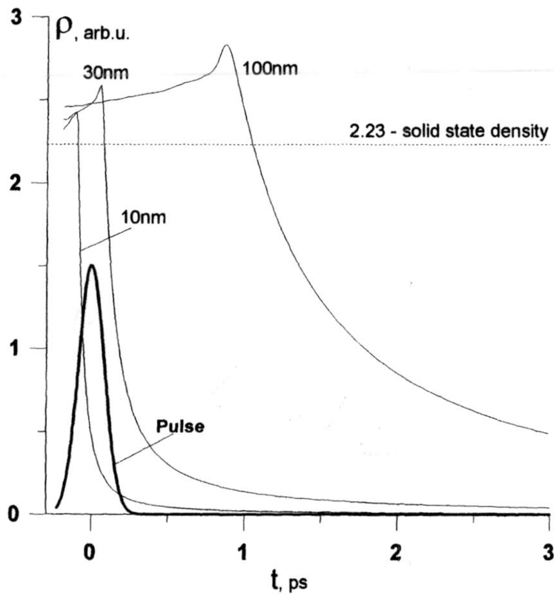

Figure 2. Peak density temporal dependence for carbon UTF, ($I \sim 2\times10^{15}$W/cm^2, λ=616nm, $\tau \sim$200fs) with different UTF depths.

According to the calculated data, when target's depth drops below 100nm, there starts a rapid increase in peak electron temperature and mean ionisation degree (fig. 1b). It should be

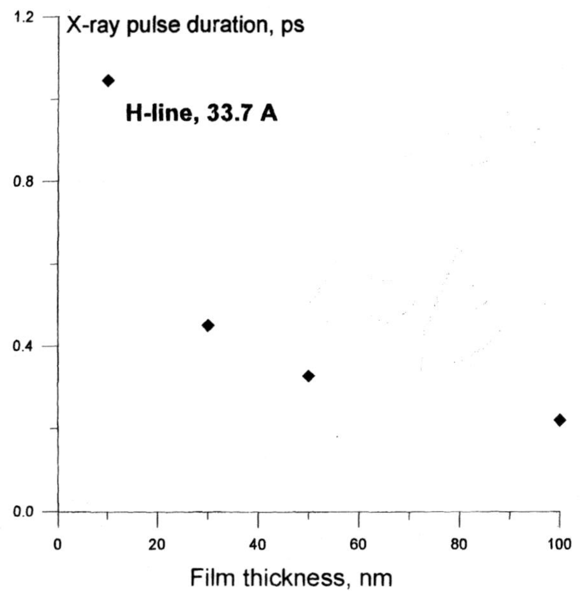

Figure 3. X-ray pulse generated by carbon FLP duration dependence on UTF thickness.

expected to be so since according to estimates made earlier at these depths target thickness drop below the heat wave propagation length, thus causing increase in specific energy per atom in heated layer. It, in turn, leads to change in ratio of partial inputs of H- and He-like ions in x-ray yield (fig.1c).

The latter fact bears a lot of significance - for it means that by adjusting target properties we can control ratio of some spectral components, including those in harder x-ray region. Thus, if at UTF depth of $\sim 10^3$nm conversion efficiency into Ly_α (line of H-like ion, λ=33.7A) to one into He-like ion line (λ=40.2A) ratio is $\eta_H/\eta_{He}\sim 0.2$, then with UTF depth being ~ 50nm it is already ~ 0.5 and becomes ~ 3.3 at the depth of ~ 10nm.

"OVERHEATING" TECHNIQUE APPLIED TO UTFS OF HEAVIER ELEMENTS

Due to practical demands as well as because of general interest, it is desirable to obtain harder x-ray quanta. To do this, natural solution is to move up the atomic number Z, choosing heavier elements as target material. On the example of Ni UTF we show below, that general tendency described above for the case of "light" carbon atoms, still applies to heavier Ni, allowing considerable increase in hard x-ray yield with the help of UTF targets.

Skin layer depth for Ni according to (1) is of the order of ~ 6.7nm, with typical heat travel depth on the timescale of laser pulse duration (200fs) ~ 3.5nm for $\sim 10^{15}$W/cm^2 and ~ 12.7nm for $\sim 10^{16}$W/cm^2. These estimates show, that there's no use in trying to "overheat" Ni UTF with intensities of the order of 10^{15}W/cm^2 since non-linear heat wave propagation depth shall be shorter than the skin layer. Hence in following simulations we assumed intensity to be 10^{16}W/cm^2.

With heavy ions spectra being rather complicated due to considerable energy states separation with different values of electrons orbital quantum numbers and with different ions full moment values, it is often feasible to approximate radiative properties of such an ion with those of an equivalent hydrogen-like ion[19].

To do that one has to introduce notion of an outer electron motion in Coulomb field of some effective charge Z_e. Effective charge is a function of ionisation degree and can be found from:

$$E_0 = \frac{Ry \cdot Z_e^2}{n_0^2},$$

where E_0 - ionisation potential of ground state and n_0 - principal quantum number of the outer electron in ground state. Transition energies and velocities can be calculated according to scale ratios:

$$E = Z_e^2 \cdot E_H,$$

$$A = Z_e^4 \cdot A_H,$$

where E_H and A_H are, respectively, energy and velocity of appropriate transition in hydrogen atom. According to estimates[19] thus approximated "hydrogen-like" transition energy coincides with "centre of gravity" of a related real ion multiplet with 20% precision.

Such approximation is well known and often yields surprisingly precise results when calculating radiative properties and is widely used in practice[19].

According to[19] hydrogen-like approximation can be applied to simulation of plasma emission in spectral range:

$$E_c \pm \frac{1}{2} \cdot \Delta E \ , \text{if}$$

$$E_c / \Delta E \ \leq 5 \ .$$

In our calculations we used a scheme based on the one described above. We considered transitions from first three excited states ($n=n_0+1$, $n=n_0+2$ and $n=n_0+3$) to ground ($n=n_0$) state. We neglected transitions between excited states since typical transition energies are comparable with ground state separation energy by orbital quantum number.

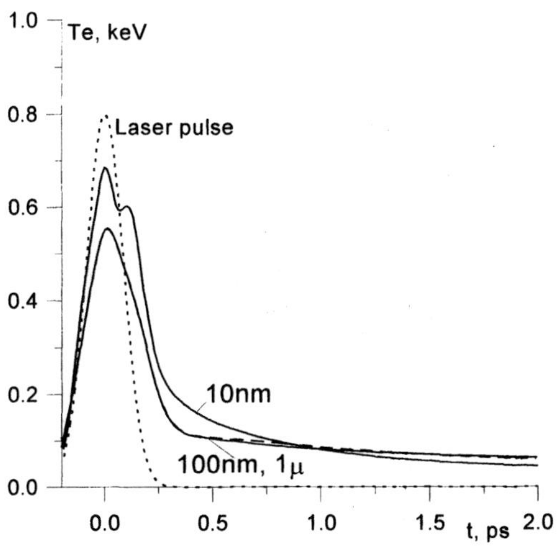

Figure 4. Temporal behaviour of peak electron temperature for Ni-UTF targets of varying thickness; $I=10^{16}W/cm^2$, $\tau=200fs$.

Fig. 4 displays dynamics of maximum (across the spatial coordinate) electron temperature calculated for a number of Ni-UTFs varying in thickness. As it is obvious from these dependencies, there's no distinct change in plasma temperature until UTF thickness of 50nm (being about 4 times larger than the estimated heat wave propagation depth), although 10nm thick UTF drives top temperature up from 554eV to 684eV, thus indicating occurrence of plasma overheating.

Simulation results concerning spectral properties of x-ray emission are presented on fig.5. It displays spectra of Ni targets of various thickness calculated for intensity of $10^{16}W/cm^2$. Comparison of these spectra proves the statement made above, on feasibility of applying UTF technique to targets with larger atomic numbers in order to enhance conversion into harder x-ray quanta. As it can be seen from the figure, changing UTF depth from 50 to 10nm drives the yield of x-ray quanta with E~7keV up by ~3 orders of magnitude constituting major increase otherwise unattainable without serious modifications to igniting laser beam itself.

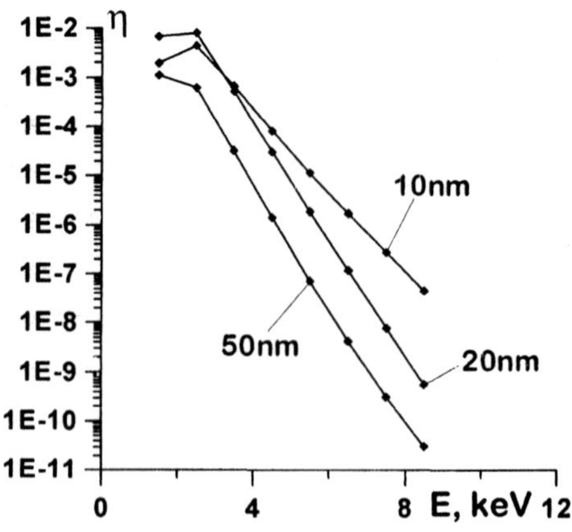

Figure 5. Simulated spectra of Ni UTF irradiated with 10^{16}W/cm^2 for various film thicknesses.

This also is in agreement with estimates made earlier according to (1) and (3): with intensity being 10^{16}W/cm^2 heat wave propagation depth is ~12.7nm, thus target depth of 50nm significantly exceeds it, while 10nm one is already ~0.78, leading to a notable overheating.

CONCLUSION

It has been shown that modifications to optical and thermal properties of near-surface target layer allows one to control interaction regime of powerful ultra short laser pulse with the target as well as relative velocities of various processes, occurring in plasma and hence, parameters of emerging FLP and its x-ray emission. Freely suspended ultra thin films allow effective control over femtosecond plasma characteristics and make it possible to design femtosecond x-ray source over wide range of x-ray quanta energy.

Usage of UTF with thickness of the order of laser skin-layer in plasma as target suppresses heat "leakage" out of the region heated directly by laser radiation of moderate intensity I~$10^{15} \div 10^{16}$W/cm^2. UTF of carbon are effective source of ultra short pulses both in the "water window" (region of water transparency) optimal for microscopy of living cells[6].

In particular, we have shown that in case of carbon target, 30nm thick one can lead to threefold increase in x-ray yield in the region of "water window" (2-4nm), and 10nm target allows to achieve electron temperature ~ 700eV and ionisation ~5.2. If applied to resonant mirrors focusing technique described in[7] it can create a powerful x-ray source in the region of "water window" with H-like ion 3.37nm characteristic line, a tempting perspective for biological microscopy requiring high spatial and temporal resolution.

It is also shown that to produce harder x-ray quanta source one should increase atomic number of target's material In case of Ni UTF three orders of magnitude in conversion efficiency for quanta energy ~ 10 keV can be gained with passing from 50nm to 10nm thick UTFs.

This work was done under the grant from Russian Foundation for Basic Research (#96-02-19146a).

REFERENCES

1. S.A.Akhmanov, *Itogi nauki i tekhniki, Seria "Sovremennie problemi lasernoi fyziki"*, Vol.4, 19, VINITI, Moscow (1991).
2. M.Murnane, H.Kapteyn, M.Rosen, R.Falcone, *Science*, Vol.251, 531 (1991).
3. L.Davis, E.G.Gamaly, et al., *Quantum electronics*, 19 (4), 317 (1992).
4. V.T.Platonenko , *Laser Physics*, 2, 852 (1992).
5. P.Gibbon, R.Forster, *Plasma Physics Control. Fusion*, 38, 769 (1996).
6. A.B.Savel'ev, M.S.Dzhidzhoev, V.M.Gordienko, A.P.Tarasevitch in: *Femtochemistry. Ultrafast chemical and physical processes in molecular systems. Lausanne, Switzerland, September 4-8, 1995*, Ed.:M.Cherqui, World Scientific, Singapore, 675-679 (1996).
7. S.A.Akhmanov, et al., *Bulletin of Russian Acad. Of Sci., Physics*, Vol.56, 9, 1372-1380 (1992).
8. D.Kuhlke, U.Herpers, D.von der Linde, *Appl.Phys.Lett.*, 50, 1785 (1987).
9. J.Workman, A.Maksimchuk, X.Liu, et.al., *Phys.Rev.Lett.*, 75(12), 2324 (1995).
10. J.F.Pelletier, M.Chaker, J.C. Kiefer, *Appl.Phys.Lett.*, 69(15), 2172 (1996).
11. U.Teubner, W.Theobald, C.Wulker, *J.Phys. B: At. Mol. Opt. Phys.*, 29, 4333 (1996).
12. U.Teubner, D.Altenbernd, E.Forster, *Digest of intern. symp. on applications of high field and short wavelngth sources, Santa Fe, New Mexico, March 19-22* (1997)
13. R.Habner, W.Theobald, R.Sauerbrey et al., Ibid.
14. H.W.K.Tom, O.R.Wood, G.D.Aumiller, M.D.Rosen, *Springer series in Chem. Phys.*, Vol.53 (1990).
15. R.V.Volkov, V.M.Gordienko, M.S.Dzhidzhoev, et al., *Quantum Electronics*, 24 (12) (1997).
16. V.G.Babaev, R.V.Volkov, V.M.Gordienko, et al., *Quantum Electronics*, 24, 271 (1997).
17. V.G.Babaev, M.S.Dzhidzhoev, V.M.Gordienko, et al., in[12].
18. M.S.Dzhidzhoev, V.M.Gordienko, V.V.Kolchin, et.al., *JOSA B*, 13, 143 (1996).
19. V.V.Kolchin, V.T.Platonenko, *Laser Physics*, 3, 849 (1993).
20. V.L.Ginsburg, *Propagation of Electromagnetic Waves in Plasmas*, Nauka, Moscow (1967).

BARRIER-SUPPRESSION IONIZATION OF COMPLEX ATOMS AND DIATOMIC MOLECULES

Vladimir P. Krainov

Moscow Institute of Physics and Technology
141700 Dolgoprudny, Moscow Region, Russia

INTRODUCTION

A theory of tunneling ionization of atoms and atomic ions by a strong low frequency electromagnetic field was developed in detailed by Perelomov et al [1] and later by Ammosov et al [2]. This approach is based on the concept of a quasi-stationary external field producing tunneling ejection of valence electrons. A complex atom or an atomic ion is considered in the framework of the quantum defect method, and its wave function is an asymptotic wave function of an atomic electron at large distances from the atomic core. The angular and energy distributions of ejected electrons in tunneling ionization were obtained with exponential accuracy for linear and circular polarization of the electromagnetic field by Corkum et al [3] and by Delone and Krainov [4].

These results are valid for tunneling ionization when the field strength is small compared with the barrier-suppression field strength [5]. A generalization of this approach to barrier-suppression ionization for ground state of the hydrogen atom was made by Krainov and Shokri [6]. This theory starts with the Keldysh-Faisal-Reiss approach (see book of Delone and Krainov [7] for details) and in addition takes into account the Coulomb correction in the Volkov wave fucntion of the final continuum state of the ejected electron. However, the results [6] are referred to only for the ground state of the hydrogen atom and for the field of circular polarization. The Coulomb correction was taken into account in the framework of the strong-field approximation [8] by Reiss and Krainov [9] for ionization of the ground state of the hydrogen atom by circularly polarized radiation, and by Krainov [10] for ionization of complex atoms and atomic ions by linearly and circularly polarized radiation, using the semiclassical perturbation theory for Coulomb potential. The last approach is applicable also to the case of molecules and molecular ions.

Of course the expressions obtained coincide with the known tunneling results in the limit of weak low frequency fields including pre-exponential factors that are important in practice because of the strong influence of the potential of the atomic core or ionic core.

Thus, in this report we generalize the theory to the case of arbitrary complex atoms, atomic ions and diatomic molecular ions and obtain simple analytical formulas for energy and angular distributions of the ejected electrons in the barrier-suppression ionization.

Everywhere we use the atomic system of units where electron mass, Planck constant and electron charge are equal to unity.

BARRIER-SUPPRESSION IONIZATION BY CIRCULARLY POLARIZED RADIATION

Now we consider the barrier-suppression ionization of complex atoms and atomic ions by a circularly polarized electromagnetic field (see paper of Krainov [10] for details). We suggest everywhere below that the radiation frequency ω is small compared with the binding energy of the atomic system; however, the amplitude of the atomic field strength F is of order of the barrier-suppression field strength, so that we can not apply any tunneling expressions. The Keldysh adiabaticity parameter is supposed to be small, i.e.

$$\gamma = \frac{\omega\sqrt{2E_i}}{F} <<1$$

in our approach. Here E_i is the energy of the initial atomic or ionic state.

The angular and energy distribution of ejected electrons is given by the expression

$$\frac{dw}{d\Omega dN} = \frac{Z\omega D^2}{4(\pi n)^2 (2F)^{1/3}} Ai^2 \left[\frac{(Z/n)^2 + (F\psi/\omega)^2 + (\omega\delta E/F)^2}{(2F)^{2/3}} \right].$$

$$(1)$$

Here the next notations are introduced: Z is the charge of atomic or ionic core, n is the effective principal quantum number, so that the binding energy of the considered state is determined by the Rydberg formula $E_i = - Z^2/2n^2$. N is the number of absorbed photons. The constant D is determined by

$$D = \left(\frac{4eZ}{Fn^4} \right)^n.$$

$Ai(x)$ is the so-called Airy function. $\psi <<1$ is the small angle between the direction of the ejected electron and the polarization plane of the electromagnetic radiation. Finally, the energy δE can be expressed via kinetic energy E_e of the ejected electron:

$$\delta E = E_e - \frac{F^2}{2\omega^2} - \frac{Z^2}{6n^2}.$$

Of course, the most of the electrons are ejected with the kinetic energy which is equal to the oscillation energy in circularly polarized field

$$E_e \approx \frac{F^2}{2\omega^2}.$$

166

Integrating Eq. (1) over all energies of the ejected electron, we obtain the angular distribution of ejected electrons at the barrier-suppression ionization by a circularly polarized radiation:

$$\frac{dw(angular)}{d\Omega} = \frac{ZFD^2}{2\omega(\pi n)^2} \int\limits_0^\infty Ai^2\left(u + x^2\right)dx. \qquad (2)$$

Here the following notation is introduced:

$$u \equiv \frac{(Z/n)^2 + (F\psi/\omega)^2}{(2F)^{2/3}}.$$

Integrating Eq. (1) over all angles of the ejected electron, we obtain the energy spectrum of ejected electrons at the barrier-suppression ionization by a circularly polarized radiation:

$$\frac{dw(E_e)}{dN} = \frac{Z(\omega D)^2}{\pi n^2 F} \int\limits_0^\infty Ai^2\left(v + x^2\right)dx. \qquad (3)$$

Here the following notation is introduced:

$$v = \frac{(Z/n)^2 + (\omega\delta E/F)^2}{(2F)^{2/3}}.$$

Finally, integrating Eq. (3) over energies of ejected electrons, or Eq. (2) over angles of ejected electrons (this produces the same result), we can find the barrier-suppression ionization rate of the atomic system by a circularly polarized field

$$w_{BSI}(k) = \frac{Z(2F)^{1/3}D^2}{2n^2}\left[\left(\frac{d}{dk}Ai(k)\right)^2 - kAi^2(k)\right]. \qquad (4)$$

Here we have introduced the notation

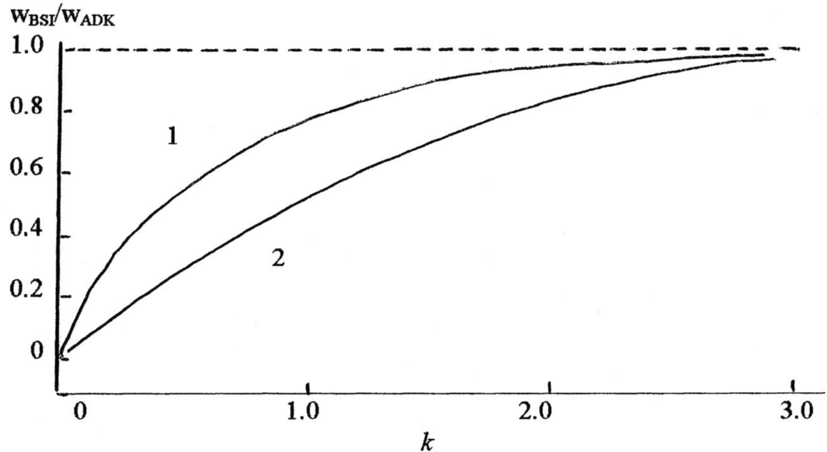

Fig. 1. The ratio of barrier-suppression ionization rate by tunneling ionization rate. Curve 1 is corresponding to circular polarization, the curve 2 - to linear polarization. Tunneling limit is achieved at the values of $k \gg 1$.

$$ k \equiv \frac{Z^2}{n^2(2F)^{2/3}}. $$

These results are valid for initial s - state. The ionization rate of p - state is 3 times larger than of s - state.

The ratio of the barrier-suppression ionization rate, Eq. (4), to the tunneling ionization rate given by known expressions of ADK approach [2] is shown in Fig. 1 as a function of the parameter k (curve 1) .

It is seen that barrier-suppression ionization rate is less than the extrapolation of tunneling ionization rate for barrier-suppression fields.

BARRIER-SUPPRESSION IONIZATION BY LINEARLY POLARIZED RADIATION

It was found [10] that barrier-suppression ionization by a linearly polarized field is determined also by an Airy function but with another argument from that for the case of circular polarization.

The energy and angular distribution of the ejected electrons in the barrier-suppression ionization by a linearly polarized field is determined by

$$ \frac{dw}{d\Omega dN} = \frac{Z(\omega pD)^2}{(\pi n)^2 (2F)^{4/3}} Ai^2 \left[\frac{\dfrac{Z^2}{n^2} + p_\perp^2 + \dfrac{nFp_\parallel^2 \gamma^3}{3Z\omega}}{(2F)^{2/3}} \right]. \tag{5} $$

Here $p_\parallel$ is the component of electron momentum along the polarization axis, and $p_\perp$ is the component of electron momentum in the perpendicular direction. The differential of the number of absorbed photons is $dN = (p_\parallel/\omega)dp_\parallel$.

Integrating this expression of the energies and angles of the ejected electrons, we obtain the barrier-suppression ionization rate by a linearly polarized field:

$$ w_{BSI}(k) = \frac{4\sqrt{3}FD^2}{\pi n(2F)^{1/3}} \int_0^\infty Ai^2\left(k + x^2\right)x^2 dx. \tag{6} $$

The value of k has been given above.

In the limiting case of tunneling ionization it reduces to known tunneling ionization rate as it should be. The ratio of barrier-suppression ionization rate, Eq. (6) by tunneling ionization rate is shown in Fig. 1 (curve 2). It is seen that barrier-suppression ionization by a linearly polarized field is smaller than barrier-suppression ionization rate by a circularly polarized field.

CONCLUSION

Simple analytical expressions are given for energy and angular distributions and also for ionization rates in the barrier-suppression ionization of complex atoms and atomic ions. These expressions for ionization rates reduce correctly to the known tunneling results in the limiting case of a relatively weak laser field compared to the barrier-suppression field. Both the linear and circular polarizations of the field are considered. The atomic, or molecular system is described in the framework of the quantum defect method, so for practical applications only energy of the state of this system is needed in our approach. The conclusion is made that barrier-suppression ionization rates are smaller than tunneling ionization rates at the same field strength.

It follows from above results that angular and energy distributions do not permit the cases of tunneling and barrier-suppression ionization to be distinguished from each other. Probably the most effective way to observe barrier-suppression ionization is to measure total ionization rates. In the case of barrier-suppression ionization the dependence of the ionization rate on the field intensity is less that in the tunneling ionization. The next way to observe barrier-suppression ionization is to determine the ratio of the ionization rate by linearly polarized radiation to that by circularly polarized radiation. We see from Fig. 1 that the slope of the curve for circular polarization is greater than the slope of curve for linear polarization.

Acknowledgments.

The author acknowledges financial support from research grant N 96-02-18299 from the Russian Council of Fundamental Researches. The authors thanks N.B. Delone and other participants of Moscow seminar on multiphoton processes for discussions and useful advices.

REFERENCES

1. A.M. Perelomov, V.S. Popov, and M.V. Terent'ev, Tunneling ionization of atoms by low-frequency electromagnetic field, *Sov. Phys. - JETP* 23:924 (1966).
2. M.V. Ammosov, N.B. Delone, and V.P. Krainov, Tunnel ionization of complex atoms and atomic ions
by an alternating electromagnetic field, *Sov. Phys. - JETP* 64:1191 (1986).
3. P.B. Corkum, N.H. Burnett, and F. Brunel, Above-threshold ionization in the long-wavelength limit, *Phys. Rev. Lett.* 62:1259 (1989).
4. N.B. Delone, and V.P. Krainov, Energy and angular electron spectra for the tunnel ionization of atoms by strong low-frequency radiation, *J. Opt. Soc. Am.* B 8:1207 (1991).
5. S. Augst, D.D. Meyerhofer, D. Strickland, and S.L. Chin, Laser ionization of noble gases by Coulomb
barrier suppression, *J. Opt. Soc. Am.* B 8: 858 (1991).
6. V.P. Krainov, and B. Shokri, Energy and angular distributions of electrons in above-barrier ionization of atoms by strong low-frequency radiation, *Sov. Phys. - JETP* 80:657 (1995).
7. N.B. Delone, and V.P. Krainov, *Multiphoton Processes in Atoms,* Springer, Berlin-Heidelberg (1994).
8. H.R. Reiss, Theoretical methods in quantum optics: S-matrix and Keldysh techniques for strong-field processes, *Prog. Quantum Electron.* 16:1 (1992).
9. H.R. Reiss, and V.P. Krainov, Approximation for a Coulomb-Volkov solution in strong fields, *Phys. Rev.* A 50:R910 (1994).
10. V.P. Krainov, Ionization rates and energy and angular distributions at the barrier-suppression ionization of complex atoms and atomic ions, *J. Opt. Soc. Am.* B14: 425 (1997).

RELATIVISTICALLY SELF-GUIDED LASER-WAKEFIELD ACCELERATION

R. Wagner, S.-Y. Chen, A. Maksimchuk, and D. Umstadter

Center for Ultrafast Optical Science
University of Michigan
Ann Arbor, Michigan 48109

INTRODUCTION

Since the concept of laser-plasma acceleration was introduced in 1979,[1] laser technology has rapidly progressed to an extent[2,3] that these accelerators have been demonstrated in many laboratories throughout the world. There is significant interest in these accelerators because of the incredible accelerating gradients of plasma waves, which have been measured to be four orders of magnitudes greater than conventional technologies ($E \geq 200$ GV/m).[4] One of the many applications these accelerators may have is the ability to create extremely short pulses of x-ray radiation on a small scale, table-top synchrotrons, that are synchronized with the laser. We have created a low-emittance beam of relativistic electrons ($\varepsilon < 1 \; \pi$ mm mrad) with over 1 nC of total charge in a picosecond bunch.[5,6] Our most recent work has shown that we can accelerate electrons beyond the previous limit of natural diffraction, and therefore more fully exploit the incredible accelerating fields. In this paper, we will review the current state of our experimental self-modulated laser-plasma accelerator.

The accelerating fields in conventional accelerators are limited by the breakdown on the walls of the accelerators. Since plasma-based accelerators are already fully ionized, this is not a limitation. In the plasma-based accelerators, a laser pulse creates a longitudinal density

Applications of High Field and Short Wavelength Sources
Edited by DiMauro *et al.*, Plenum Press, New York, 1998

wave through the so-called ponderomotive force. This density wave gives rise to regions of excess electrons (net negative charge) and regions with a shortage of electrons (net positive charge due to the background ions). These longitudinal variations in charge create an electric field that propagates with a phase velocity is equal to the group velocity of the laser pulse creating the plasma wave and can quickly accelerate electrons to relativistic speeds. The magnitude of the accelerating electric field is given by $E\,[\text{V/cm}] \approx 0.96\alpha\sqrt{n_e}\,[\text{cm}^{-3}]$, where n_e is the plasma density and $\alpha = \delta n_e/n_e$ is the amplitude of the plasma wave. For typical plasma densities and wave amplitudes in our self-modulated experiments, this accelerating field is of the order of 1 GeV/cm (assuming $n_e \approx 10^{19}$ cm^{-3} and $\alpha \approx 30\%$).

We have chosen to explore the self-modulated laser wakefield accelerator[7,8,9] where an intense, short pulse laser is focused into an underdense plasma with the laser pulse duration, τ, much longer than the plasma wave period, τ_p (where $\tau_p = 2\pi/\omega_p$, $\omega_p = \sqrt{4\pi e^2 n_e/\gamma m_e}$, e is the electron charge, n_e is the electron density, m_e is the electron rest mass, and γ is the relativistic factor of the electrons quivering transversely in the laser pulse). The relativistic factor, γ, is related to the normalized vector potential, a_o, of the laser pulse by $\gamma = \sqrt{1+a_o^2}$, where $a_o = \mathcal{W}_{os}/c = eE/m_o\omega c = 8.5\times10^{-10}\lambda[\mu\text{m}]I^{1/2}[\text{W/cm}^2]$. Under these conditions, the laser pulse will self-modulate and break up into a series of shorter laser pulses through the forward Raman scattering instability. This modulation of the laser pulse enhances the plasma wave and further modulates the laser pulse, creating large plasma waves that can accelerate electrons. This instability requires extremely high power lasers such that the laser power exceeds the critical power for relativistic self-focusing.

In Gaussian optics, the laser remains focused for a length given by the Rayleigh range, $Z_R = \pi r_o^2/\lambda$, but this tends to be rather short (100's of microns) for the laser systems used in these experiments. However, relativistic effects in the plasma, such as relativistic self-focusing, can be used to extend the length of the focal region well beyond the natural diffraction limit. The index of refraction in a plasma is given by $n = \sqrt{1-\omega_p^2/\omega_o^2}$, where ω_o is the laser frequency, and since the plasma frequency depends on the intensity profile of the laser for ultraintense pulses, the index of refraction will be larger in the center where the intensity is highest. The plasma will then act like a positive lens and continue to focus the laser through the plasma. This effect has a power threshold given by $P_c \approx 17 n_c/n_e$ [GW] where n_c is the critical density (when $\omega_p = \omega_o$).[10] Using this technique, we can break the natural diffraction barrier and exploit the large accelerating fields of the self-modulated wakefields.

EXPERIMENT

In our experiments at the Center for Ultrafast Optical Science, we used a hybrid Ti:sapphire–Nd:glass chirped-pulse-amplification laser system. This laser is capable of producing a 3 J, 400 fs pulse at 1.053 μm in a 43 mm beam. We focused this beam with a *f*/4 off-axis parabolic mirror to vacuum spot size of $r_o = 8.5\,\mu m$ (1/e^2), which corresponds to a vacuum focal intensity of 4×10^{18} W/cm^2. This laser pulse was focused onto a supersonic Helium gas jet with a long flat-top interaction region (750 μm) and a sharp gradient (250 μm)—see figure 1. The gas density produced by the gas jet was found to vary linearly with backing pressure up to a maximum density of 3.6×10^{19} cm^{-3} at 1000 PSI, which corresponds to a critical power of 470 GW for our wavelength. The design of the gas jet was found to be critical in producing the plasma wave and the relativistic electron beam—the sharper the density gradient, the "better" the electron beam (more electrons, lower thresholds, and higher wave amplitudes). Since the plasma was created by the foot of the intense laser pulse tunnel ionizing the gas, this ionization can lead to defocusing of the main part of the laser pulse. By steepening the gas jet profile, this effect can be minimized, and more of the laser energy delivered on target.

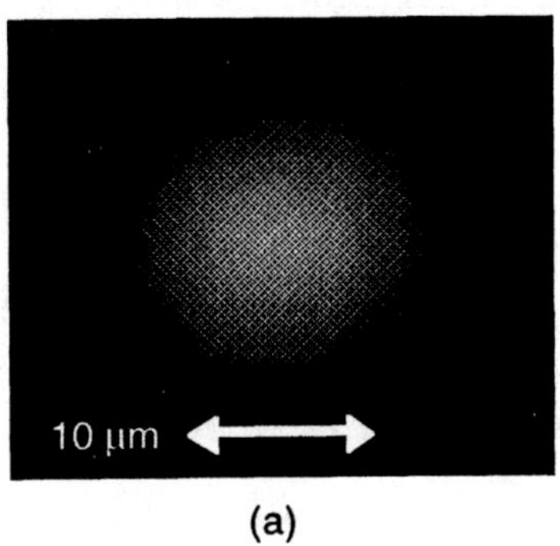

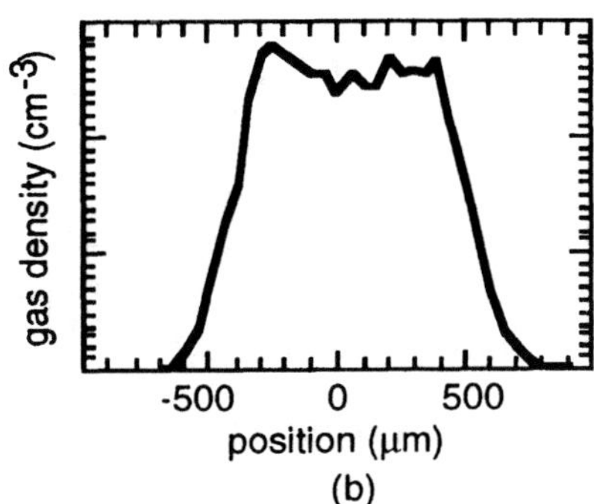

Figure 1. The image to the left (a) shows the vacuum focal image obtained with an equivalent plane imaging system. The arrow shows the scale, and the 1/e^2 radius was measured to be 8.5 μm. The graph at the right (b) shows the density profile for the supersonic gas jet used in these experiments. Note the sharp gradient and long flat-topped region. This data was obtained by measuring the recombination emission with a low power alignment beam focused above the gas jet.

Plasma Wave Characterization

The plasma wave was characterized using collective collinear Thomson scattering. A portion of the main pulse was split off, frequency doubled with a type-I KDP crystal, and recombined to form a collinear, temporally adjustable probe. This probe (ω_o, k_o) coupled to the plasma wave (ω_p, k_p) to produce scattered sidebands shifted by the plasma frequency ($\omega_o \pm \omega_p$, $k_o \pm k_p$). We were able to determine the amplitude of the plasma wave by measuring the amplitudes of the scattered sidebands,[4] and the maximum amplitude measured was 40%—corresponding to an accelerating field of 2 GeV/cm. The temporal duration of the plasma wave was found to be 1.5 ps (FWHM) by varying the delay of the collinear Thomson probe and mapping the amplitude.[5] The growth and decay rates were also found from this measurement to be $\gamma_g = 4$ ps^{-1} and $\gamma_d = 1.9$ ps^{-1}, respectively. We also observed that the plasma frequency decreased from 3.3×10^{14} rad/s to 2.5×10^{14} rad/s, as the laser power increased, and that this scaled with the relativistic correction to the plasma frequency due to changing mass of the electrons.

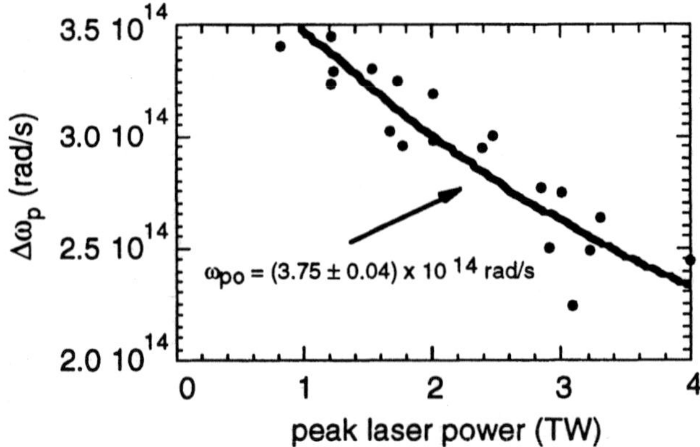

Figure 2. The plasma frequency detunes as the laser power increases due to the relativistic quiver motion of the electrons in the plasma. The points are the measured values of the plasma frequency and the curve represents the fit using the detuning equation with the free parameter ω_{po}. This detuning was only observed when we also observed other relativistic effects, such as self-focusing.

An imaging system, with 15 μm resolution, was set up to measure up the length of the plasma wave from the 90° Thomson scattered light. For our focusing conditions, the natural diffraction length is given by $Z_R = \pi r_o^2 / \lambda = 215$ μm, and the dependence of the plasma channel length was studied by varying the laser power and gas density. The gas density was fixed at 3.6×10^{19} cm^{-3}, and the laser power was slowly increased. At lower laser powers ($<2.6 P_c$), the channel length is considerably shorter (~125 μm) than the diffraction length. However, as the laser power increases, the channel length extends first to 250 μm at $3.9 P_c$ before finally reaching the maximum length of 850 μm at $9.1 P_c$. Note that the maximum length of the plasma channel was limited only by the length of the gas jet itself. The same trend was observed when the gas density was varied for a fixed laser power. In this case, the channel was limited to 250 μm at 400 PSI ($3.2 P_c$ and 1.4×10^{19} cm^{-3}) before extending to 750 μm at 800 PSI ($7.0 P_c$ and 2.9×10^{19} cm^{-3}). Transverse lineouts of the plasma channel show that the width of the channel remains unchanged throughout the entire length, indicating that the images are of an extended channel and not simply due to increased emission from larger laser powers. The length of the plasma channel depends only on the ratio of the laser power to the critical power, indicating that the mechanism for extending the plasma channel is relativistic self-focusing.

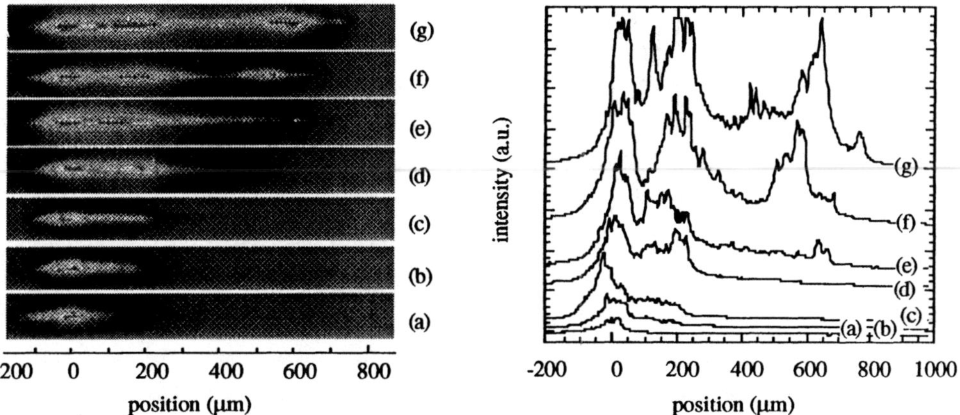

Figure 3. Sidescattered images (left) and lineouts (right) of the plasma channel at various laser powers. Note that the laser is incident from the left and the electron density is 3.6×10^{19} cm^{-3}. The laser powers are $P/P_c =$ (a) 1.6, (b) 2.6, (c) 3.9, (d) 5.5, (e) 7.2, (f) 8.4, and (g) 9.1. Note: the curves have been displaced vertically for ease of viewing.

Electron Beam Characterization

Once the plasma wave was characterized, we looked at the self-trapped, relativistic electrons from this self-modulated laser wakefield accelerator. Two independent diagnostics, a Faraday cup and a scintillator/PMT, were used to measure the total number of electrons accelerated and were found to be self-consistent. There was a sharp threshold (see figure 4) in production at the critical power for relativistic self-focusing, and, in the best conditions, over 10^{10} electrons (or 1 nC) were detected in the forward direction.

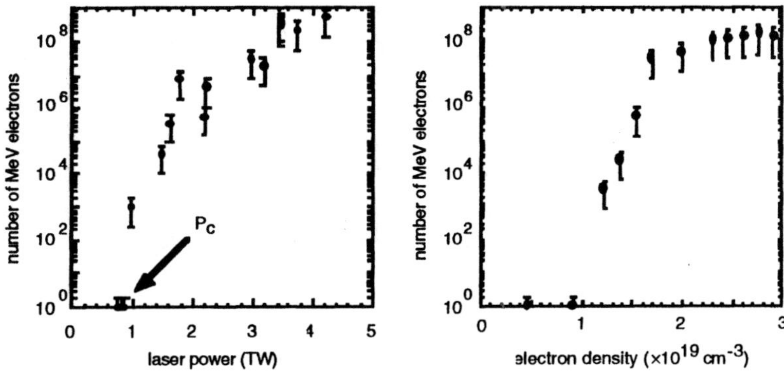

Figure 4. There was a sharp threshold in electron production at the critical power for relativistic self-focusing. In the graph on the left, the laser power was varied and the electron density was fixed at 2.3×10^{19} cm^{-3}, whereas in the graph on the right, the electron density was varied and the laser power fixed at 1.3 TW.

With this number of relativistic electrons, there was enough to take a picture of the electrons using a scintillating film (LANEX). The scintillating film was placed behind an aluminum foil shield and imaged with a 12-bit CCD. This shielding blocked both the incident laser light and electrons with energy less than 100 keV, and the images in figure 5 represent electrons with energy from 100 keV to 3 MeV (this was found from spectroscopy shown later). From these measurements, the transverse emittance of the electron beam can be calculated $\left(\varepsilon = r_o \theta_{\mathrm{HWHM}}\right)$, assuming that source size of the electron beam is equal to the laser focal spot size. The emittance was found to decrease from 0.8 π mm mrad to 0.4 π mm mrad as the length of the plasma channel increased, and this is consistent with the electrons accelerating down a longer plasma channel. This minimum emittance was found to be consistent with the space charge divergence limit, assuming 10^9 electrons at 1 MeV in a 1 ps bunch.[11]

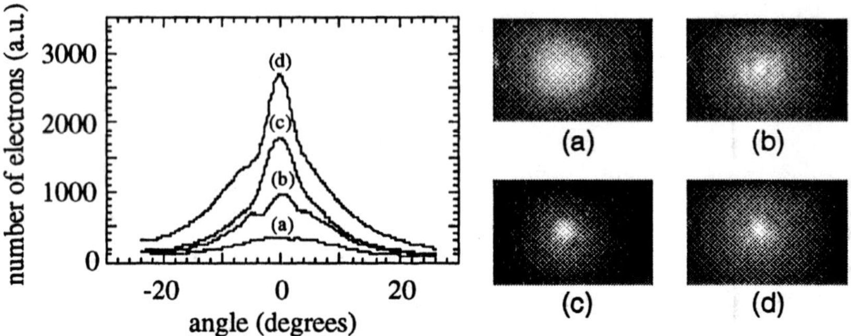

Figure 5. The divergence of the electron beam varies as a function of laser power due to the increased acceleration length from relativistic self-focusing. The various curves and images show this dependence on the laser power where $P/P_c =$ (a) 3.4, (b) 5.0, (c) 6.0, and (d) 7.5.

The energy of the electrons was measured using a 60° sector dipole magnet that bends the electrons onto a LANEX screen which was then imaged by a 12-bit CCD camera. This spectrometer could only measure electrons with kinetic energy from 0.8–2.3 MeV, and in this range, the energy spectrum of the electrons was found to only depend on the length of the plasma channel. When the plasma channel was less than 250 μm ($< 6P_c$), the slope of the energy spectrum was fixed and followed a functional form of $\exp(-\gamma)$, which corresponds to a temperature of 500 keV. However, when the plasma channel extended to 750 μm ($> 6P_c$), the slope abruptly changed to follow $\exp(-0.67\gamma)$, which corresponds to a temperature of 750 keV. This same behavior was observed if the laser power was held fixed and the plasma density varied, which indicated that accelerating length determined the temperature of the accelerated electrons.

CONCLUSION

From these experiments, we have shown that relativistic self-focusing had a pronounced effect on the electrons accelerated by this self-modulated laser wakefield. The divergence of

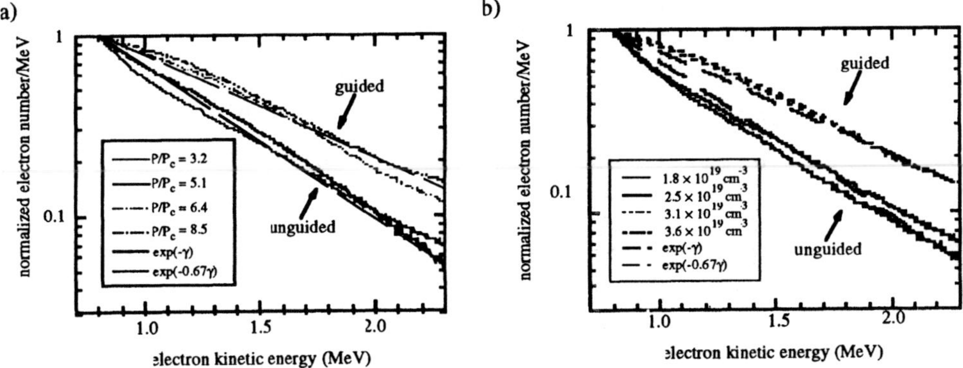

Figure 6. The electron kinetic energy spectrum depends on the accelerating length of the electrons. When the laser is unguided (<250 μm), the electron temperature is fixed at 500 keV, however when the laser is guided fro 750 μm, the temperature abruptly increases to 750 keV. Graph (a) shows the energy spectra for a fixed electron density of 3.6×10^{19} cm^{-3} and varying laser power, whereas graph (b) shows the energy spectra for fixed laser power of $P/P_c \approx 6.4$ TW and varying electron density.

the electron beam decreased to the space-charge limit and the total energy increased as the length of the plasma channel increased. We were able to extend the length of the channel to over four times the Rayleigh length by a factor of seven over the low power length, ultimately limited by the length of the gas jet itself. We were able to produce a beam of broadband, MeV electrons with a total charge of 1 nC in a 1.5 ps bunch with a space-charge limited transverse emittance. By breaking the natural diffraction limit, we have shown that incredible accelerating gradients of laser-plasma accelerators can be fully exploited to produce the next generation of electron accelerators.

The authors would like to acknowledge financial support from DOE/LLNL subcontract B307953, NSF PHY 972661, and NSF STC PHY 8920108.

REFERENCES

1. T. Tajima and J. Dawson, "Laser electron accelerator," *Phys. Rev. Lett.* 43:267 (1979).
2. G. Mourou and D. Umstadter, "Development and applications of compact high-intensity lasers," *Phys. Fluids B* 4:2315 (1992).
3. M. D. Perry and G. Mourou, "Terawatt to petawatt subpicosecond lasers," *Science* 264:917 (1994).
4. D. Umstadter, S.-Y. Chen, A. Maksimchuk, G. Mourou, and R. Wagner, "Nonlinear optics in relativistic plasmas and laser wake field acceleration of electrons," *Science* 273:472 (1996).

5. S. P. LeBlanc, M. C. Downer, R. Wagner, S.-Y. Chen, A. Maksimchuk, G. Mourou, and D. Umstadter, "Temporal characterization of a self-modulated laser wakefield," *Phys. Rev. Lett.* 77:5381 (1996).

6. R. Wagner, S.-Y. Chen, A. Maksimchuk, and D. Umstadter, "Electron acceleration by a laser wakefield in a relativistically self-guided channel," *Phys. Rev. Lett.* 78:3125 (1997).

7. P. Sprangle, E. Esarey, J. Krall, and G. Joyce, "Propagation and guiding of intense laser pulses in plasmas," *Phys. Rev. Lett.* 69:2200 (1992).

8. T. M. Antonsen and P. Mora, "Self-focusing and Raman scattering of laser pulses in tenuous plasmas," *Phys. Rev. Lett.* 69:2204 (1992).

9. N. E. Andreev, L. M. Gorbunov, V. I. Kirsanov, A. A. Pogosova, and R. R. Ramazashvili, "Resonant excitation of wakefields by a laser pulse in a plasma," *JETP Lett.* 55:571 (1992).

10. C. Max, J. Arons, and A. B. Langdon, "Self-modulation and self-focusing of electromagnetic waves in plasmas," *Phys. Rev. Lett.* 33:209 (1974).

11. S. Humphries, *Charged Particle Beams*, Wiley, New York (1990).

A SINGLE-BEAM, PONDEROMOTIVE-OPTICAL TRAP FOR ENERGETIC FREE ELECTRONS

J.L. Chaloupka[a], T.J. Kessler, and D.D. Meyerhofer[a,b]

University of Rochester
Laboratory for Laser Energetics
250 East River Road
Rochester, NY 14607
[a]Also Department of Physics and Astronomy
[b]Also Department of Mechanical Engineering

INTRODUCTION

In 1957, it was shown that charged particles interacting with an oscillating electro-magnetic field will experience a force proportional to the gradient of the field intensity in the direction of decreasing intensity.[1] This so-called ponderomotive force results in the rapid expulsion of free electrons from a high-intensity Gaussian laser focus.[2] If, however, the gradient of the intensity is made to point inwards, this force could be used to confine rather than repel. This was known for the case of RF-fields in 1958,[3] and was extended to laser fields in 1966.[4] Since that time, laser peak intensities have seen a dramatic rise.[5] This has led to the observation of a vast array of non-linear light-matter interactions, but the accompanying increase in the magnitude of the ponderomotive force has made the observation of some effects difficult. In this paper, we re-visit the idea of ponderomotive trapping, but with a specific goal in mind: the confinement of electrons in a high-field region, with the expectation of observing second-harmonic generation from their relativistic oscillations.[6] We present a novel scheme to generate a single-beam, three-dimensional trap, along with simulated electron trajectories that demonstrate confinement in a high-field region. Also, we present images of the trapping beam generated with a high-power laser. It is important to note that there exist other schemes of laser-based trapping.[7,8] These more complicated methods would trap electrons in only two dimensions, while our single-beam trap offers three dimensions of confinement. To our knowledge, this is the only single-beam, three-dimensional trap for energetic electrons to be proposed, and the only trap of any kind to be generated with a high-power laser.

Applications of High Field and Short Wavelength Sources
Edited by DiMauro *et al.*, Plenum Press, New York, 1998

SIMULATED ELECTRON TRAJECTORIES - I

A simple computer code has been written to model the cycle-averaged motion of electrons in arbitrary intensity distributions. To motivate the need for a trapping beam, the simulation is first run using a Gaussian beam profile. The characteristics of our T^3 laser[9] ($\lambda = 1.05$ μm, $I_{PEAK} = 10^{18}$ W/cm^2, $\tau_{FWHM} = 2.0$ ps, $w_0 = 5.0$ μm) and electrons from Ar^{3+} (BSI[10] threshold intensity equal to 1.2×10^{15} W/cm^2) were used. Figure 1a shows an image of the focal region along the radial and axial dimensions. The "x" marks the point of origin of the electron. Figure 1b shows the distance of the electron from the laser axis as a function of time, and figure 1c shows the intensity that the electron experiences as a function of time. At the peak of the pulse, the electron is ~$10w_0$ from the laser axis. It experiences a peak intensity of two orders of magnitude below the peak intensity of the laser. In a sense, the extra intensity in the laser is wasted, as it only serves to ionize a larger volume of atoms.

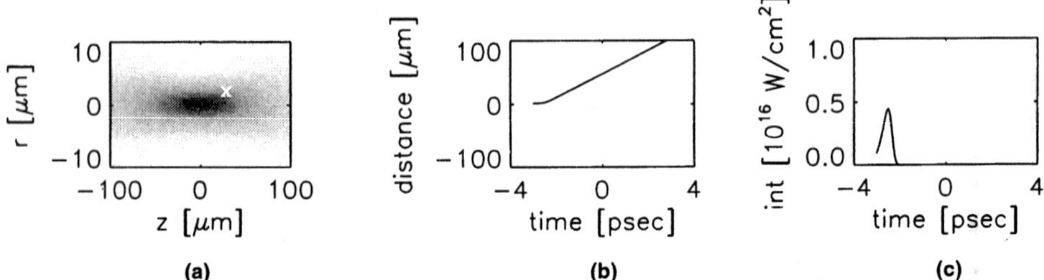

Figure 1. The simulated trajectory of an electron from Ar^{3+} in a Gaussian laser focus: (a) Image of the focus and the electron's point of origin. (b) The electron's distance from the laser axis as a function of time. (c) The intensity that the electron experiences as a function of time.

ALTERING THE LASER FOCUS

As is well known, altering the near-field amplitude of a laser beam has little effect on the centrally-peaked structure of the focal spot. In fact, this can be advantageous if one requires an altered beam in the near-field and a centrally-peaked one in the focal plane.[11] However, to yield deep far-field amplitude modulations, one must modulate the phase of the beam in the near-field.[12] We have found that a remarkably simple phase plate produces the desired "trapping" focus. The phase plate consists of only two regions: an inner disk and an outer annulus. The disk adds a π-phase to the central portion of the beam, while the outer region remains unshifted. The light from these two regions interferes at the center of the focal spot. If half of the incident electric field amplitudes are shifted and half are not, then there will be complete destructive interference at the center of the focus. This will occur for a Gaussian beam if the disk diameter (d_π) equals 1.65w (where w is the $1/e^2$ (in intensity) radius of the incident beam). Figure 2a shows the calculated focal region for a Gaussian beam incident on a $d_\pi = 1.65w$ phase plate as a function of the radial and axial coordinates. In this and all subsequent figures, w_0 and z_0 refer to the characteristics of the unaltered beam ($1/e^2$ radius (in intensity and at best focus) and Rayleigh range, respectively). The contour lines are labeled in terms of the unaltered beam's peak intensity. The contour line at 8.3% is the boundary of the region of complete trapping. This region has a volume of ~$2w_0^2 z_0$.

The focal region described above represents an ideal "dark" trap. However, we are interested in confining electrons in a high-intensity region, not in a region of zero intensity. By simply changing the size of the π-disk, this can be achieved. If more than half of the incident field is shifted, then the destructive interference at the center of the focus will not be complete. Figure 2b shows the trapping region generated with a $d_\pi = 2.20w$ phase plate. The region of complete trapping is surrounded by the 23.3% contour line. The intensity at the center of the trap is 17% of the unaltered beam's peak intensity. This region has a volume of $\sim w_o^2 z_o$.

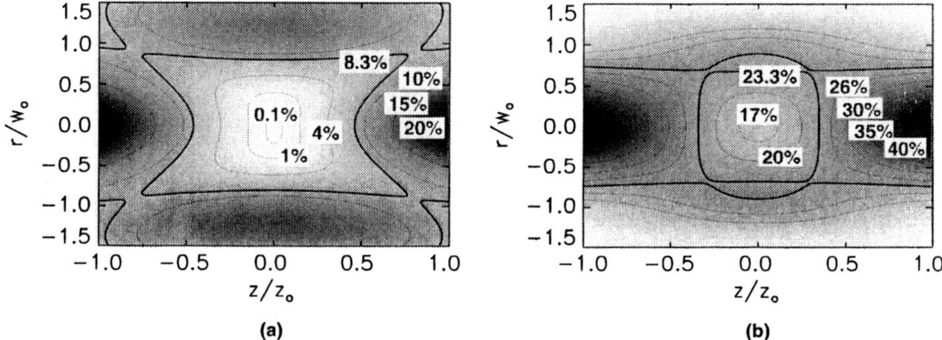

Figure 2. Calculated trapping focal regions, with intensities labeled in terms of the unaltered beam's peak intensity: (a) The focal region generated with a $d_\pi = 1.65w$ phase plate. The contour line at 8.3% encloses the region of complete trapping, with a volume of $\sim 2w_o^2 z_o$. (b) The focal region generated with a $d_\pi = 2.20w$ phase plate. The contour line at 23.3% encloses the region of complete trapping, with a volume of $\sim w_o^2 z_o$.

SIMULATED ELECTRON TRAJECTORIES - II

For the same laser conditions as before, an electron's motion is modeled in the trapping focal regions. Figure 3 shows the results of the simulation for an electron interacting with the $d_\pi = 1.65w$ trap. The electron is confined to within w_o of the laser axis from T_{BIRTH} (the time at which it is released into the field by ionization, or "born," where time equals zero at

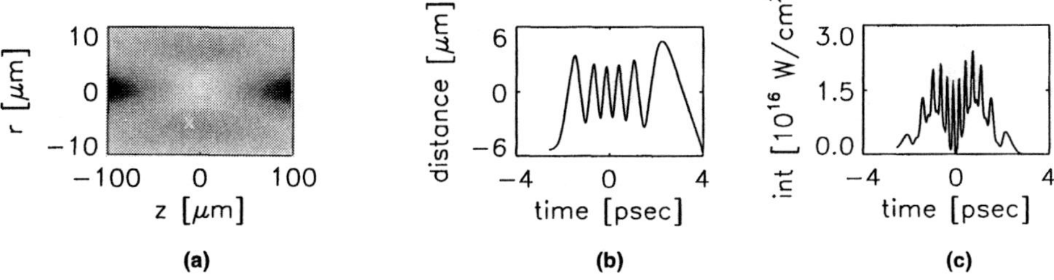

Figure 3. The simulated trajectory of an electron from Ar^{3+} in a trapping laser focus, generated with a $d_\pi = 1.65w$ phase plate: (a) Image of the focus and the electron's point of origin. (b) The electron's distance from the laser axis as a function of time. (c) The intensity that the electron experiences as a function of time.

the peak of the laser pulse) to $|T_{BIRTH}|$. Even though the electron is not confined in a high-field region, it experiences a higher peak intensity, for a longer period of time, than does the electron interacting with the centrally-peaked, Gaussian focus. Figure 4 shows the results of the simulation for the electron interacting with the $d_\pi = 2.20w$ trap. Again, the electron is confined for a time equal to $2\times|T_{BIRTH}|$. But now, the peak intensity that the electron experiences is equal to $\sim2\times10^{17}$ W/cm^2. Compared to the experienced intensity for the Gaussian case, the electron now interacts with a field forty times as intense for four times as long.

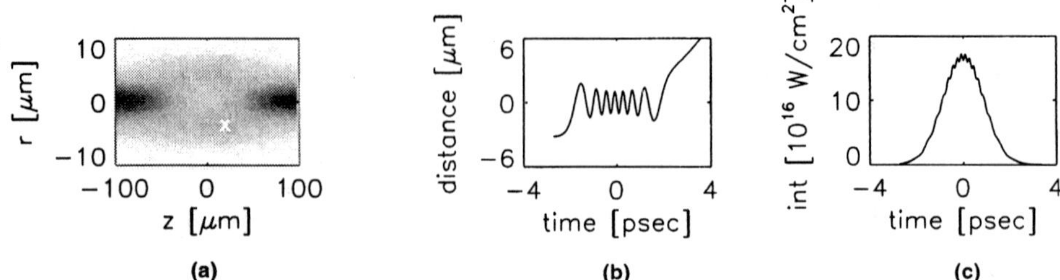

Figure 4. The simulated trajectory of an electron from Ar^{3+} in a trapping laser focus, generated with a $d_\pi = 2.20w$ phase plate: (a) Image of the focus and the electron's point of origin. (b) The electron's distance from the laser axis as a function of time. (c) The intensity that the electron experiences as a function of time.

EXPERIMENTAL ARRANGEMENT

Although the phase plates described above can be created using photo-lithographic methods, we have developed a novel segmented half-wave plate arrangement to create the altered focus. A half-wave plate retards light polarized along one axis by π with respect to light polarized along the orthogonal axis. If part of an incident beam travels as a fast wave, while the rest travels as a slow wave, a π-phase shift will be introduced between the two portions. To accomplish this with the desired geometry, a 4-cm disk was cut from the center of an 8-cm mica half-wave plate. The disk and annulus were mounted on cross-hairs and aligned to the laser beam, as shown in figure 5. By rotating the inner disk by 90° with respect to the outer annulus, the o-axis of one section will be oriented parallel to the e-axis of

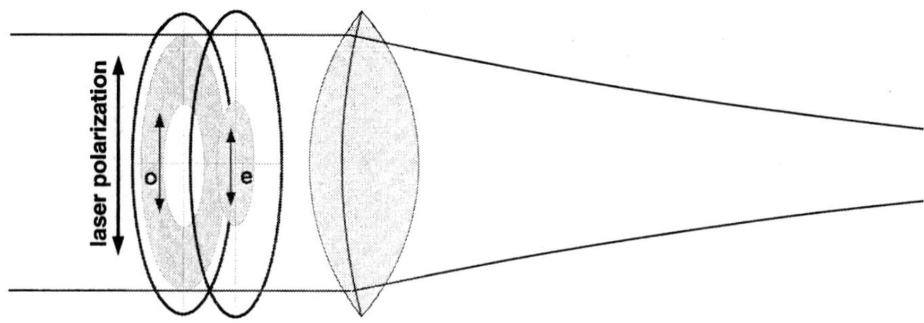

Figure 5. The experimental configuration for the generation of a trapping focal region, consisting of the segmented half-wave plate and the focusing lens.

the other. This introduces a π-phase shift to the inner portion of the beam. Such a scheme has several advantages. Instead of making a precise custom coating on an optical flat, an off-the-shelf wave plate is altered with a single, circular cut. Also, by rotating the two wave-plate pieces as a unit, the polarization of the laser can be rotated without introducing a separate optic. But most importantly, the depth of the trapping region can be tuned by rotating the inner disk with respect to the outer annulus. As the inner disk is rotated, the central portion of the beam has its polarization rotated, putting part of its energy into the non-interfering orthogonal polarization. In the focal region, the destructive interference is no longer complete and the minimum begins to fill in. An extreme case of this is to rotate the inner disk back to its original position, which creates an ordinary, centrally-peaked focus.

Figure 6 shows the imaged focal region when the wave-plate arrangement is in place with the inner disk rotated by 90° with respect to the outer annulus. Figure 6a shows the laser spot at best focus, and figures 6b and 6c show the lineouts along the y- and x-directions, respectively. Figure 6d shows the lineout at r=0 along the z-direction. The axes are labeled in terms of the experimentally determined waists, Rayleigh range and peak intensity ($x_{o\,exp}$, $y_{o\,exp}$, $z_{o\,exp}$, $I_{o\,exp}$) of the unaltered beam. This focal region exhibits a minimum along all three dimensions. These images were taken with the laser firing at moderate energies (~40 mJ). With the final, single-pass amplifier firing, there was little change in the shape of the focal spot.

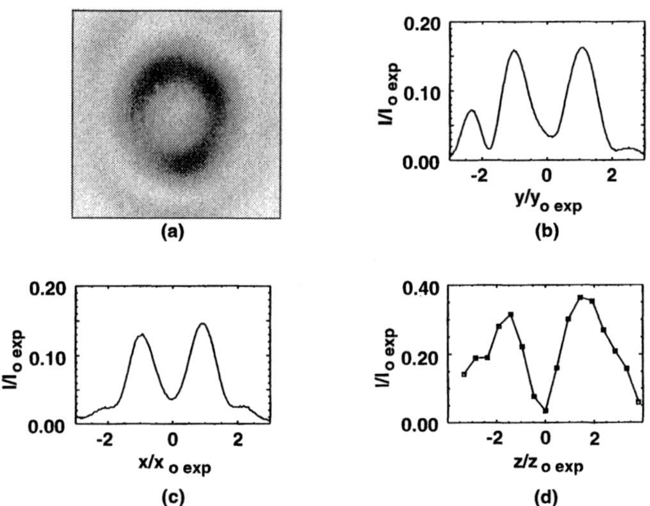

Figure 6. Experimental results: (a) Image of the altered beam at best focus. (b) Lineout in the y-direction at best focus. (c) Lineout in the x-direction at best focus. (d) Intensity at r=0 as a function of z. All axes are labeled in terms of the experimentally determined waists, Rayleigh range and peak intensity ($x_{o\,exp}$, $y_{o\,exp}$, $z_{o\,exp}$, $I_{o\,exp}$).

CONCLUSION

Achieving the intensities at which an electron's quiver motion becomes relativistic (~10^{18} W/cm^2 for $\lambda = 1$ μm) has become increasingly common-place. Unfortunately, to observe some aspects of this relativistic motion (such as harmonic generation), it is necessary to control the rapid acceleration of the electrons out of the high-field region. We have developed a single-beam trap that will confine electrons in an intense region in three-dimensions, forcing them to interact with the strong driving field. Simulations show that electrons born within the trapping region will be confined for a time greater than the pulse width, and will experience high intensities. This trap has been generated with a high-power laser using a novel segmented wave-plate approach. Experiments to verify the trapping of free electrons are underway.

This work is supported by the National Science Foundation, with additional support by the Office of Inertial Confinement Fusion, U.S. Department of Energy (DOE), under Cooperative Agreement DE-FC03-92SF19460, and by the University of Rochester. The support of the DOE does not constitute an endorsement by the DOE of the views expressed in the paper.

REFERENCES

1. H.A.H. Boot and R.B.R.-S.-Harvie, Nature 180:1187 (1957).
2. C.I. Moore, J.P. Knauer, and D.D. Meyerhofer, Phys. Rev. Lett. 74:2439 (1995).
3. V. Gaponov and M.A. Miller, J. Exptl. Theoret. Phys. (U.S.S.R.) 34:242 (1958).
4. N.J. Phillips and J.J. Sanderson, Phys. Lett. 21:533 (1966).
5. M.D. Perry and G. Mourou, Science 264:917 (1994).
6. J.L. Chaloupka, Y. Fisher, T.J. Kessler, and D.D. Meyerhofer, Opt. Lett. 22:1021 (1997).
7. C.I. Moore, J. Mod. Opt. 39:2171 (1992).
8. U. Mohideen, H.W.K. Tom, R.R. Freeman, J. Bokor, P.H. Bucksbaum, J. Opt. Soc. Am. B 9:2190 (1992).
9. Y.-H. Chuang, D.D. Meyerhofer, S. Augst, H. Chen, J. Peatross, and S. Uchida, J. Opt. Soc. Am. B 8:1226 (1991).
10. S. Augst, D.D. Meyerhofer, D. Strickland, and S.L. Chin, J. Opt. Soc. Am. B 8:858 (1991).
11. J. Peatross, J.L. Chaloupka, and D.D. Meyerhofer, Opt. Lett. 19:942 (1994).
12. L.W. Casperson, Opt. Quantum Electron. 10:483 (1978).

ULTRAFAST DIFFRACTION FROM RYDBERG WAVE PACKETS USING HIGH HARMONICS

Kenneth J. Schafer[1] and Jeffrey L. Krause[2]

[1]Department of Physics and Astronomy
Louisiana State University
Baton Rouge, LA 70803-4001

[2]Quantum Theory Project
University of Florida
Gainesville, FL 32611-8435

INTRODUCTION

Rydberg wave packets have been the object of intense experimental and theoretical study over the past decade due to their many fascinating semiclassical and quantum characteristics. Long-lived wave packets localized in both the radial and angular dimensions have now been produced in alkali metal atoms.[1-4] The methods of quantum control can also be applied to calculate the optimally shaped laser pulse which creates a wave packet having a desired distribution in phase space at a specified "target" time.[5] Such wave packets can take the form, for instance, of "Schrödinger cat" states,[6] or coherent superpositions of macroscopically distinct electronic states.

An important issue in Rydberg wave packet research is their detection. Normally the wave packet is localized in an interesting distribution only when it is far from the ion core, in regions close to the turning point of the Coulomb potential, where the electron is nearly free. Since free electrons do not absorb or emit optical photons, the wave packet distributions are difficult to detect with traditional pump-probe methods. Several alternative methods, each sensitive to different properties of the wave packet, have been implemented, including wave packet interferometry,[7] half-cycle pulses[8] and "atomic streak cameras".[9]

Applications of High Field and Short Wavelength Sources
Edited by DiMauro *et al.*, Plenum Press, New York, 1998

Rydberg packets typically have length scales of 10-1000 nm and orbital periods of several picoseconds. In this paper we present calculations to demonstrate that high order harmonics from a high repetition rate TW-class Ti:Sapphire laser interacting with a nonlinear medium can serve as an appropriate light source for ultrafast diffraction from Rydberg' wave packets. The high harmonics from such a source are easily shorter than 100 fs,[10] and a wide range of wavelengths from 8-80 nm are available, making high harmonics an ideal light source for time-dependent diffraction from transient nanostructures.

The measurement of electron distributions in atoms by gas phase diffraction dates back to the early part of this century and played a key role in establishing the experimental basis for quantum mechanics.[11] For example, early x-ray diffraction experiments by Wollan[12] were instrumental in confirming the usefulness of Hartree's self-consistent field method.[13] In the intervening years, x-ray diffraction and electron diffraction have been used to measure the electronic densities of atoms and many small molecules in their ground states. In the past few years, several schemes for using ultrafast methods to measure low-lying excited state distributions via time-dependent diffraction have been proposed.[5,14–16]

METHOD AND RESULTS

In the first Born approximation the elastic cross section for electron-photon scattering is proportional to the square of the atomic form factor, which has three contributions[16]

$$f(\vec{q}, t) = n_0 f_0(\vec{q}) + n_{wp} f_{wp}(\vec{q}, t) + f_{c.t.}. \tag{1}$$

Here, n_0 and n_{wp} are the populations of the ground state and the wave packet, and f_0 and f_{wp} are the normalized ground state and wave packet form factors. The cross term $f_{c.t.}$, which results from the product of the ground state amplitude and the wave packet, oscillates on a time scale set by the excitation laser. For XUV diffraction pulses lasting much longer than the excitation time, this rapidly oscillating cross term makes no contribution to the signal.[5,16] For a wave packet with structure on the nanometer length scale, the optimal photon source has a wavenumber $k_0 \ll 1$ au. For these wavelengths the scattering from the ground state component is nearly isotropic. The scattered intensity, which is proportional to $|f(\vec{q}, t)|^2$, can then be written as

$$I(\vec{q}, t) \propto n_0^2 + 2n_0 n_{wp} f_{wp}(\vec{q}, t) + \mathcal{O}(n_{wp} f_{wp})^2. \tag{2}$$

This equation shows that the signal oscillates about its average value with an amplitude proportional to the population of the wave packet. This allows us to extract f_{wp}, which can then be inverted to give the spatial distribution of the wave packet.

As an example of the signal that can be expected from a Rydberg wave packet, consider first an $\ell = 1$ Gaussian wave packet located at a distance $\bar{r}$ from the nucleus, with a width σ. If $\bar{r} > \sigma$ we can approximate the form factor by

$$f_{wp}(\vec{q}) \approx e^{-q^2\sigma^2/4} \left[j_0(q\bar{r}) - 2j_2(q\bar{r}) P_2(\hat{q} \cdot \hat{\mu}) \right] \tag{3}$$

where $\hat{\mu}$ is the optical laser polarization, j_n are Bessel functions, and P_2 is the second Legendre polynomial. We see that the signal consists of a damped oscillatory pattern, with a monotonically decreasing envelope for increasing $|\vec{q}|$. The period of the oscillations depends on the distance of the center of the wave packet from the atomic core, and the damping is proportional to the width of the Gaussian target.

Fig. 1 illustrates how $f_{wp}(\vec{q})$ changes as a function of time as an $\ell = 1$ radial wave packet with an average principle quantum number of $n \approx 45$ moves between its inner and outer turning points on its first orbit. Note that this packet focuses slightly before its outer turning point. We choose $k_0 = .015\,\mathrm{au}$, which corresponds to 22.2 nm radiation. The excitation laser is polarized parallel to the z axis, the XUV pulse propagates along the y axis, and and the detector is assumed parallel to the $x - y$ plane. The scattering angle θ_q in the $z - y$ plane is defined by $q = 2k_0 \sin(\theta_q/2)$. Due to the $\ell = 1$ character of the wave packet, the diffraction pattern is not cylindrically symmetric about the y axis.

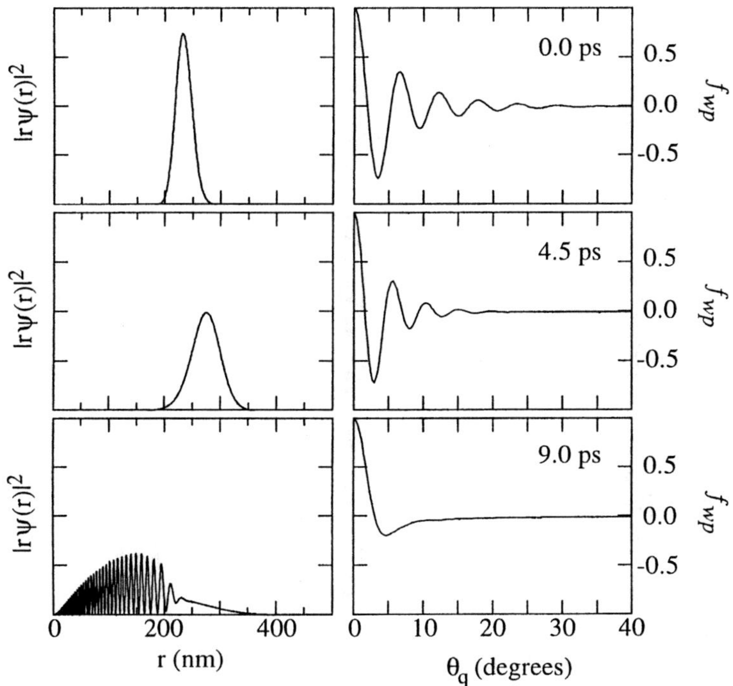

Figure 1. Rydberg wave packet probability distributions as a function of time (left panels) and the form factors $f_{wp}(\theta_q)$ calculated from these distributions (right panels). The photon wave length is 22.2 nm

The changing spatial distribution of the wave packet results in a time-dependent diffraction pattern which is clearly visible at this wavelength. We find that the approximate form factor in Eq. 3 is an excellent approximation, except when the wave packet is at its closest approach to the atomic core.

Next we consider the diffraction signal from a shaped wave packet that is programmed to assume a specified distribution at a target time. For a wave packet that can be decomposed into a sum of several simple substructures, the form factor is the sum of the form factors for each structure. Hence, we can expect to see interference effects from the oscillatory contributions of a multiple-peaked wave packet. However, these effects are present only while the wave packet is near the target time and the structures are well-developed. In addition, the partitioning of the wave packet into several narrow structures mitigates the effects of the exponential prefactor in Eq. 3 and thereby enhances the diffraction signal.

Fig. 2 illustrates the diffraction from a shaped Rydberg wave packet, consisting of a trident structure. Note the non-monotonic envelope of the form factor, which is due to the structure displayed by the packet at the target time. This feature is absent at later times when the wave packet disperses.

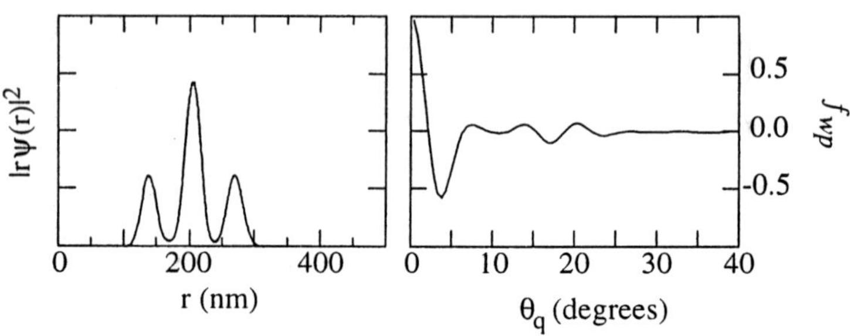

Figure 2. A three-peaked "tailored" wave packet at the target time (left panel) and its form factor $f_{wp}(\theta_q)$ (right panel). The photon wavelength is 22.2 nm

DISCUSSION

The cross section for electron-photon scattering is very small and does not benefit from coherent enhancement due to periodic structure in the sample. Rotational averaging, however, is not necessary, because the wave packet axes are aligned by the pump laser. This lack of periodic structure in the gas sample means that the entire form factor (at all values of $\vec{q}$) can be measured, and then inverted to determine the wave packet density. The feasibility of the experiment is increased greatly by the vastly different length scales in the problem (ground versus Rydberg state), and the fact that the interference signal in Eq. 2 depends primarily on the form factor, and not its square (Eq. 2). The best systems for actually producing shaped wave packets are the alkali metal atoms, since the excitation can be accomplished using optical wavelengths. In addition, for an atom with Z_c core electrons the expression for the signal is altered by replacing n_0 in Eq. 2 with $n_0 + Z_c$, thus increasing the overall signal by a factor of Z_c^2 and the interference term by Z_c. Recording a gas phase diffraction pattern such as Fig. 2 requires a very bright, short-pulse source of photons. For the wave lengths required in this experiment, the optimal source is high harmonics generation from an intense short pulse laser interacting with an atomic gas. The diffraction signal is proto the photon flux, so a high intensity source of photons is clearly a desideratum. Our estimates indicate that current sources are adequate to perform the experiment,[5] and improvements in the design and optimization of such sources are occurring rapidly.[17]

Finally we note that time-resolved electron diffraction is also an attractive detection scheme, because the electron-electron scattering cross section is much higher than the electron-photon cross section. However, for wave packets with a nanometer length scale, the electron energies required would be extremely small, making this method impractical to implement in the laboratory.

References

[1] J. Parker and C. R. Stroud, Jr., Coherence and decay of Rydberg wave packets, *Phys. Rev. Lett.* 56:716 (1986).

[2] A. ten Wolde, L. D. Noordam, A. Lagendijk, and H. B. van Linden van den Heuvell. Observation of radially localized electron wave packets, *Phys. Rev. Lett.* 61:2099 (1988).

[3] J. A. Yeazell, M. Mallalieu, J. Parker, and C. R. Stroud, Jr., Classical periodic motion of atomic-electronic wave packets, *Phys. Rev. A* 40:5040 (1989).

[4] Maciej Kalinski and J. H. Eberly, New states of hydrogen in a circularly polarized electromagnetic field, *Phys. Rev. Lett.* 77:2420 (1996).

[5] J. L. Krause, K. J. Schafer, and M. Ben-Nun, Creating and detecting shaped Rydberg wave packets, *Phys. Rev. Lett.* in press (1997).

[6] M. Noel and C. R. Stroud, Jr., Young's double slit experiment within an atom, *Phys. Rev. Lett.* 75:1252 (1995); Excitation of an atomic electron to a coherent superposition of macroscopically distinct states, *Phys. Rev. Lett.* 77:1913 (1996).

[7] R. R. Jones, C. S. Raman, D. W. Schumacher, and P. H. Bucksbaum, Ramsey interference in strongly driven Rydberg systems, *Phys. Rev. Lett.* 71:2575 (1993).

[8] R. R. Jones, D. You, and P. H. Bucksbaum, Ionization of Rydberg atoms by subpicosecond half-cycle electromagnetic pulses, *Phys. Rev. Lett.* 70:1236 (1993).

[9] G. M. Lankhuijzen and L. D. Noordam, Streak-camera probing a rubidium wave packet decay in an electric field, *Phys. Rev. Lett.*, 76:1784 (1996).

[10] J. M. Schins, P. Breger, P. Agostini, R. C. Constantinescu, H. G. Muller, A. Bouhal, G. Grillon, A. Antonetti, and A. Mysyrowicz, Cross-correlation measurements of femtosecond extreme-ultraviolet high-order harmonics, *JOSA B*, 13:197 (1996).

[11] J. T. Randall. *The diffraction of X-rays and electrons by amorphous solids, liquids, and gases*, John Wiley and Sons, New York, 1934.

[12] E. O. Wollan, Scattering of x-rays from gases. *Phys. Rev.* 37:862 (1931).

[13] I. Waller and D. R. Hartree, *Proc. Roy. Soc. A* 124:119 (1929).

[14] J. C. Williamson, M. Dantus, S. B. Kim, and A. H. Zewail, Ultrafast diffraction and molecular structure, *Chem. Phys. Lett.* 196:529 (1992).

[15] T. Anderson, I. V. Tomov, and P. M. Rentzepis, A high repetition rate, picosecond hard x-ray system, and its application to time-resolved x-ray diffraction, *J. Chem. Phys.* 99:869 (1993).

[16] M. Ben-Nun, T. J. Martinez, P. M. Weber, and K. R. Wilson, Direct imaging of excited electronic states using diffraction techniques: Theoretical considerations, *Chem. Phys. Lett.* 262:405 (1996).

[17] J. Zhou, J. Peatross, M. M. Murnane, and H. C. Kapteyn, Enhanced high-harmonic generation using 25 fs laser pulses, *Phys. Rev. Lett.* 76:752 (1996).

BACKWARD AND MULTI-ECHO FIELD IONIZATION
BY INTENSE NON-ENVELOPE 'SUPERPULSES'

P. L. Shkolnikov, A. E. Kaplan, and S. F. Straub [*]

Electrical and Computer Engineering Department The Johns Hopkins University, Baltimore, MD 21218
* Permanent address: Abteilung fur Quantenphysik, Universitat Ulm, Germany

We have discovered theoretically that the ionization of a quantum well or a hydrogen-like atom by a unipolar, i. e. strongly **asymmetrical** , ultrashort electromagnetic pulse produces a highly **symmetrical** photoelectron cloud, the electrons emmitted both along the field and in the opposite direction in well-separated groups (shells) whose velocities for a quantum well reflect resonances in the ionization continuum.

Photoionization is one of the most fundamental and rich phenomena in atomic physics. The Einstein's formula, $E_k = \hbar\omega - U_0$, which connects the kinetic energy of a photoelectron, E_k, with the photon energy, $\hbar\omega$, and the electron binding energy, U_0, for relatively weak quasi-monochromatic fields, breaks down in strong fields due to multiphoton processes, ATI, tunneling and over-the-barrier ionization, etc. With all their diversity, however, all of the above mentioned effects are caused by quasi-monochromatic optical fields typical for a laser. We have recently demonstrated theoretically the feasibility of intense ultrashort (near- or sub-femtosecond), high-intensity non-oscillating pulses [1]. Such "superpulses", capable of strongly exciting or ionizing quantum systems within the time shorter than the evolution time of the system, may become an important new probing tool and a source of new effects in nonlinear optics and atomic and molecular physics, Atomic ionization by superpulses is, however, largely unexplored. Recent research on the ionization of Rydberg atoms by almost-unipolar ("half-cycle") pulses, whose duration ($\sim 0.5\,ps$) is comparable to the orbital times of the Rydberg states, however, has already revealed substantial differences with the atomic ionization by oscillating optical fields or by longer field pulses [2]. In this paper, we predict a qualitatively new effect. We show, theoretically and by computer simulation of a 3D hydrogen-like atom and a one-dimensional quantum well, that the photoelectron cloud produced by a highly asymmetrical -- unipolar -- superpulse, which is modeled by a Dirac δ-function, displays a highly symmetrical spatio-temporal structure. In particular, the spatial distribution of photoelectrons immediately after the interaction with a superpulse is fully symmetrical with regard to the electric field direction; and the photoelectrons propagate in several, well-separated in space bunches (shells) in both forward and backward directions [3].

An important difference of our approach from other research on atomic ionization is our focus on the electron wavefunction in the coordinate representation. While the ionization of Rydberg atoms by an electrical field "kick" has been addressed before (see e.g. [2]; we are not aware of similar research on quantum wells), the attention has been given so far almost solely to the ionization (or survival) probability; to our knowledge, no attempts have been made to predict the spatio-temporal behavior of photoelectrons. Most recent experimental results [4] on half-cycle pulse ionization of Rydberg atoms seem to corroborate our predictions.

Schrödinger equation of a quantum system interacting with the electric field $\mathcal{E}(t)$ in the dipole approximation is

$$i\hbar\partial\Psi(t,\mathbf{r})/\partial t = [H_0(\mathbf{r}) - e\,\mathbf{r}\cdot\mathcal{E}(t)]\Psi(t,\mathbf{r}) \tag{1}$$

where H_0 is the field-free Hamiltonian, and t and $\mathbf{r}$ are time and displacement, respectively. Modeling the pulse field by the Dirac δ-function of time, $\mathcal{E}_\delta = \mathcal{E}_p\,\delta(t/\tau_p)$, where $\mathcal{E}_p$ is the amplitude of the finite-width pulse, $|\mathcal{E}(t)| \leq \mathcal{E}_p$, and τ_p is the width of the effective rectangular pulse, $\tau_p \equiv \int_{-\infty}^{\infty}\mathcal{E}(t)dt/\mathcal{E}_p$, one obtains a simple relation between the pre-pulse ($t = 0-$) wavefunction ψ_0, and that immediately after the pulse as (see e.g. [5])

$$\Psi(0+,\mathbf{r}) = \psi_0(\mathbf{r})\,exp(i\mathbf{p}_\delta\cdot\mathbf{r}/\hbar), \tag{2}$$

where $\mathbf{p}_\delta = e\mathcal{E}_p\tau_p = \int_{-\infty}^{\infty} e\mathcal{E}(t)dt$ is the classical (equal to the quantum average) momentum transfer due to the δ-kick. Eq. (2) is obtained by neglecting H_0 in Eq. (1), which constitutes the impulse approximation [5]. As follows from Eq. (2), the probability distribution is not altered by the δ-kick, $|\Psi(0+,\mathbf{r})|^2 = |\psi_0(\mathbf{r})|^2$, since the system has no time to move.

The evolution of the system after the pulse is governed by the field-free Schrödinger equation, with $\Psi(0+,\mathbf{r})$ as the initial function. The solution of this equation is most straightforwardly expressed through the full orthonormal set of eigenfunctions of the field-free Hamiltonian as

$$\Psi(t,\mathbf{r}) = \Psi_{bound} + \Psi_{ion}, \quad \Psi_{bound} = \sum b_j\psi_j(\mathbf{r})e^{-iE_jt/\hbar}, \quad b_j = \int d\mathbf{r}\psi_j^*(\mathbf{r})\,e^{i\mathbf{p}_\delta\cdot\mathbf{r}/\hbar}\,\psi_0(\mathbf{r}) \tag{3}$$

Note that for weak δ-kicks, $b_j \approx i\mathbf{p}_\delta\mathbf{d}_{1,j}/\hbar$, where $\mathbf{d}_{1,j} = \int\psi_0(\mathbf{r})\mathbf{r}\psi_j^*(\mathbf{r})d\mathbf{r}$ is the dipole moment of the transition between the initial and the j-th eigenstates. Of primary interest to us is the "ionized part" of the total wavefunction, Ψ_{ion} (see below), that comprises the eigenfunctions of the continuum and, therefore, describes the motion of photoelectrons. We investigate in detail two fundamental quantum systems: a 1D quantum well (QW) and a 3D hydrogen atom (HA). Much of the surprising symmetry of the superpulse ionization, however, is a general phenomenon unrelated to particular potentials.

(i) Immediately after a highly **asymmetrical** (unipolar) kick, the spatial distribution of photoelectrons, $|\Psi_{ion}(0+,\mathbf{r})|^2$, is **fully symmetrical** with regard to the direction of the kick, and this holds for any real and inversion-symmetrical H_0. Indeed, bound eigenfunctions of such H_0 can always be chosen real and of definite parity, with the ground-state wavefunction being even; assume ψ_0 even. Separating in Eq. (3) (at $t = 0+$) sums over odd and even states and taking into account that b_j are real (imaginary) for even (odd) ψ_j, one can see that $\text{Re}[\Psi_{bound}(0+,\mathbf{r})]$ is an even function of $\mathbf{r}$, while $\text{Im}[\Psi_{bound}(0+,\mathbf{r})]$ is odd. The same apparently holds for $\Psi(0+,\mathbf{r})$, and, therefore, for $\Psi_{ion}(0+,\mathbf{r})$; the symmetry of $|\Psi_{ion}(0+,\mathbf{r})|$ follows. Computer simulations based on the Eqs. (4), (9) below at $t = 0$ confirm this (Fig. 1). Moreover, Fig. 1a reveals a strikingly ordered structure of $|\Psi_{ion}(0+,x)|$ for QW, which contains pronounced, equally and symmetrically spaced peaks in the amount of the number of bound states for this well plus one; we will address this structure in detail elsewhere. Here, we just point out that the location of the highest peaks reflects in some sense the "size" of the potential in H_0: the borders of a QW, and the radial extent of the ground state of a HA.

(ii) For weak-kick ($p_\delta\, a/\hbar \ll 1$, where a is a characteristic size of the system) ionization of a 1D system, the photoelectron propagation after the kick is approximately symmetrical with regard to the field direction.

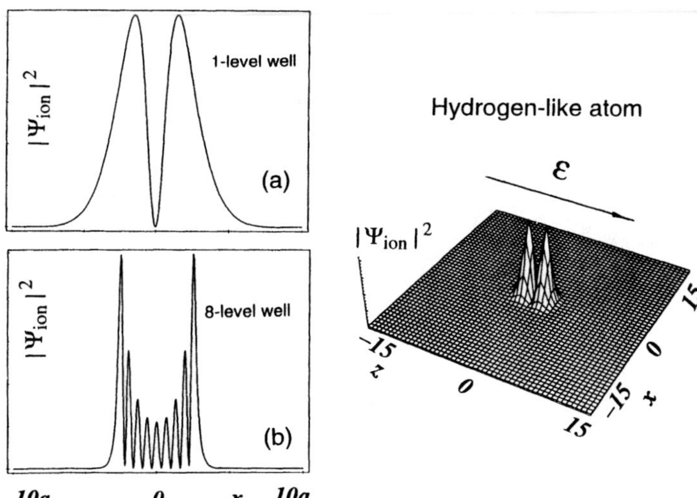

Fig. 1. Symmetrical, ordered structure of the photoelectron cloud immediately after the δ-kick ionization of: (a) 1-level (top) and 8-level (bottom) QW; (b) hydrogen-like atom. On Fig. 1b, the distances in x and z axes are in atomic units.

Indeed, the continuos spectrum of a 1D inversion-symmetrical Hamiltonian is double-degenerate. We may choose the eigenfunctions $\psi_\pm(E,x)$ for a given energy E as $\psi_-(E,x) = \psi_+(E, -x)$, so that the respective spatial distributions mirror each other. Then

$$\Psi_{ion}(t,x) = U_0^{-1} \int_0^\infty dE\, \psi_E(x)\, e^{-iEt/\hbar}\ , \quad \psi_E(x) = b_+\psi_+ + b_-\psi_-\ , \quad b_\pm = \int dx\ \Psi(0+,x)\, \psi_\pm^*(E,x) \quad (4)$$

For weak kicks, therefore,

$$b_+(E) \approx -b_-(E) \approx i(p_\delta/\hbar)\, d_{j,E}\ , \quad |b_+(E)|^2 \approx |b_-(E)|^2 \tag{5}$$

This means, in particular, that the photoelectron spectra for weak kicks are almost identical to the spectrum of the dipole matrix elements of the field-free Hamiltonian, and that photoelectrons born by a weak kick are equally likely to propagate in both directions. The approximate *spectral* symmetry, Eq. (5), combined with the above choice of the eigenfunctions, leads to the approximate *spatial* symmetry of the expanding photoelectron cloud.

To move further, we turn to particular quantum systems.

(i) One-dimensional quantum well (QW). A 1D quantum well in the x-axis is defined by the potential $U = -U_0 = const < 0$ at $|x| \leq a$ and $= 0$ elsewhere. We choose ψ_+ as a (non-normalized) plane wave that propagates from the left to the right, so that there are both incident and reflected wave on the left of the well, and only one (transmitted) right to it [6]. Hence, $\psi_+ = e^{ixk_E} + A\, e^{-ixk_E}$, at $x < -a$; $\psi_+ = B_+ exp(ixk_U) + B_- exp(-ixk_U)$, at $|x| \leq a$, and $\psi_+ = C exp(ik_E x)$, at $x > a$, where $k_E = \sqrt{2mE}/\hbar$, and $k_U = \sqrt{2m(E+U_0)}/\hbar$, or $k_U^2 = k_0^2 + k_E^2$, where $k_0 = \sqrt{2mU_0}/\hbar$. Similarly, $\psi_-(x) = \psi_+(-x)$ propagates from the right to the left. The constants A, $B_\pm$ and C are found from the continuity of $\psi_\pm(x)$ and $d\psi_\pm/dx$ at the boundaries $x = \pm a$ as $A = 2ik_0^2 sin(2ak_U)exp(-2iak_E)D^{-1}$, $B_\pm = 2k_E$

$(k_U \pm k_E)exp[-ia(k_E \pm k_U)]D^{-1}$, and $C = 4k_E k_U exp(-2iak_E)D^{-1}$, where $D = (k_U + k_E)^2 exp(-2iak_U) - (k_U - k_E)^2 \ exp(2iak_U)$; $|A|^2 + |C|^2 = 1$. The QW is transparent, $|C|^2 = 1$, if $ak_U = n\pi/2$, n integer. The respective resonant energies are:

$$E_n = -U_0 + n^2\pi^2\hbar^2/(8ma^2), \quad n \geq S \equiv 2ak_0/\pi \tag{6}$$

These resonances, in effect "continuum shadows" of QW, are at the core of the effects considered here. E_n are equal to the positive eigenenergies of an infinitely-deep well of the size $2a$ (same as the size of the original QW), whose walls begin at the energy $-U_0$. This infinite well emulates the structure of the ionization continuum of the finite QW, due to the following easily verifiable selection rule: $|d_{1,E_n}|^2$ is equal to zero for even n and reaches local maxima at approximately $E_{n\ odd}$. (It is worth noting that the very first maximum of $|d_{1,E}|^2$ does not correspond to any integer n.) Weak δ-kicks "map" this structure into the energy spectrum of photoelectrons, which, in particular, does not contain electrons energies $\sim E_{n\ even}$; stronger kicks shift the the maxima and zeroes in the energy spectrum (a Stark shift), while maintaining the overall structure. Regardless of the magnitude of the momentum transfer, however, δ-pulse translates the structured ionization continuum into a spatial stratification of the photoelectron cloud. This phenomenon may be called *coherent ionization*, since all the photoelectron spectral components have a trans-spectral coherency of their phases, due to the same initial (immediately after the kick) phase; the result is a well-ordered photoelectron cloud. Fig. 2, calculated based on Eq. (4), displays "multiple ionization echoes" -- a sequence of well-separated moving peaks, which describe

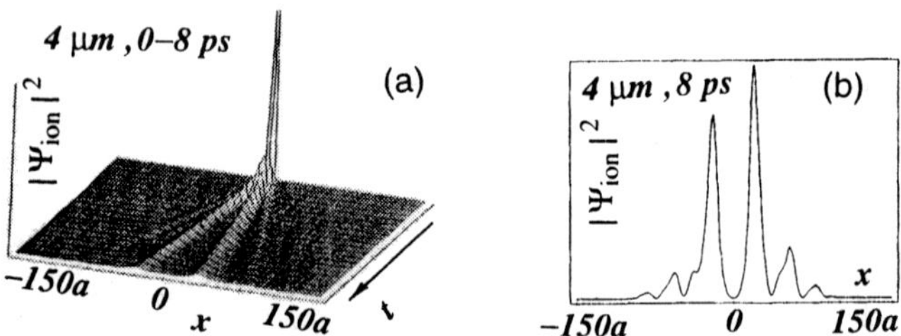

Fig. 2. Backward and multiple photoelectron "echoes" in the δ-kick ionization of a QW: (a) evolution in time; (b) at a particular time. The size of the spatial window and the time elapsed after the kick are given for $U_0 = 100\ mV$, $a = 70\ \text{Å}$.

the electron bunches propagating both in the direction of the pulse field and in the opposite direction. The constant velocities of the peaks are equal to $\sqrt{2E_{max}/m}$, where E_{max} are the energies for which the photoelectron energy spectra $|b_\pm(E)|^2$ reach local maxima, while the peak divides propagate with the velocities that correspond to zeroes in those spectra. Moreover, the relative heights of the "echoes" reflect the relative amplitudes $|b_\pm(E^{j,max})|^2$ at the respective maxima. For weak kicks, as follows from Eq. (5) and the selection rule, these relative amplitudes are proportional to the square modulus of the dipole matrix element for odd resonances in Eq. (6); it is easy to show that $|d_{1,Eodd}|^2 \sim (|E_0| + E)^{-4}E^{-1/2}$ where E_0 is the ground-level energy.

(ii) 3D hydrogen/hydrogen-like atom (HA). The amplitudes $b(\mathbf{k})$ between the ground state and continuum positive-energy eigenfunctions of HA, $\psi^-(\mathbf{k},\mathbf{r})$, were calculated for the needs of the theory of atomic collisions long ago [7,8]; for an arbitrary initial state see [9]. Based on Ref. [8], we write Ψ_{ion} for HA as follows (in atomic units):

$$\Psi_{ion}(t,\mathbf{r}) = \int d\mathbf{k}\chi(\mathbf{k})\psi^-(\mathbf{k},\mathbf{r})exp(-ik^2t/2), \tag{7}$$

$$\psi^-(\mathbf{k},\mathbf{r}) = (2\pi)^{-3/2}\Gamma(1+i/k)e^{\pi/2k}e^{i\mathbf{ks}}\ _1F_1(-ik^{-1}, 1, -i\mathbf{kr}-ikr),$$

$$b(\mathbf{k}) = \frac{16}{(2\pi)^{3/2}} e^{\pi/2k}\Gamma(1-\frac{i}{k})\left[\frac{1+(\mathbf{q}-\mathbf{k})^2}{1+q^2-k^2-2ik}\right]^{i/k} \frac{q^2-\mathbf{q}\mathbf{k}(1+ik^{-1})}{[1+(\mathbf{q}-\mathbf{k})^2]^2\,(1+q^2-k^2-2ik)} \quad (8)$$

Here Γ is the gamma-function, and $_1F_1$ is the confluent hypergeometrical function. Substituting χ and ψ^- into Ψ_{ion}, we obtain:

$$\Psi_{ion}(t,\mathbf{s}) = \frac{4}{\pi^2}\int d\mathbf{k}\,\frac{\exp(-ik^2t/2)}{k\,[1-\exp(-2\pi/k)]}\,exp\,(-i\theta/k)\,exp\,[-k^{-1}\tan^{-1}(\frac{2k}{1+q^2-k^2})] \quad (9)$$

$$\times\,\frac{q^2-\mathbf{q}\mathbf{k}(1+ik^{-1})}{[1+(\mathbf{q}-\mathbf{k})^2]^2\,(1+q^2-k^2-2ik)}\,exp\,(i\mathbf{k}\mathbf{r})\,_1F_1(-ik^{-1},1,-i\mathbf{k}\mathbf{r}-ikr)\,,$$

$$\theta \equiv \tfrac{1}{2}\ln[1+(q-k)^2]\,[1+(q+k)^2]-\ln[1+(\mathbf{q}-\mathbf{k})^2]$$

The photoelectron spatial distribution calculated using Eq. (9) for a weak kick, Fig. 3, shows apparent qualitative similarity with Fig. 2: approximate symmetry with regard to the field direction and pronounced maxima (shells) in the photoelectron distribution.

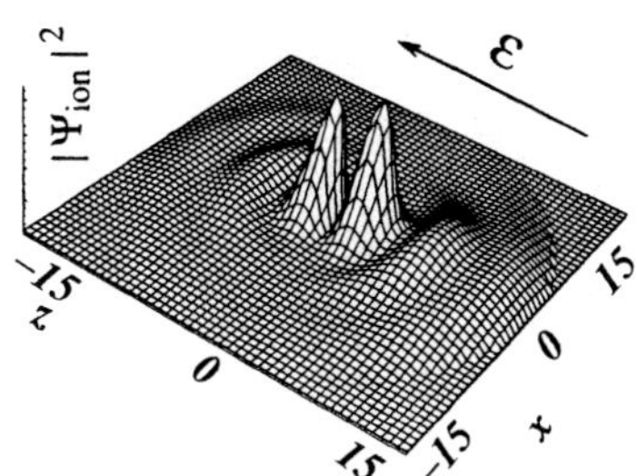

Fig. 3. Backward and multiple photoelectron "echoes" in the δ-kick ionization of a hydrogen-like atom; $t=10\,a.u.$ after the kick.

Now, however, the relationship between these shells and the photoelectron energy spectra is not straightforward, in particular because a shell does not correspond to a bunch of electrons with close velocities, as is the case for QW. We will address this problem elsewhere. For the δ-ionization of the ground-state hydrogen atom, the required pulse duration $\ll 1\,a.u. \sim 10^{-17}\,sec$ is apparently beyond reach of present experimental techniques. (It could become possible in the future, though, with the methods proposed in [1].)

So far, we have not discussed the actual numbers behind our assumption that a pulse can be treated as a δ-kick, and, therefore, the experimental feasibility of the predicted effects. The effects considered in this paper for HA could, however, be observed in the ionization of (hydrogen-like) Rydberg atoms by existing sub-picosecond half-cycle pulses. In fact, a recent experimental paper [4] reports observations that are not inconsistent with the existence of ionization echoes predicted in this paper. At the same time, pulses that emulate a δ-kick for a ground-state ionization of a shallow QW could be relative long, which makes experiments with superpulse ionization of QW feasible in near future. Our results for finite-width pulse ionization, to be published elsewhere, show that many features of the δ-ionization might be visible even if the pulse duration is as long as the characteristic QW time $h/2U_0$.

In conclusion, we have theoretically demonstrated the feasibility of a qualitatively new quantum effect, backward and multi-echo field ionization, in the interaction of superpulses with atoms and quantum wells.

This work is supported by AFOSR.

REFERENCES

[1] A. E. Kaplan, Phys. Rev. Lett. 73, 1243 (1994); A. E. Kaplan and P. L. Shkolnikov, Phys. Rev. Lett. 75, 2316 (1995); A. E. Kaplan, S. F. Straub, and P. L. Shkolnikov, Opt. Lett. 22, 405 (1997).

[2] R. R. Jones, D. You, and P. H. Bucksbaum, Phys. Rev. Lett. 70, 1236 (1993); R. R. Jones and P. H. Bucksbaum, Comments At. Mol. Phys. 30, 347 (1995).

[3] P. L. Shkolnikov and A. E. Kaplan, in *Quantum Electronics and Laser Science Conference* , V. 12, 1997 OSA Technical Digest Series (OSA, Washington, D.C., 1997), p.31.

[4] R. B. Vrijen, G. M. Lanhuijzen, and L. D. Noordam, Phys. Rev. Lett. 79, 617 (1997).

[5] C. O. Reinhold, M. Miellse, H. Shao, and J. Burgdorfer, J. Phys. B 26, L659 (1993); M. D. Girardeau, K. G. Kim, and C. C. Widmayer, Phys. Rev. A 46, 5932 (1992).

[6] D. K. Ferry, *Quantum Mechanics* (IOP Publ. Ltd., 1995), Ch. 2.5.

[7] H. A. Bethe, Ann. Phys., Lpz., 5, 325-400 (1930); H. S. W. Massey and C. B. O. Mohr, Proc. R. Soc. A 140, 613 (1933).

[8] A. R. Holt, J. Phys. B 2, 1209 (1969); C. O. Reinhold and J. Burgdorfer, J. Phys. B 26, 3101 (1993).

[9] Dž. Belkič, J. Phys. B 17, 3629 (1984).

DECONVOLVED IONIZATION PROBABILITIES FOR STRONG FIELD IONIZATION OF XENON

Mark A. Walker, Peter Hansch, and Linn D. Van Woerkom

Department of Physics
The Ohio State University
174 W. 18th. Ave.
Columbus, OH 43210

INTRODUCTION

Using a novel experimental technique, Intensity Selective Scanning[1,2], we have collected time-of-flight mass spectra of ions produced by strong field laser ionization. The hallmark of this technique is that it eliminates a persistent problem inherent in traditional experiments: the complex spatial distribution of intensities present in the ionizing pulse. As a consequence, it becomes possible to extract ionization yields as a function of intensity directly from the experimental data, removing the complication of deducing them from a spatially averaged signal. We will present results for xenon ionized by laser pulses at intensities ranging from 10^{13} to 4×10^{14} W/cm^2.

We recorded time-of-flight mass spectra of xenon using a 1 kHz Ti:Sapphire laser system, generating 120 fs, 800 nm output pulses with an energy of 0.70 mJ per pulse. The absolute peak intensity at the minimum beam waist was held fixed while spectra over the range were recorded via the ISS technique, which involves limiting the volume observed by placing an aperture at the end of the spectrometer's flight tube near the interaction region. Ions are detected only from positions in the laser focus which are in the direct line of sight of the detector.

When ion yields are measured without such a restriction, there is no way to resolve the signals originating in different regions of the gaussian focus,

$$I(I_0, r, z) = \frac{I_0}{1 + (z/z_0)^2} \exp\left(-\frac{2r^2}{\omega_o^2\left(1 + (z/z_0)^2\right)}\right).$$

Consequently, the experiment will yield a spatially integrated signal S given by

$$S(I_0) = \int N(I(\mathbf{r}))d^3r = 2\pi \int_{-\infty}^{\infty} dz \int_0^{\infty} N(I_0, r, z)r dr$$

Applications of High Field and Short Wavelength Sources
Edited by DiMauro *et al.*, Plenum Press, New York, 1998

Where N(I) is the density of ions produced by a pulse of temporal peak intensity I. Because of the intensity distribution, the signal can also be represented as an integral in intensity space,

$$S(I_0) = \int N(I, I_0)\left(\frac{\partial V}{\partial I}(I, I_0)\right) dI$$

The dependence upon all but the overall peak intensity is integrated away. In traditional experiments this parameter is varied to produce curves of ion yield vs. peak intensity. These must then be interpreted in the context of the averaging they contain in order to understand the physics underlying the data. Subtle intensity dependent phenomena cannot be distinguished in these cumulative signals.

By restricting the spatial extent of the region from which the observed signal originates, in these experiments to a 500μm wide rectangular "slice" along the longitudinal axis of the focus, we were able to resolve the contributions to the signal from different points along this dimension. This effectively reduced the usually complicated spatial distribution of intensities responsible for producing ions to a one-dimensional radial distribution about the axis. This reduced dimensionality provides an additional experimental degree of freedom, which ultimately makes it possible to deconvolve the signal. The ionization yields as a function of intensity could thus be recovered directly from the experimental data.

DECONVOLUTION PROCEDURE

As mentioned previously, the limited extraction volume which provides the ISS signal contains intensities whose spatial variation can be described as a function of a single variable, the radial coordinate r. The z dependence is fixed and the measured quantity is an average over this transverse distribution:

$$S(z) = 2\pi\Delta z \int_0^\infty rdrN(I(r, z))$$

The signal is a function of the z position of the pinhole, or equivalently of the on-axis intensity I_{0L} at that position. This allows the intensity behavior to be mapped while holding the overall intensity fixed.

The reduction of the signal to a one-dimensional integral over r, which is in direct correspondence with I, allows for a simple change of variables within the integral:

$$|rdr| = \frac{\omega_0^2}{4}\left(\frac{I_0}{I_{0L}}\right)\frac{dI}{I}$$

$$S(z) = \frac{\pi z_0 \omega_0^2 \Delta z}{2}\int_0^{I_{0L}}\frac{dI}{I}N(I)$$

This form can be inverted by applying the Liebniz rule

$$\frac{d}{dx}\int_{\alpha(x)}^{\beta(x)} F(x, y)dy = \int_{\alpha(x)}^{\beta(x)}\frac{\partial F}{\partial x}(x, y)dy + F(x, \beta)\frac{\partial \beta}{\partial x} - F(x, \alpha)\frac{\partial \alpha}{\partial x}$$

to obtain

$$N(I_{0L}(z)) \propto \left[\frac{I_{0L}(z)}{dI_{0L}(z)/dz}\right]\frac{d}{dz}\left[I_{0L}(z)S(z)\right]$$

198

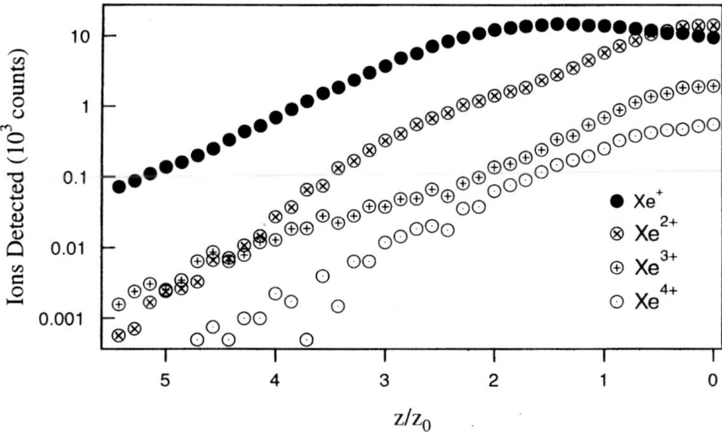

Figure 1. Multiple ionization data for xenon taken using the ISS method. Charge states up to the fourth were observed in the intensity range of the experiment. Data have been corrected for detection sensitivity by the method described in the text.

This allows the behavior of N(I) to be mapped at the values of I_{0L} corresponding to the z values at which our data is collected. This requires making several measurements, each at different z positions, and each with sufficient statistics to maintian good signal to noise, especially since the inversion procedure involves differentiating the signal. Fortunately we are able to do so given the high repetition rate of the laser system. In order to calibrate I_{0L} vs z, we can examine photoelectron spectra which contain resonance peaks which appear at very specific intensities. We also should have some reassurance that our assumption of a Gaussian focus is accurate, so we have performed imaging measurements of the focus to corroborate this as well as our intensity calibration.

RESULTS AND DISCUSSION

Intensity dependent ion signals for xenon are shown in Figures 1 and 2. Figure 1 shows the data in their original form as a function of z. Figure 2 shows these same data as a

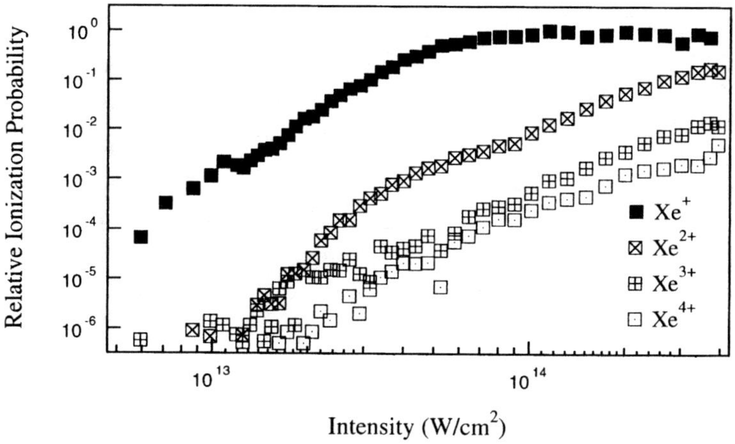

Figure 2. Deconvolved ionization data for xenon. These represent the result of applying the inversion procedure to the data of Figure 1.

199

function of intensity after inversion using the procedure described above. These curves represent $N_i(I)$, the relative number of ions of the ith charge state (i=1 to 4) per unit volume produced at intensity I. The number of singly charged ions saturates around 7×10^{13} W/cm^2 and does not increase beyond this point. This is expected, since the probability for ionization has a maximum value of unity. The slope of the curve becomes slightly negative at the highest intensities with the onset of depletion by double ionization. The occurrence of non-sequential double ionization (NSDI) is clearly evident in the Xe^{2+} curve, which shows significant double ionization before the Xe^+ saturation intensity, as well as a change in slope near 9×10^{13} W/cm^2. Here the dominant process changes from nonsequential to sequential ionization. The signal to noise for the Xe^{3+} and Xe^{4+} curves is oor, but they are shown here for completeness.

Since all geometric dependencies have been removed from the data, it is possible to calibrate the ion yields for relative detection efficiencies. The sum of the different charge states is required to be unity for intensities above the saturation intensity. It should be possible, in fact, to absolutely calibrate the detection efficiency using data with better statistics due to the probability for ionization having an absolute maximum of unity.

CONCLUSION

We have shown that the results of ISS experiments, because of their limited extraction volume, can be inverted easily to exhibit their original intensity dependence. We have provided an example of this in our ion yield data, but the technique is not limited to this type of experiment. It could be applied for example to look at intensity dependences of features in photoelectron spectra, or to study ionization and dissociation of molecules. As the study of high intensity phenomena expands, the possibility of recovering data without averaging effects provides the opportunity to obtain a clearer picture of the underlying physics involved in these processes.

REFERENCES

1. P. Hansch and L. D. Van Woerkom, Opt. Lett. **21**, 1286 (1996)
2. P. Hansch, M. A. Walker and L. D. Van Woerkom, Phys. Rev. A **54**, R2559 (1996)

PRECISION TESTS OF LASER-TUNNELING IONIZATION MODELS

B. Buerke[a], J. P. Knauer, S. J. McNaught[b], and D. D. Meyerhofer[a,b]

University of Rochester Laboratory for Laser Energetics
250 East River Road
Rochester, NY 14623

[a]Also Dept. of Physics and Astronomy
[b]Also Dept. of Mechanical Engineering

INTRODUCTION

The energy and number distributions of photoelectrons observed outside the ionizing field can give insight into the ionization process in a strong laser field. The physics of this process can be considered to occur in a two-step process known as the "simpleman's model."[1] First, the electron escapes the atom by tunneling, and second, it interacts with the external field. If the time it takes the electron to tunnel out of the atom is much shorter than the laser period, then ionization occurs in the adiabatic limit (the ionization potential I_p is greater than the photon energy ω), and static field calculations of the ionization rate are valid.[1] In the experiments described below, the Keldysh adiabaticity parameter $\gamma = (I_p/2U_p)^{1/2} \ll 1$,[2] where I_p is the ionization potential of the atom and $U_p = F^2/2\omega^2$ is the ponderomotive energy, indicating that ionization occurs in the tunneling regime. (Here the electric field F is proportional to the square root of the laser intensity.) Once liberated, the electron will gain a drift momentum which is determined by the phase of the laser field at the time of ionization. If the polarization is linear, then the drift momentum will be zero unless the external electric field is not at its peak value. The drift momentum will be nonzero and directed perpendicular to the polarization vector if the laser is elliptically polarized. If the laser is circularly polarized, the drift due to initial conditions will give the electron a directed kinetic energy of about $2U_p$ when ponderomotive effects are included. We have used a 2-ps, 1-μm laser with intensities up to 1×10^{18} W/cm^2 and two electron spectrometers to test the static tunneling models.

MEASURING THE TUNNELING RATE FOR HYDROGENIC HELIUM

The ionization of hydrogenic atoms under quasistatic conditions is ideal for testing strong-field ionization theories. Theoretical analysis of ionization is greatly simplified by the absence of multielectron effects. Further simplification is possible under quasistatic conditions,

Applications of High Field and Short Wavelength Sources
Edited by DiMauro *et al.*, Plenum Press, New York, 1998

where the time it takes for the electron to escape the atom is much shorter than the laser period. When the field is quasistatic, multiphoton resonance effects are eliminated and atomic structure becomes unimportant.

Quasistatic ionization occurs in one of two ways. The first is classical, where the electron escapes over the Coulomb barrier, with an ionization threshold given by[3]

$$F_{classical} = I_p^2/4Z = Z^3/16n^4.$$ (1)

The field F and ionization potential I_p are in atomic units. The second is tunneling ionization, where the electron tunnels through the barrier, described by the well-known semiclassical rate[4]

$$w = 8I_p\left(\frac{(2I_p)^{3/2}}{F}\right)\exp\left(-\frac{2(2I_p)^{3/2}}{3F}\right).$$ (2)

Both of these models have been successful in describing the ionization of noble gases in the optical regime.[3]

The experiment uses circularly polarized light, with $\omega = 0.043$, to ionize hydrogenic helium and determine the ionization rate to high precision. The large ionization potential of He^{1+} ($I_p = 2.0$) ensures that the atom, unlike hydrogen for optical frequencies, ionizes under quasistatic conditions. Circular polarization suppresses resonant excitation[5] and keeps the field magnitude constant, allowing comparisons with static-field tunneling theories.

The rate is determined by measuring the energies of electrons ejected from the high-intensity focus. According to the two-step model discussed above, each electron acquires an initial drift energy and an additional ponderomotive energy while traversing the focus. For circular polarization the total available energy is $2U_p$. This energy provides a direct measure of the laser intensity acting on each electron at the moment it is released from the atom.

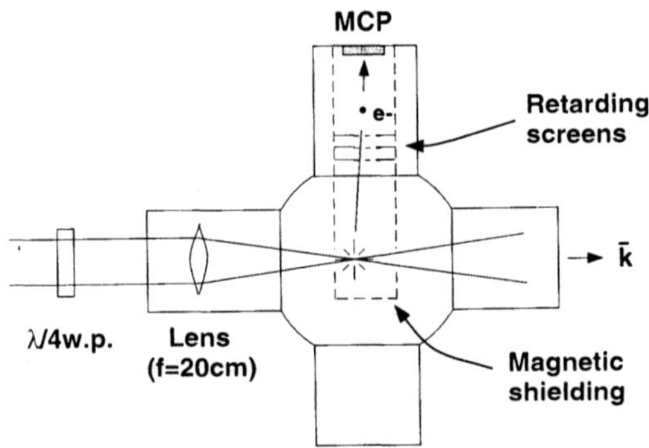

Figure 1. Schematic of vacuum chamber showing laser focus, spectrometer, and detector.

To obtain a precise calibration between the ionization intensity and electron energy, we performed a Monte Carlo simulation of the electron motion in the focus using fully relativistic equations of motion and accounting for the laser's measured spatial and temporal characteristics. The electron energy spectra are Gaussian in shape--due primarily to the temporal pulse--with a peak and width that depend on the ionization model used in the simulation. For the classical model of Eq. 1, where the electrons are produced at a single intensity, the distribution for our energy range is narrowly peaked at 97.5% of $2U_p$. For tunneling ionization models, where the

202

electron can be produced over a range of intensities, the distribution is wider and the peak energy gained is actually slightly more than expected ($E = 2.03U_p$).

Helium is ionized by a circularly polarized, 2.0-ps laser pulse with a peak intensity of 1.5×10^{17} W/cm^2. A retarding-potential spectrometer is used to measure the energy of the ejected electrons, as shown in Fig. 1. Electrons with sufficient energy pass through the screens and are detected by a microchannel plate. The output is digitized and integrated for each shot, and these values are binned into 5% energy windows and averaged together after adjusting for variations in peak intensity and pressure.

A detailed retarding-potential spectrum for hydrogenic helium is shown in Fig. 2. The vertical scale is the estimated number of electrons produced in the focus. The solid line through the data points is the best error-function fit to the data, corresponding to a Gaussian with a peak at 1927 ± 40 eV and a $1/e$ half-width of 740 ± 50 eV. In addition to the averaging described above, the data points have been adjusted to account for the measured dependence of the MCP efficiency at higher energies. The systematic error, due mostly to the absolute uncertainty in the pulsewidth, is identical to the random error listed above.

Two conclusions can immediately be drawn from these results. First, the energy gained by the electrons is far less than the energy available at the peak, since at 1.5×10^{17} W/cm^2 the total energy is $2U_p = 31$ keV. The observed energies represent the intensities at ionization rather than the intensity profile of the focus. Second, the atom ionizes well below the classical over-the-barrier prediction of Eq. 1, which gives $2U_p = 3630$ eV. Previous experiments[3] were consistent with both tunneling and classical ionization. Here the latter is completely eliminated.

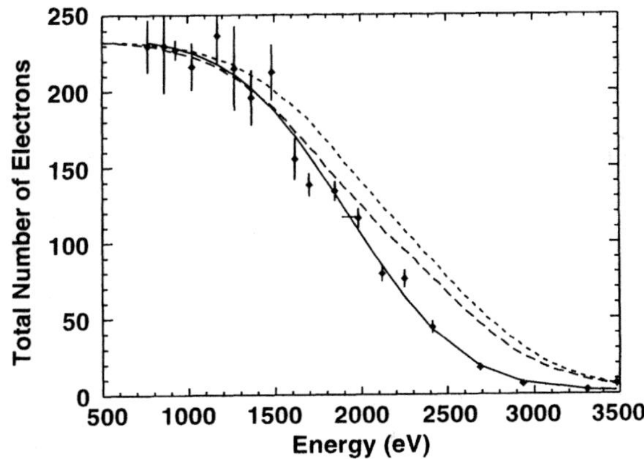

Figure 2. Energy distribution of electrons ejected during the ionization of He^{1+} with a peak intensity of 1.5×10^{15} W/cm^2 and a pressure of 5×10^6 T. The solid curve is a best fit to the data, the dashed curve gives the semiclassical tunneling prediction, and the dotted curve is the semiclassical theory with the DC Stark shift.

The best agreement between data and theory is obtained for the tunneling formula in Eq. 2. The result, shown in Fig. 2, gives a Gaussian distribution with a peak of 2060 eV and $1/e$ half-width of 859 eV. The relative widths for data ($38.4 \pm 2.6\%$) and theory (41.7%) are in reasonable agreement, confirming that the atom ionizes by tunneling. The primary advantage of Eq. 2 is that it accounts for the long-range nature of the Coulomb potential, which is responsible for the field-dependent factor in front of the exponential and greatly increases the ionization rate. Our data confirm the importance of this effect.

It should be noted, however, that the semiclassical theory fails to predict the correct peak, giving an energy too high by three standard deviations. Moreover, the theory ignores DC Stark shift of the ground state, which effectively increases the ionization potential and reduces the rate. This effect worsens the discrepancy between theory and experiment. Based on the intensity-to-energy calibration, the data imply that the most probable ionization intensity for hydrogenic helium for a 2.0-ps pulsewidth is $9.12 \pm 0.19 \times 10^{15}$ W/cm^2, equivalent to an ionization rate 1.59 times the semiclassical rate and 2.72 times the DC Stark-shifted rate. At present, no known theory of ionization is adequate to fully explain the experimental results.

MEASUREMENT OF PHOTOELECTRON INITIAL CONDITIONS

The tunneling model of ionization can also be tested by measuring the angular distribution of ejected electrons in the plane of polarization, as this distribution is a function of the electron's initial drift momentum. In the second experiment, the electron's average initial drift momentum in the long-pulse tunneling limit has been measured and shown to agree with the drift momentum predicted by the Ammosov-Delone-Krainov (ADK) tunneling model.[4,7] An elliptically polarized laser was used to ionize neon gas at low density and the high-energy photoelectron spectra was measured as a function the azimuthal angle φ (relative to the major axis of the polarization ellipse) using a magnetic spectrometer. These azimuthal distributions were measured at an energy and forward angle θ where the maximum signal was observed from electrons born in the creation of the charge states Ne^{6+} through Ne^{8+}.[6] For linear polarization, electrons with energies up to 30 keV were observed,[7] and electrons with energies up to 70 keV were observed for an elliptic polarization of $\varepsilon = 0.7$.[8] Here ε is defined to be the amplitude of the minor component of the electric field divided by that of the major component. For linearly polarized light, a significant asymmetry in the electron number in the plane perpendicular to the propagation direction was observed, with twice as many electrons detected along the direction of polarization as perpendicular to it.[7] For elliptically polarized light, the asymmetry reversed, and many more electrons were observed along the minor axis of the electric field ellipse than along the major axis.[8] Fig. 3 shows the azimuthal spectrum for electrons injected into an elliptically polarized field in the creation of Ne^{8+} ions; about seven times as many electrons were observed along the minor axis of the polarization ellipse as along the major axis.

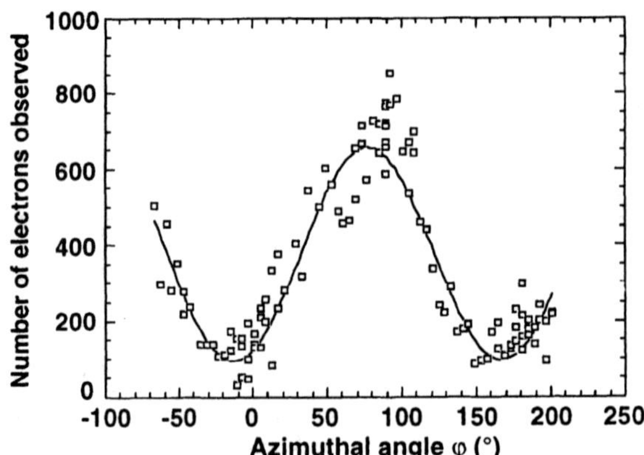

Figure 3. Measured azimuthal distribution of electrons liberated in the creation of Ne^{8+} in an elliptically polarized field ($\varepsilon = 0.7$).

This asymmetry can be explained by the electron initial conditions at the time of ionization. The ADK theory predicts that the ionization rate falls off exponentially as the amplitude of the electric field,[9] and differs from Eq. 2 only in its pre-exponential factor. Fig. 4 shows the ADK ionization rate $w(\eta)$ as a function of laser phase η for laser ellipticities of $\varepsilon = 0$, $\varepsilon = 0.7$, $\varepsilon = 0.9$, and $\varepsilon = 1.0$, calculated for electrons born in the creation of Ne^{8+} ions with the peak field held constant. Note that the distribution has a significant width for linear

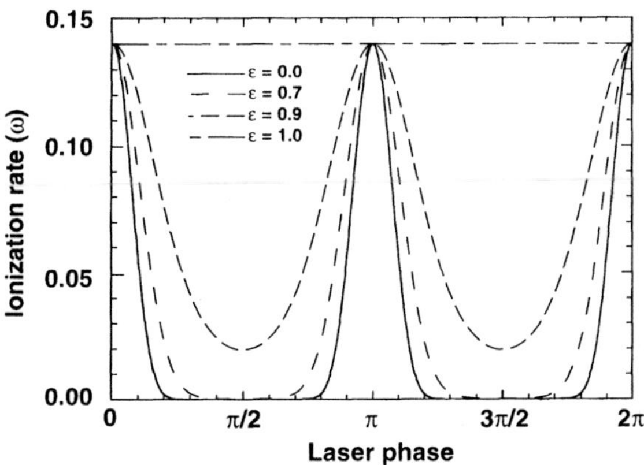

Figure 4. Ionization rate $w(\eta)$ as a function of phase within the optical cycle for various ellipticities, given in units of the laser frequency ω. The parameters used are those for electrons produced in the creation of Ne^{8+} ions.

polarization, changes little as ε increases to 0.7, still has a large modulation at 0.9, and finally becomes constant for circular (ε =1.0) polarization, since in the last case the field magnitude is a constant function of phase. Several mechanisms give rise to electron drift once it is freed from the atom at the optical phase determined by the distributions shown in Fig. 4. The electron will gain a total drift $p_d = p_k(\eta 0) + p_F(\eta 0) + p_{pond}$, where $p_k(\eta 0)$ is the initial kinetic momentum, $p_F(\eta 0)$ is the field momentum, and p_{pond} is any drift due to ponderomotive effects. The field momentum $p_F(\eta 0)$ is determined by the laser phase at ionization $\eta 0$ and is given for a plane-wave field by

$$p_F\left(\eta_0\right) = \frac{F}{\omega\sqrt{\varepsilon^2 + 1}}\left(\hat{x}\sin\eta_0 + \varepsilon\hat{y}\cos\eta_0\right). \tag{3}$$

Note that when ε = 0, the y component of $p_F(\eta 0)$ is zero and that there is no x component of $p_F(\eta 0)$ when $\eta 0 = 0$, π, 2π, etc.

Fig. 5 shows the root-mean-square field momentum amplitudes $p_{F,x}$ and $p_{F,y}$ as a function of laser ellipticity. At the top of the figure, polar plots of the resultant angular distribution when ponderomotive effects are accounted for are shown for ellipticities of ε = 0.0, ε = 0.3, ε = 0.7, and ε = 1.0. Here the parameters for electrons born in the creation of Ne^{8+} ions are used. The features of the angular distribution of photoelectrons as predicted by the ADK model can be described in terms of the main features of Fig. 5. If $p_{F,y}$ is larger than $p_{F,x}$, more electrons will be ejected in the y direction than in the x direction, and vice versa. Eq. 3 shows that for a linearly polarized field (ε = 0), $p_{F,y} = 0$, so more electrons will appear along the x direction. The resultant 2:1 angular distribution calculated by a Monte Carlo computer simulation of the experiment is shown at the top of Fig. 5, with ponderomotive effects included. This result agrees with the experimental data.[7] As ε increases, the magnitude of the y component of the average drift approaches that of the x component, producing an isotropic angular distribution when ε = 0.3. At ε = 0.7 there is the greatest difference between $p_{F,x}$ and $p_{F,y}$, so the asymmetry is maximized. When ponderomotive effects are accounted for, this gives a 1:7 asymmetry ratio, which also agrees with the experimental data.[8] As the ellipticity increases beyond 0.7, the probability of ionization as a function of phase $\eta 0$ becomes uniform enough that there is less of a preferred direction for $p_F(\eta 0)$, when at ε = 1 the angular distribution again becomes isotropic.

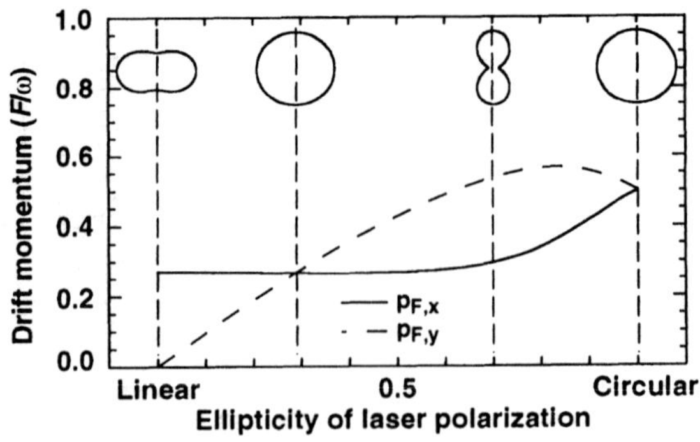

Figure 5. The values of the rms drift momenta $p_{F,x}$ and $p_{F,y}$ as a function of ellipticity using the ADK ionization rate.[9] At the top of the figure, polar plots of the resultant angular distribution when ponderomotive effects are accounted for are shown for ellipticities of $\varepsilon = 0.0$, $\varepsilon = 0.3$, $\varepsilon = 0.7$, and $\varepsilon = 1.0$.

The rms value of the field momentum in both directions was calculated in the model assuming that the initial *kinetic* momentum $p_k(\eta_0)$ was zero. Because the experimental data agreed with the values of $p_{F,x}$ and $p_{F,y}$ predicted by the ADK model shows that $p_k(\eta_0)$ must be negligible and allows an upper limit to be placed on its amplitude. In the case of linear polarization, the initial kinetic momentum is limited by the data to only 0.2% of the average quiver energy U_p of the electron in the field, which for the 22 keV electrons observed is only about 45 eV. This is the most precise experimental limit ever placed on the electron initial conditions in this regime known to the authors.

This work is supported by the National Science Foundation. Additional support was provided by the U. S. Department of Energy Office of Inertial Confinement Fusion under Cooperative Agreement No. DE-FC03-92SF19460, the University of Rochester, and the New York State Energy Research and Development Authority. The support of DOE does not constitute an endorsement by DOE of the views expressed in this article.

REFERENCES

1. P. B. Corkum, *Phys. Rev. Lett.* **71**: 1994 (1993).
2. L. V. Keldysh, *Sov. Phys. JETP* **20**: 1307 (1965).
3. S. Augst, D. D. Meyerhofer, D. Strickland, and S.L. Chin, *J. Opt. Soc. Am. B* **8**: 858 (1991).
4. A. M. Perelomov, V. S. Popov, and M. V. Terent'ev, *Sov. Phys. JETP* **23**: 924 (1966).
5. P. H. Bucksbaum, L. D. Van Woerkom, R. R. Freeman, and D. W. Schumacher, *Phys. Rev. A* **41**: 4149 (1990).
6. C. I. Moore, J. P. Knauer, and D. D. Meyerhofer, *Phys. Rev. Lett.* **74**: 2439 (1995).
7. S. J. McNaught, J. P. Knauer, and D. D. Meyerhofer, *Phys. Rev. Lett.* **78**: 626 (1997).
8. S. J. McNaught, J. P. Knauer, and D. D. Meyerhofer, *Laser Physics* **7**: 712 (1997).
9. M. V. Ammosov, N. B. Delone, and V. P. Krainov, *Sov. Phys. JETP* **64**: 1191 (1986).

SELF-GUIDING WITHOUT FOCUSING NONLINEARITY: LEAKING MODE SELF-EFFECT DUE TO FIELD-INDUCED SATURABLE IONIZATION

A.M.Sergeev,[1] M.Lontano,[2] M.D.Chernobrovtseva,[1] and A.V.Kim[1]

[1]Institute of Applied Physics of the Russian Academy of Sciences
46 Ulyanov st., 603600 Nizhny Novgorod, Russia
[2]Istituto di Fisica del Plasma, Consiglio Nazionale delle Ricerche
EURATOM-ENEA-CNR Association, 20133 Milano, Italy

INTRODUCTION

Creation of elongated plasma structures for guiding of powerful ultra-short laser pulses is a challenging scientific problem and an important application in the area of superstrong field interaction with matter. Self-channeling regimes recently observed in experiments[1-4] have been attributed to the focusing nature of relativistic and Kerr nonlinearities that cause an increase of the refractive index and deviation of light rays toward stronger field regions. In the case of gas ionization at the axis of a Kerr-effect induced waveguide[3-5] the influence of focusing nonlinearity should be especially strong since it is not only to prevent the divergence of rays due to linear diffraction but also to balance refraction of radiation from the axis, that is caused by emerging plasma. Experimentally observed extra-long waveguides produced by few mJ, 100 fs laser pulses at ionization of atmospheric air[3,4] have been interpreted as plasma structures having a core where the ionization nonlinearity prevails and an outer cladding where the dominating Kerr nonlinearity generates opposite-in-sign positive variations of the refractive index and hence keeps the radiation from divergence.

In this report we demonstrate that the saturable ionization nonlinearity alone, without any focusing nonlinearity, is a sufficient mechanism for self-channeling of an ultrashort laser pulse. At first sight, this statement looks absurd since in accord with a common concept a nonlinearity with a growing dependence of the refractive index on the field intensity is needed for the self-guiding effect. The idea of self-guiding at defocusing ionization nonlinearity consists in the following. Owing to a strong dependence of the field ionization rate on the field intensity a laser pulse can produce a plasma distribution that is smooth near the axis and sharply bounded at the periphery of the cross-section (a plasma filament with sharp boundaries). This distribution in spite of a negative variation of the refractive index at the axis can guide an electromagnetic wave in the form of a leaking mode with exponentially small losses over the distances of many free-space Rayleigh lengths. As distinct from the common self-guiding effect where the field localization is achieved due to the total internal reflection at the periphery of a guide, in this case the quasi-localization is obtained due to a strong reflection of the trapped wave from the plasma boundary that is sharp as compared to the transverse scale (transverse wavelength) of this wave. Hence, the leakage losses are an inherent feature of the plasma waveguide though this factor may have only a minor contribution to the overall wave dissipation as compared for example to the ionization losses. Note that the leaking wave radiation has been measured in the recent experiment,[4] however the authors have interpreted the observed channeling as a self-effect due to the Kerr nonlinearity.

BASIC EQUATIONS

Formation of a channel with a quasi-rectangular radial plasma profile can be facilitated in the case of saturation of the ionization, which is typical for this nonlinearity and corresponds to the complete depletion of one or several electronic states in atoms. The result of saturation is a flattening of the plasma profile near the axis and a corresponding decrease of refraction in the center of the channel where the main part of the laser energy is propagated. This regime seems easier to be implemented in single-species gases at not so high pressures when the saturation can be reached before the free electron concentration becomes too large and leaves no chance to balance the strong refraction.

A combination of two factors, a sharp dependence of the ionization rate and a strong saturation, allows us to compose a simple analytical model for the self-guiding.[6] Assuming the electron concentration to be saturated at a level N_0 everywhere inside the induced plasma waveguide with the radius a, so that $(k_o a)^{-2} << N_o/N_{cr} << 1$, we obtain for the spatial decrement h of the self-trapped leaking mode the following expression

$$h \approx \frac{6}{k_o^2 a^3 \sqrt{N_o/N_{cr}}},$$

(1)

where $k_o = \omega/c$ and $N_{cr} = m\omega^2/4\pi e^2$. For a quite powerful laser pulse focused on the gas in a spot with the size a_0 we expect that the ionized region is wider than the radiation beam, $a > a_0$. The ratio of the leakage distance $z^* = 1/h$ to the free-space Rayleigh length z_F can be presented in the form

$$\frac{z^*}{z_F} = \frac{k_o a^3}{a_o^2} \sqrt{\frac{N_o}{N_{cr}}}.$$

(2)

This expression demonstrates the physical requirements for the long distance channeling $\left(z^*/z_F >> 1\right)$.

We have also performed a detailed computational study of the new self-guiding effect. For the optical field with the scalar complex amplitude E in the paraxial approximation we have used the equation

$$2ik_o \frac{\partial E}{\partial z} + \frac{\partial^2 E}{\partial x^2} + \frac{\partial^2 E}{\partial y^2} - \frac{4\pi e^2 N}{mc^2} E = 0$$

(3)

that includes the factors of diffraction in the transverse (x, y) directions, ionization nonlinearity. The electron density N has been governed by a simple dynamical equation

208

$$\frac{\partial N}{\partial \tau} = (N_o - N) f(|E|) \tag{4}$$

that describes the ionization with the field-dependent rate $f(|E|)$, saturated at the level N_o. Here N_o is the gas density. The time $\tau = t - z/V_{gr}$ has been counted from the pulse arrival at a given point along the propagation path z. Below we present some results in the case of tunneling ionization with the well known dependence of the ionization frequency on the field amplitude[7]

$$f(|E|) = 4\gamma \left| \frac{E_a}{E} \right|^{1/2} \exp\left(-\frac{E_a}{|E|} \right). \tag{5}$$

The values of the atomic field E_a, and the frequency γ may range in wide intervals depending on the concrete kind of the ionized species.

SIMULATION RESULTS

Here we present the results obtained for $\gamma = 10^5/\tau_0$, $N_o = 25 \cdot (k_o a_F)^2 N_{cr}$, and the collimated incident laser pulse with the Gaussian temporal and transverse distributions of intensity $|E|^2 (z = 0, x, y, \tau) = E_o^2 \cdot E_a^2 \exp\left[-(x^2 + y^2) - \tau^2 \right]$, $E_o = 1.5$.

For numerical study we used the following dimensionless variables: $z/z_F \rightarrow z$, $\vec{r}_\perp/a_F \rightarrow \vec{r}_\perp$, $\tau/\tau_o \rightarrow \tau$, $N/(k_o a_F)^2 N_{cr} \rightarrow N$, where $z_F = k_o a_F^2$ is the vacuum Rayleigh length, and τ_0 is the laser pulse duration.

Figure 1 presents the maximum intensity of the pulse and the field intensity in the middle point of the pulse along the propagation distance z. In this picture the self-trapping of the ionizing laser pulse in the induced long plasma channel over 6 Rayleigh length is distinctly seen. Our simulation points to that the maximum of the laser intensity displaces to the back part of the laser pulse due to stronger beam spreading the leading part. So, here laser pulse is shortening and, therefore, energy that belongs to the back part guides for a longer distance. As you may see in figure 2, in transverse directions laser intensity is well localized. This is due to sharp profile of the plasma channel density.

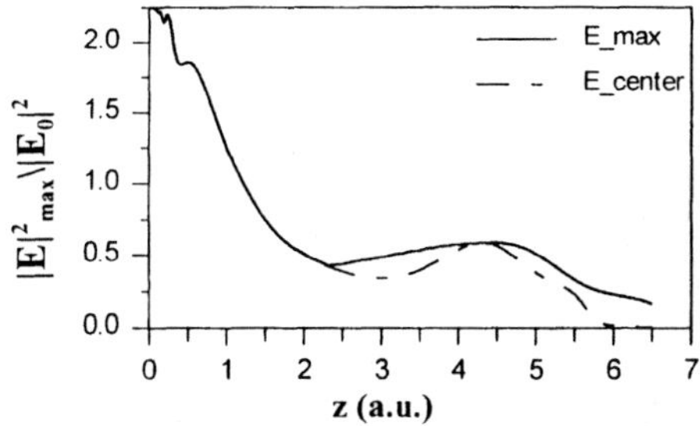

Figure 1. Field intensity at the axis of the channel in the middle of the temporal profile of the pulse as a function of propagation distance z measured in Rayleigh lengths.

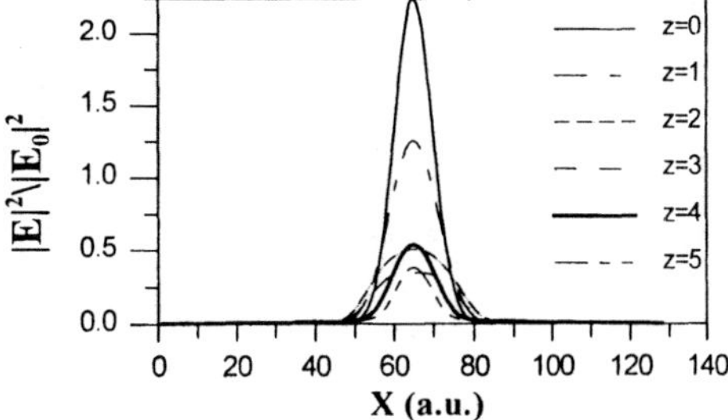

Figure 2. Transverse distribution of the field intensity in the plasma channel for different propagation distances z measured in Rayleigh lengths.

The spatial distribution of plasma channel after the laser pulse passage is shown in figure 3. Almost homogeneous long plasma channel with sharp boundaries was created which has good quality, for example, for wake-field plasma wave experiments with ultrashort laser pulses.

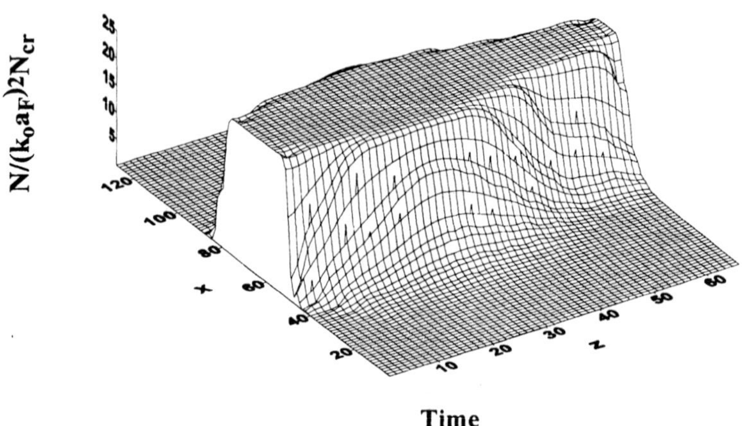

Figure 3. Transverse distribution of plasma density in the channel for different distances z measured in Rayleigh lengths. Here Rayleigh length equals 10.

The self-guiding effect under the investigation is characterized by several remarkable features. First of all it concerns the form of the plasma filament. Due to the decrease of the intensity caused by the wave leakage, the area of the cross-section, occupied by the field capable of strong ionization, is gradually narrowing. As a result with receding from the boundary the plasma filament becomes thinner (see Fig. 3) and takes ultimately the form of a sharpened needle. Hence, the decrease of the energy transmitted through the nonlinear guide is accompanied by a narrowing of the channel itself, which is rather unusual for self-guiding waves.

Another important feature that can be proved directly in an experimental observation is a specific transverse distribution of the frequency shift acquired by the guided wave. It is well known that any group element of the wave producing ionization experiences a blue shift of the frequency.[8,9] In the central part of the channel (near the axis), a strong blue shift is acquired only at the leading front of the pulse due to gas ionization to the saturation level whereas the rest major part of the pulse propagates at a fixed frequency in the pre-formed plasma. On the contrary, at the periphery of the channel the radiation increases the frequency due to a gradual ionization over the time of the full pulse duration. If one evaluates the frequency shift averaged over the pulse at different positions from the axis, a frequency gradient directed toward the channel periphery will be readily seen. This effect is even more pronounced if we take into account that short-scale (i.e. high frequency) components trapped in the channel have a greater leakage coefficient at reflection from the sharp plasma boundary as compared to the large-scale transverse components. These two factors results in a remarkable effect: the radiation frequency averaged over the pulse at the axis of the channel is decreasing in spite of the strong ionization whereas the radiation propagating at the channel periphery is essentially blue-shifted, as seen in figure 4, where the radiation frequency shift averaged over the whole profile at each cross-section point are shown for different distances from the gas boundary.

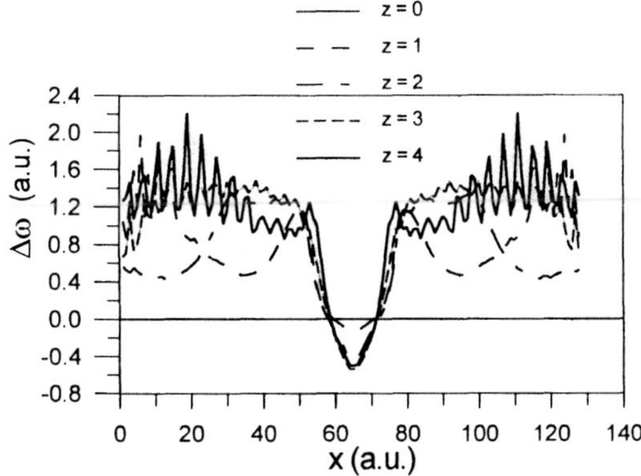

Figure 4. Radiation frequency shift averaged over the temporal pulse profile at different z measured in Rayleigh lengths. Here parameters for calculation are taken as $\gamma = 1.2 \cdot 10^4/\tau_0$, $N_O = 9.5 \cdot (k_O a_F)^2 N_{cr}$, E_0 =0.4, others are the same as in Fig. 1 – 3.

CONCLUSION

We have proposed and described an unusual opportunity for self-guiding of an ultrashort laser pulse without a focusing nonlinearity. This effect is attributed to the specific properties of ionization nonlinearity in the superstrong optical fields and is promising for creating long distance plasma channels for applications.

ACKNOWLEDGMENT

The authors acknowledge support for this work from the Russian Basic Research Foundation.

REFERENCES

1. A. Sullivan, H. Hamster, S.P.Gordon et al., Propagation of intense, ultrashort laser pulses in plasmas, *Opt.Lett.*, 19:1544 (1994).
2. A.B. Borisov et al., *JOSA B*, 11:1941 (1994).
3. A. Braun G. Korn, X. Liu et al.,Self-channeling of high-peak-power femtosecond laser pulses in air, *Opt.Lett.*, 20:73 (1995).
4. E.T.J. Nibbering, P.F. Curley, G. Grillon et al., Conical emission from self-guided femtosecond pulses in air, *Opt.Lett.*, 21:62 (1996).
5. D. Anderson, A.V. Kim, M. Lisak et al., Self-sustained plasma waveguide structures produced by ionizing laser radiation in a dense gas, *Phys.Rev.E* 52:4564 (1995).
6. A.A. Babine, A.V. Kim, A.M. Kiselev et al., Interaction of superstrong laser fields with matter: effects and applications, *Izv.VUSov Radiofizika*, 39:713 (1996).
7. N.B. Delone and V.P. Krainov, Multiphoton Processes in Atoms, Springer, Berlin (1994).
8. W.M. Wood, G.B. Focht, and M.C.Downer, Tight focusing and blue shifting of millijoule femtosecond pulses from a conical axicon amplifier, *Opt. Lett.*, 13:984 (1988).
9. V.B. Gildenburg, A.V. Kim, and A.M. Sergeev, Possibility of sharp increase in the frequency of the radiation of ionizing laser pulse in gas, *JETP Lett.*, 51:104 (1990).

CLASSICAL THEORY OF NONLINEAR COMPTON SCATTERING AND THREE-DIMENSIONAL PONDEROMOTIVE SCATTERING

F.V. Hartemann[1,2], E.C. Landahl[1,2], J.R. Van Meter[1,2], A.L. Troha[1,2], A.K. Kerman[1,3], and N.C.Luhmann, Jr.[2]

[1]Institute for Laser Science and Applications, LLNL, Livermore, CA 94550
[2]Department of Applied Science, UC, Davis, CA 95616
[3]Department of Physics, MIT, Cambridge, MA 02139

INTRODUCTION

The physics of laser-electron interactions changes dramatically at relativistic intensities, where the transverse momentum of the charge, measured in electron units, exceeds one. Three fundamental processes are known to occur in this regime : nonlinear ponderomotive[1] and Compton[2,3] scattering, and high intensity Kapitza-Dirac scattering[4,5]. These vacuum interactions correspond to the following geometries : collinear propagation, head-on collision, and electron diffraction in a laser standing wave, respectively.

Detailed knowledge of the three-dimensional (3D) electromagnetic field distribution of the focusing laser wave is required to study these phenomena and properly model experimental results. For example, two ultrahigh intensity relativistic electron scattering experiments are currently underway at SLAC and CEA. In the first case, nonlinear (multiphoton) Compton backscattering is investigated using the SLAC 50 GeV beam and a tightly focused TW-class laser[6] ; at CEA, low energy electrons are accelerated by a TW laser[7]. In both instances, the 3D nature of the focused laser pulse is an essential feature of the experiment and must be described accurately to interpret the resulting data.

The purpose of this paper is to present a comprehensive theoretical and numerical description of the relativistic dynamics of a charged particle interacting with an external electromagnetic field propagating *in vacuo*. To accurately describe the focusing and diffraction of the drive laser wave in vacuum, the paraxial propagator approach is used, where the mass shell condition (vacuum dispersion relation) is approximated by a quadratic Taylor expansion in the 4-wavenumber[8]. This approach is extremely accurate for any realizable laser focus, and yields analytical expressions for the fields. In addition, the gauge condition is satisfied exactly everywhere, thus yielding a proper treatment of the axial electromagnetic field components due to wavefront curvature. The electron phase is used as the independent variable, thus allowing for particle tracking over an arbitrarily large number of Rayleigh

Applications of High Field and Short Wavelength Sources
Edited by DiMauro *et al.*, Plenum Press, New York, 1998

ranges, independently of the nonlinear slippage and relativistic Doppler shift due to radiation pressure[3]. The scattering of radiation by the accelerated electron is also reviewed, in the case of plane electromagnetic waves of arbitrary intensity, and radiative corrections are briefly discussed within the context of the classical Dirac-Lorentz equation[10].

PLANE WAVE DYNAMICS, CANONICAL INVARIANTS

The electron 4-velocity and 4-momentum are defined as

$$u_\mu = \frac{dx_\mu}{d\tau} \equiv \gamma\,(1, \beta) \quad , \quad p_\mu = m_0 c\, u_\mu \quad , \quad u_\mu u^\mu = -1, \tag{1}$$

where τ is the proper time along the dimensionless electron world line $x_\mu(\tau)$, where distances are measured in units of the laser wavelength, $\frac{c}{\omega_0}$. In the absence of radiative corrections[10], the energy-momentum transfer equations are driven by the Lorentz force

$$\frac{du_\mu}{d\tau} = -\,(\partial_\mu A_\nu - \partial_\nu A_\mu)\, u^\nu. \tag{2}$$

For plane waves, the 4-vector potential of the laser wave is given in units of $\frac{m_0 c}{e}$ by

$$A_\mu(\phi) = [0, \mathbf{A}(\phi)] \quad , \quad \phi = k_\mu x^\mu(\tau), \tag{3}$$

where ϕ is the relativistically invariant phase of the traveling wave along the electron trajectory. Choosing $k_\mu \equiv (1, 0, 0, 1)$, with the wave propagating in the z-direction, we have

$$\frac{d\phi}{d\tau} = \gamma - u_z = q, \tag{4}$$

which defines the light-cone variable q, and the 4-momentum transfer equations read

$$\frac{d\mathbf{u}_\perp}{d\tau} = q\, \frac{d\mathbf{A}_\perp}{d\phi}(\phi), \tag{5}$$

$$\frac{du_z}{d\tau} = \frac{d\gamma}{d\tau} = \mathbf{u}_\perp \bullet \frac{d\mathbf{A}_\perp}{d\phi}(\phi). \tag{6}$$

Equation (6) shows that q is invariant : $q = q_0 = \gamma_0\,(1 - \beta_0)$; additionally, Eq. (5) is readily integrated to yield the well-known transverse momentum invariant

$$\mathbf{u}_\perp(\tau) = \mathbf{A}_\perp(\phi), \tag{7}$$

and the energy and axial momentum are immediately obtained using the fact that the 4-velocity is a unit 4-vector :

$$u_z(\tau) = \gamma_0 \left[\beta_0 + \frac{\mathbf{A}_\perp^2(\phi)}{2}\,(1 + \beta_0) \right], \tag{8}$$

$$\gamma(\tau) = \gamma_0 \left[1 + \frac{A_\perp^2(\phi)}{2} (1 + \beta_0) \right]. \tag{9}$$

The quadratic dependence of the energy and axial momentum on the 4-vector potential, measured in electron units, distinguishes the relativistic scattering regime, where $A_\perp \geq 1$. In this regime, the ponderomotive force dominates the electron dynamics, yielding nonlinear slippage and Doppler shifts [1,3]. Equation (9) also provides a scaling for the maximum energy in a plane wave : $\frac{\gamma^*}{\gamma_0} \cong A_\perp^2$, for relativistic electrons. Finally, the electron position is given by $\mathbf{x}(\phi) = \frac{1}{q_0} \int_{-\infty}^{\phi} \mathbf{u}(\psi) \, d\psi$.

COMPTON BACKSCATTERING, NONLINEAR SPECTRA

We now focus on the spectral characteristics of the radiation scattered by the accelerated charge. As discussed by Jackson[11], the distribution of energy radiated per unit solid angle, per unit frequency can be derived by considering the instantaneous radiated power, as described by the Larmor formula, and applying Parsival's theorem to obtain

$$\frac{d^2N(\varpi, \mathbf{n})}{d\varpi \, d\Omega} = \frac{\alpha}{4\pi^2} \, \varpi \left| \int_{-\infty}^{+\infty} \mathbf{n} \times (\mathbf{n} \times \boldsymbol{\beta}) \, \exp\left[i \, (\varpi t - \mathbf{n} \bullet \mathbf{x}) \right] dt \right|^2, \tag{10}$$

where $\varpi = \frac{\omega}{\omega_0}$ is the normalized frequency, and $\alpha = \frac{1}{137.036}$ is the fine structure constant. The quantity in Eq. (10) corresponds to the average radiated photon number.

Using the phase as the independent variable, we now have

$$\frac{d^2N(\varpi, \mathbf{n})}{d\varpi \, d\Omega} = \frac{\alpha}{4\pi^2} \, \frac{\varpi}{q_0^2} \left| \int_{-\infty}^{+\infty} \mathbf{n} \times [\mathbf{n} \times \mathbf{u}(\phi)] \, \exp\left\{ i \, \varpi \, [\phi + z(\phi) - \mathbf{n} \bullet \mathbf{x}(\phi)] \right\} d\phi \right|^2. \tag{11}$$

Here, we have used the plane wave invariance of $q = q_0$.

The most interesting case is the backscattered radiation, where most of the power is radiated and where we obtain the maximum relativistic Doppler upshift. In this case,

$$\mathbf{n} = -\hat{z}, \quad z(\phi) - \mathbf{n} \bullet \mathbf{x}(\phi) = 2 \, z(\phi), \quad \hat{z} \times [\hat{z} \times \mathbf{u}(\phi)] = -\mathbf{A}_\perp(\phi),$$

and Eq. (11) can now be recast in a manifestly covariant way, to read

$$\frac{d^2N(\varpi, \mathbf{n})}{d\varpi \, d\Omega} = \frac{\alpha}{4\pi^2} \, \chi \left| \int_{-\infty}^{+\infty} \mathbf{A}_\perp(\phi) \, \exp\left\{ i \, \chi \left[\phi + \int_{-\infty}^{\phi} A_\perp^2(\psi) \, d\psi \right] \right\} d\phi \right|^2, \tag{12}$$

where we have introduced the normalized Doppler-shifted frequency $\chi = \varpi \left(\dfrac{1 + \beta_0}{1 - \beta_0} \right)$.

The functional dependence of the spectrum is now independent of β_0, which only sets the frequency scale. This fact is not surprising, as it results directly from covariance : by changing the reference frame in which the scattering is viewed, one can vary the sign of β_0 and go continuously from the FEL geometry to the laser acceleration geometry. For the FEL, the laser frequency is Doppler-upshifted, while it is downshifted in the second case. In both cases, the normalized vector potential and the average photon number are conserved as they are Lorentz invariants.

Having derived the expression for the nonlinear backscattered light spectrum for arbitrary polarizations and intensities, we will now focus on one important case : circularly polarized plane waves. In this case, the dimensionless 4-vector potential can be expressed as $\mathbf{A}_\perp(\phi) = A_0\, g(\phi)\, [\hat{x}\, \sin\phi + \hat{y}\, \cos\phi]$, which implies that the magnitude of the 4-vector potential varies adiabatically as the pulse intensity envelope : $A_\mu A^\mu = \mathbf{A}_\perp^2(\phi) = A_0^2\, g^2(\phi)$. A simple physical model for the pulse envelope is given by a hyperbolic secant, namely $g(\phi) = \cosh^{-1}\left(\dfrac{\phi}{\eta}\right)$, so that the electron's axial position can be determined analytically. We then have

$$\int_{-\infty}^{\phi} \mathbf{A}_\perp^2(\psi)\, d\psi = \int_{-\infty}^{\phi} \frac{A_0^2}{\cosh^2\left(\frac{\psi}{\eta}\right)}\, d\psi = A_0^2\, \eta\left[1 + \tanh\left(\frac{\psi}{\eta}\right)\right], \tag{13}$$

and the nonlinear backscattered spectrum is now proportional to

$$\chi \left| A_0\, e^{i\,\chi\, A_0^2\, \eta} \int_{-\infty}^{+\infty} \frac{\hat{x}\,\sin\phi + \hat{y}\,\cos\phi}{\cosh(\phi/\eta)}\, \exp\left\{i\,\chi\,\eta\left[\frac{\phi}{\eta} + A_0^2\,\tanh\left(\frac{\phi}{\eta}\right)\right]\right\}\, d\phi \right|^2. \tag{14}$$

This Fourier transform can be evaluated analytically by performing two changes of variable[3], with the result that

$$\frac{d^2 N(\varpi, -\hat{z})}{d\varpi\, d\Omega} = \frac{\alpha}{8}\, A_0^2\, \eta^2\, \chi \left\{ \left| \frac{\Phi\left(\mu_-, 1, 2\, i\, A_0^2\, \chi\, \eta\right)}{\cosh\left[\frac{\pi}{2}\, \eta\, (\chi - 1)\right]} \right|^2 + \left| \frac{\Phi\left(\mu_+, 1, 2\, i\, A_0^2\, \chi\, \eta\right)}{\cosh\left[\frac{\pi}{2}\, \eta\, (\chi + 1)\right]} \right|^2 \right\}. \tag{15}$$

Here, Φ is the degenerate (confluent) hypergeometric function, and $\mu_\pm = \frac{1}{2}\,[1 + i\,\eta\,(\chi \pm 1)]$. The behavior of the nonlinear spectral function is illustrated in Fig. 1 for $\eta = 5$, and different values of A_0. Within this context, the onset of nonlinear relativistic spectral effects corresponds to a situation where the electron phase ϕ and the axial Doppler shift $\int_{-\infty}^{\phi} \mathbf{A}_\perp^2(\psi)\, d\psi$ become comparable because of the intense radiation pressure of the drive laser pulse.

For linear polarization, the nonlinear effects are even stronger because there is an additional modulation of the axial velocity at the second harmonic of the laser, resulting from the $\mathbf{v} \times \mathbf{B}$ ponderomotive (radiation pressure) force.

THREE-DIMENSIONAL DYNAMICS, PONDEROMOTIVE SCATTERING

The 3D behavior of the laser electromagnetic field propagating in vacuum is now described within the context of the paraxial propagator formalism. The wave equation governing the 4-vector potential is $[\partial_\nu\, \partial^\nu]\, A_\mu + \mu_0\, j_\mu = 0$, where we have introduced the 4-gradient operator, defined as $\partial_\mu \equiv \dfrac{\partial}{\partial x^\mu} \equiv \left(\dfrac{1}{c}\, \partial_t, \nabla\right)$, and the 4-current density $j_\mu \equiv (c\rho, \mathbf{j})$. In addition, it is important to note that the 4-potential must satisfy the Lorentz gauge condition $\partial_\mu\, A^\mu = 0$.

In vacuum (no sources), a general solution to the wave equation can be constructed as a Fourier superposition of wavepackets of the form

$$A_\mu(x_\nu) = \frac{1}{(2\pi)^2} \iiiint \widetilde{A}_\mu(k_\nu)\, \exp(i\, k_\nu\, x^\nu)\, d^4 k_\lambda, \tag{16}$$

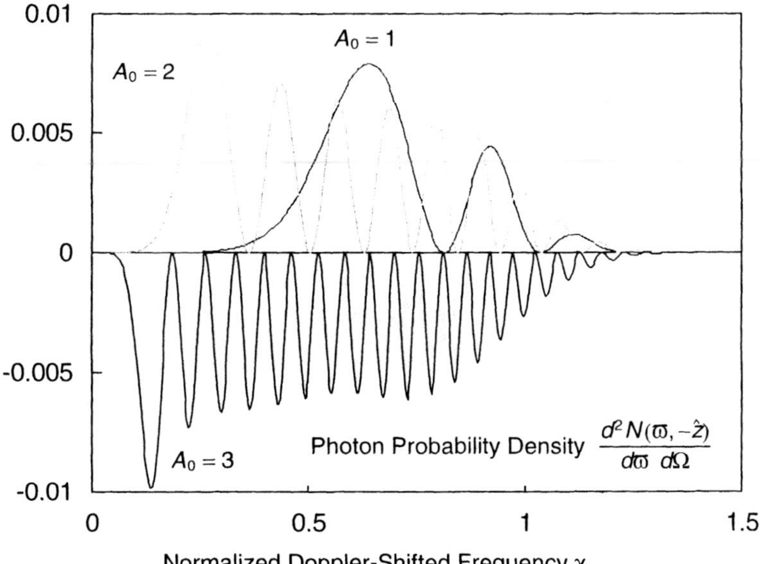

Fig. 1 Nonlinear Compton backscattered spectrum for a circularly polarized $cosh^{-2}$ envelope and different values of A_0 ; $\eta = 5$.

where the 4-wavenumber k_μ satisfies the vacuum dispersion relation, which is also the mass shell condition for the photon field : $\hbar^2\,(k_\mu\,k^\mu) = 0$.

When the laser pulse characteristics are defined at focus ($z = 0$), we can obtain the electromagnetic field distribution in any given z-plane by performing the following integral

$$A_\mu(x,y,z,t) = \frac{1}{(2\pi)^{3/2}} \iiint \tilde{A}_\mu(\mathbf{k}_\perp,\omega,z=0)\ \exp\left[i\left(\omega t - k_x x - k_y y - \sqrt{\frac{\omega^2}{c^2} - \mathbf{k}_\perp^2}\ z\right)\right] d^2\mathbf{k}_\perp\, d\omega,$$

$$(17)$$

where we have introduced the focal spectral density $\tilde{A}_\mu(\mathbf{k}_\perp,\omega,z=0)$.

The physics of this solution can be understood as follows : the temporal evolution of each wavepacket is described by the frequency spectrum, while the transverse profile of the laser wave is expressed as an integral over a continuous spectrum of transverse vacuum eigenmodes. The dispersion relation indicates how each transverse component of the wavepacket propagates, thus yielding wavefront curvature and diffraction.

In vacuum, the gauge condition reduces to $\nabla \bullet \mathbf{A} = 0$, and can be satisfied by requiring that $\mathbf{A} = \nabla \times \mathbf{G}$. For a linearly polarized Gaussian-elliptical focus, $\mathbf{G}$ takes the form

$$\mathbf{G}(x_\mu) = \hat{y}\, G_y(x_\mu) \quad,\quad G_y(x,y,z=0,t) = \frac{A_0}{k_0}\, \exp\left[-\left(\frac{x}{w_{0x}}\right)^2 - \left(\frac{y}{w_{0y}}\right)^2\right] g(t), \qquad (18)$$

where w_{0x} refers to the beam waist along the x-axis and w_{0y} refers to the beam waist along the y-axis, A_0 is the amplitude of the vector potential at focus, $k_0 = \dfrac{\omega_0}{c} = \dfrac{2\pi}{\lambda_0}$ corresponds to the central laser wavelength, and $g(t)$ is the temporal variation of the laser wave, which can be arbitrary. The propagation integral is now approximated by

$$G_y(x,y,z,t) = \frac{1}{(2\pi)^{3/2}} \iiint \tilde{G}_y(\mathbf{k}_\perp,z=0,\omega)\ \exp\left\{i\left[\omega t - k_x x - k_y y - \left(\frac{\omega}{c} - \frac{k_\perp^2}{2k_0}\right)z\right]\right\} d^2\mathbf{k}_\perp\, d\omega,$$

$$(19)$$

217

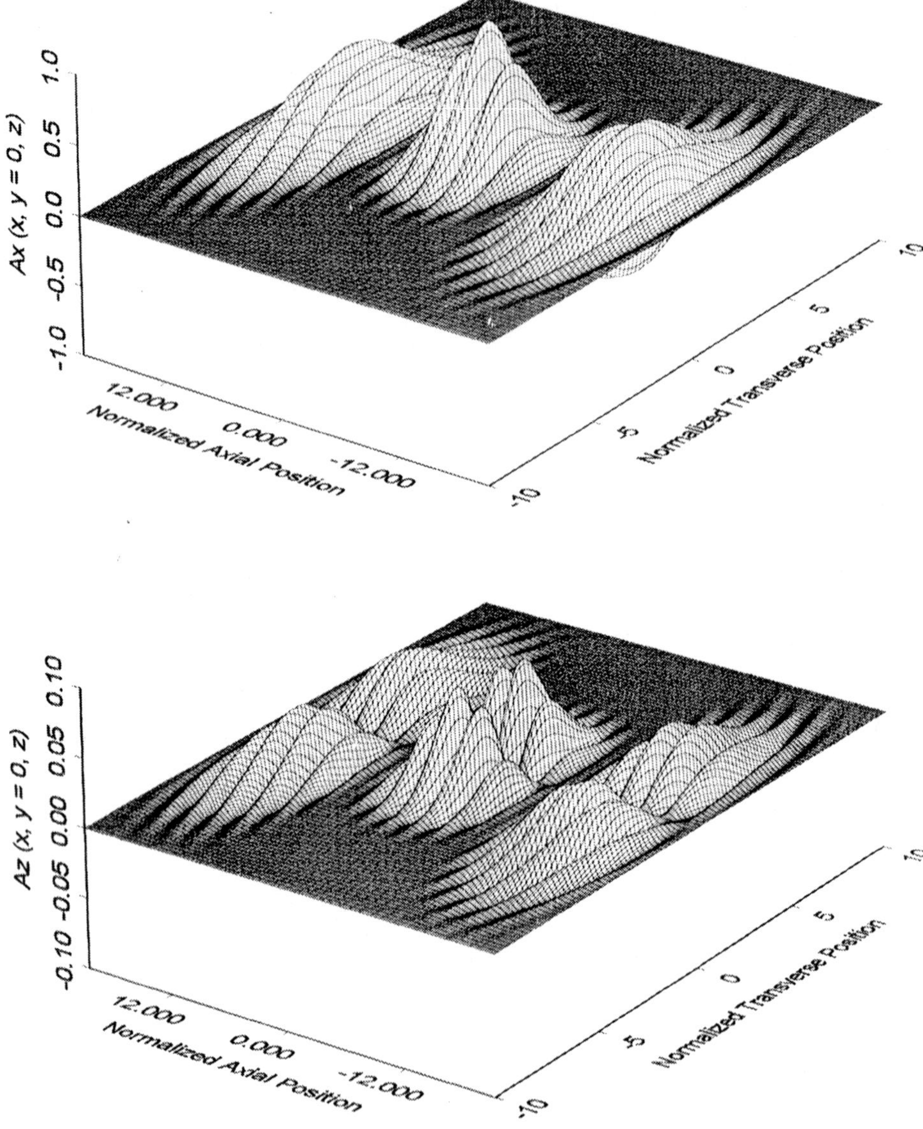

Fig. 2 Transverse and axial vector potential components, in the plane of polarization, at three different times. The pulses are 6 cycles long and f: 3.

where the square root factor has been Taylor expanded to second order around $\omega = \omega_0$ and $\mathbf{k}_\perp = 0$. It is clear that the exact and Taylor expanded axial phase differ only for large values of the transverse wavenumber, where the spectral density is vanishingly small. The integrals over $d^2\mathbf{k}_\perp$ are readily obtained, as they correspond to complex Gaussians :

$$G_y(x,y,z,t) = \frac{A_0}{k_0}\, g(\frac{\phi}{\omega_0}) \left[1 + \left(\frac{z}{z_{0x}}\right)^2\right]^{-\frac{1}{4}} \left[1 + \left(\frac{z}{z_{0y}}\right)^2\right]^{-\frac{1}{4}} \exp\left[-\left(\frac{x}{w_x}\right)^2 - \left(\frac{y}{w_y}\right)^2\right]$$

$$\times \exp\left\{ i\left[\frac{1}{2}\arctan\left(\frac{z}{z_{0x}}\right) + \frac{1}{2}\arctan\left(\frac{z}{z_{0y}}\right) - \frac{z}{z_{0x}}\left(\frac{x}{w_x}\right)^2 - \frac{z}{z_{0y}}\left(\frac{y}{w_y}\right)^2\right]\right\}, \qquad (20)$$

where $w_{x,y}(z) = w_{0x,y}\sqrt{1 + \left(\frac{z}{z_{0x,y}}\right)^2}$ and $z_{0x,y} = \frac{1}{2}k_0 w_{0x,y}^2$ is the Rayleigh range for each f.

The vector potential of the focusing wave is shown in Fig. 2. The fields are then derived from this vector potential. The algorithm developed to model the dynamics of a charge interacting with the 3D laser fields employs the second order Runge-Kutta method and uses the axial electron phase as the integration variable to handle the nonlinear slippage and the relativistic Doppler shift.

For a laser focus with extremely large f-numbers, excellent agreement is found between the numerical results and the theoretical analytic expressions obtained for plane wave dynamics ; for smaller values of f, scattering is obtained. To obtain efficient scattering, the electron must be seeded far from focus, so that by the time it has slipped into the nonlinear temporal phase of the pulse, the focus is reached and the electron interacts with the spatio-temporal maximum of the laser wave. High energy scattering occurs for intermediate values of the f-numbers : for low values, the focus is too tight and the electron scatters away too early ; for very large values, we recover the plane wave interaction. The fact that, for higher initial injection energies, efficient scattering requires larger values of the f-number is not

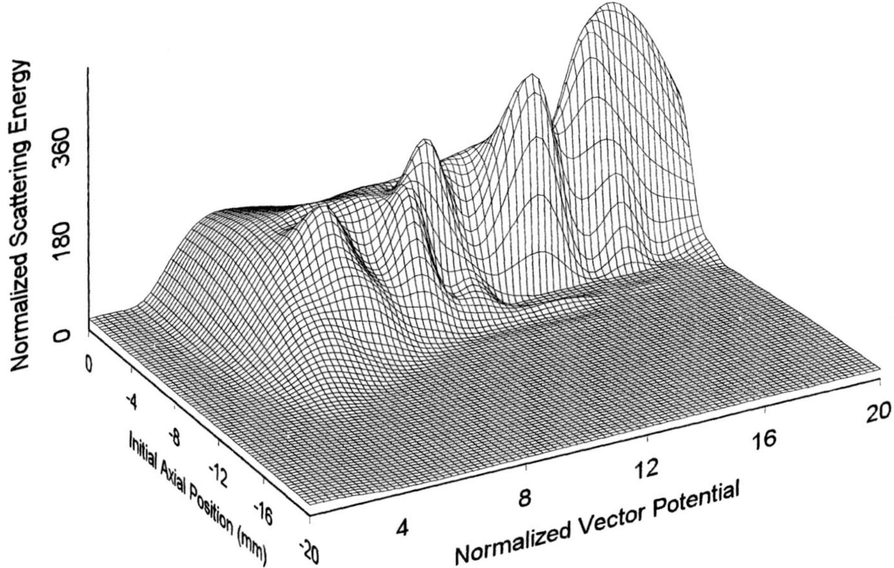

Fig. 3 Scan of A_0 and initial z, for a constant pulse energy of 20 J. $\lambda_0 = 1$ μm, FWHM 20 fs, initial energy 10 MeV, $x_i = y_i = 0$.

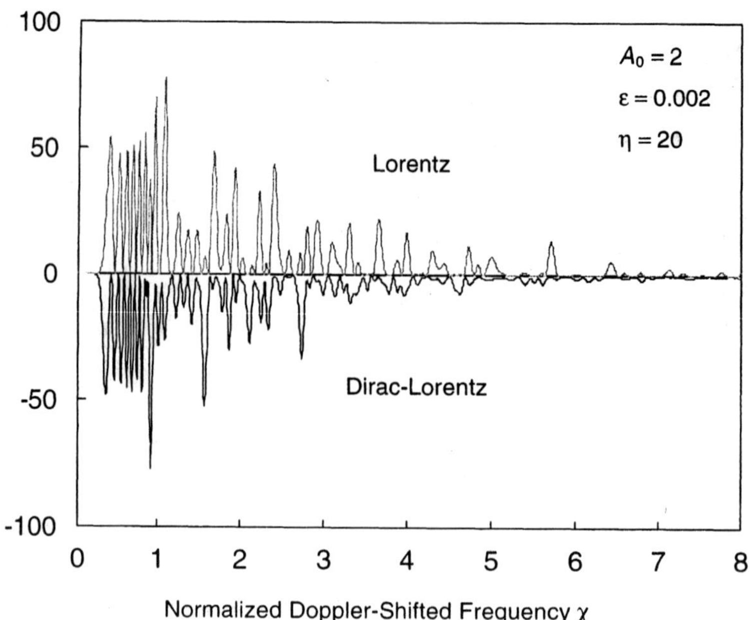

Normalized Doppler-Shifted Frequency χ

Fig. 4 Lorentz and Dirac-Lorentz backscattered spectra for a linearly polarized pulse. The electron initial energy is 48 GeV in the lab frame, where $\lambda = 1$ μm.

surprising, since the transverse electron excursion scales like $q^{-1} \cong q_0^{-1} = \gamma_0(1 + \beta_0)$. The largest scattering energies are achieved for $f_x > f_y$, which is due to the linear polarization.

To assess the feasibility of laser acceleration based on this scattering process, the total energy in the laser pulse is obtained by integrating the Poynting vector flux over the focal spot and the pulse duration to obtain

$$\frac{W}{m_0 c^2} = \frac{3\pi}{32} A_0^2 \frac{w_{0x}}{\lambda_0} \frac{w_{0y}}{\lambda_0} \frac{c \, \Delta t}{r_0}, \qquad (21)$$

where $r_0 = 2.81785 \times 10^{-15}$ m is the classical electron radius. We then study a cylindrical focus ($w_{0x} = w_{0y} = w_0$), and maintain the pulse FWHM at 20 fs, and the product $A_0 \, w_0$ at 250 μm (constant energy of 20 J), while varying A_0 between 1 and 20, and varying the injection position between -20 mm and focus. The results are shown in Fig. 3, and clearly indicate the existence of an optimum combination of f-number and laser intensity for high energy scattering, approximately obtained for a normalized vector potential of 17.5 and f: 23. Here, there is a sharply defined acceptance range in the initial position, and the scattering energy reaches 0.25 GeV. The acceleration process occurs over 3 mm, yielding a gradient of 85 GeV/m. These laser parameters correspond to the next generation of CPA lasers[9].

RADIATIVE CORRECTIONS, DIRAC-LORENTZ ELECTRODYNAMICS

The Dirac-Lorentz equation[10] describes the covariant dynamics of a classical point electron, including the radiation reaction effects due to the electron self-interaction. In the case of nonlinear Compton scattering, the Dirac-Lorentz equation can be given as

$$a_\mu = \frac{du_\mu}{d\tau} = L_\mu + \tau_0 \left[\frac{da_\mu}{d\tau} - u_\mu \, (a_\nu \, a^\nu) \right] \quad , \quad \mathbf{L}_\perp = q \, \mathbf{E}_\perp \quad , \quad L_z = L_0 = \mathbf{u}_\perp \bullet \mathbf{E}_\perp, \quad (22)$$

where we recognize the light-cone variable $q = \gamma - u_z$, and the laser transverse electric field. $\tau_0 = 2r_0/3c = 0.626 \; 10^{-23}$ s is the Compton time-scale. Subtracting the axial component of Eq. (22) from the temporal component, we obtain an equation governing the evolution of q,

$$\frac{dq}{d\tau} = \tau_0 \left[\frac{d^2 q}{d\tau^2} - q \, (a_\nu \, a^\nu) \right]. \quad (23)$$

Noting that $\mathbf{E}_\perp = q \, \dfrac{d\mathbf{A}_\perp}{d\phi}$, we also obtain an equation governing the evolution of the canonical momentum :

$$\frac{d}{d\tau}(\mathbf{u}_\perp - \mathbf{A}_\perp) = \tau_0 \left[\frac{d^2 \mathbf{u}_\perp}{d\tau^2} - \mathbf{u}_\perp \, (a_\nu \, a^\nu) \right]. \quad (24)$$

Introducing the small parameter $\varepsilon = \omega_0 \, \tau_0$, which measures the Doppler-shifted laser wavelength in units of r_0, and using ϕ as the independent variable, Eq. (23) now reads

$$\frac{dq}{d\tau} = \varepsilon \left[\frac{d^2}{d\phi^2} \left(\frac{q^2}{2} \right) - q^2 \left(\frac{du_\mu}{d\phi} \, \frac{du^\mu}{d\phi} \right) \right]. \quad (25)$$

Since the right-hand side of Eq. (25) is at least of order ε, we can replace the terms in the brackets by their zeroth-order (Lorentz dynamics) approximation ; in this case, we obtain a simple differential equation for the light-cone variable perturbation

$$\frac{d}{d\phi} \left[\frac{1}{q(\phi)} \right] \cong \varepsilon \, A_0^2 \, g^2(\phi), \quad (26)$$

where we recognize the envelope of the circularly polarized laser pulse. Equation (26) can easily be integrated to yield

$$\frac{1}{q(\phi)} = \frac{1}{q_0} + \varepsilon \, A_0^2 \int_{-\infty}^{\phi} g^2(\psi) \, d\psi. \quad (27)$$

This equation describes the electron recoil for beam parameters similar to those of SLAC[6]. It is clear that at sufficient intensities, the relative radiative energy loss becomes significant.

Finally, the Dirac-Lorentz equation can be integrated backward in time to avoid runaways due to the electromagnetic mass renormalization[2] and the Compton backscattered spectrum is obtained by evaluating

$$\frac{d^2 N(\varpi, -\hat{z})}{d\varpi \, d\Omega} = \frac{\alpha}{4\pi^2} \, \varpi \left| \int_{-\infty}^{+\infty} \frac{\mathbf{u}_\perp(\phi)}{q(\phi)} \, \exp \left\{ i \, \varpi \left[\phi + 2 \int_{-\infty}^{\phi} \frac{u_z(\psi)}{q(\psi)} \, d\psi \right] \right\} \right|^2. \quad (28)$$

For the SLAC beam parameters, and a TW-class laser, the result is shown in Fig. 4. At this intensity $(0.22 \; \mathrm{TW}/\mu\mathrm{m}^2)$, the nonlinear Doppler shift yields complex spectra for both Lorentz and Dirac-Lorentz dynamics, but the lines are shifted, in similarity with the Lamb shift, and damped at high frequencies.

ACKNOWLEDGMENTS

This work is supported in part by DoD/AFOSR (MURI) F49620-95-1-0253, AFOSR (ATRI) F30602-94-2-001, ARO DAAHO4-95-1-0336, and LLNL/LDRD DoE W-7405-ENG-48. We wish to thank Prof. H.A. Baldis and D.T. Santa Maria for very stimulating discussions.

REFERENCES

1. F.V. Hartemann, *et al*, Phys. Rev. **E51**, 4833 (1995).
2. F.V. Hartemann and A.K. Kerman, Phys. Rev. Lett. **76**, 627 (1996).
3. F.V. Hartemann, *et al*, Phys. Rev. **E54**, 2956 (1996).
4. P.L. Kapitza and P.A.M. Dirac, Proc. Cambridge Philos. Soc. **29**, 297 (1933).
5. P.H. Bucksbaum, *et al*, Phys. Rev. **A41**, 4119 (1990).
6. C. Bula, *et al*, Phys. Rev. Lett. **76**, 3116 (1996).
7. G. Malka, E. Levebvre, and J.L. Miquel, Phys. Rev. Lett. **78**, 3314 (1997).
8. P.W. Milonni and J.H. Eberly, *Lasers* (Wiley Interscience, New York, NY 1988), Chap. 14.
9. M.D. Perry and G. Mourou, Science **264**, 917 (1994).
10. P.A.M. Dirac, Proc. R. Soc. London Ser. **A167**, 148 (1938).
11. J.D. Jackson, *Classical Electrodynamics* (2nd edition, John Wiley and Sons, New York, NY, 1975).

THE MOLECULAR HYDROGEN B-STATE IN AN INTENSE LIGHT FIELD

H. Rottke, J. Ludwig, and W. Sandner

Max-Born-Insitut
P.O. Box 1107
D-12474 Berlin
Germany

INTRODUCTION

Molecular hydrogen is a preferred candidate for the investigation of excitation and decay of molecular systems in an intense light field (Zavriyev et al., 1990; Yang et al., 1991; Rottke et al., 1996; Shao et al., 1996; Walsh et al., 1997; Ludwig et al., 1997). In many respects H_2 behaves similar to more complex diatomics. Compared to these molecules the only two electrons which are involved in the interaction of light with H_2 simplify a theoretical approach and therefore a detailed understanding of the influence of light on molecular decay processes.

Up to date experiments which expose H_2 to intense light pulses used $X\,^1\Sigma_g^+(v=0)$ thermally populated rotational levels as initial states. We have started experiments which aim at understanding the role this initial state plays in the interaction. This is done by preparing states prior to interaction with the intense pulse. By suitably choosing the initial state we also expect that it will become possible to influence the main decay pathway. In this way specific decay mechanisms can be studied in more detail over a wider range of light intensities.

Here we present a first experiment where we specifically look into multiphoton dissociation of the first electronically excited state $B\,^1\Sigma_u^+$ in infrared (IR) laser pulses in an intensity range intermediate between perturbative and non-perturbative ($2 \times 10^{10} - 2 \times 10^{12}\,\mathrm{W/cm^2}$). Single $v=3$ ro-vibrational states are prepared by one photon excitation from the electronic ground state by short wavelength vacuum ultraviolet light.

EXPERIMENT

Tunable 106 nm vacuum ultraviolet (VUV) radiation necessary to excite the molecular hydrogen $B - X(3,0)$ band is generated by phase matched frequency tripling with Xenon as nonlinear medium. The ultraviolet input radiation for the conversion process

Applications of High Field and Short Wavelength Sources
Edited by DiMauro *et al.*, Plenum Press, New York, 1998

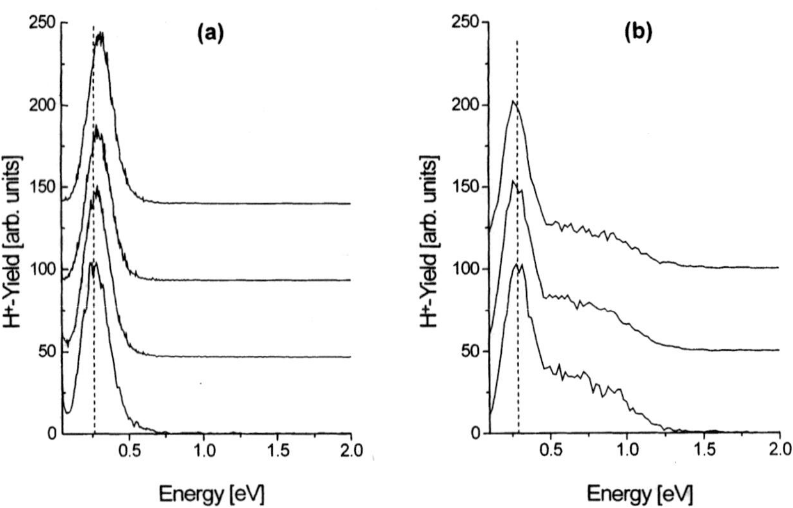

Figure 1. Kinetic energy distribution of H^+ ions from multiphoton dissociation and ionization of H_2 by (a) Nd:YAG and (b) Ti:Sapphire IR laser pulses. Initial states are (a) $B\,^1\Sigma_u^+(v=3, J=1)$ and (b) $B\,^1\Sigma_u^+(v=3, J=2)$. From bottom to top the light intensities are (a): 2.5×10^{10}, 4×10^{10}, 6×10^{10}, and $7 \times 10^{11}\,\mathrm{W/cm^2}$; (b) 4×10^{11}, 7×10^{11}, and $2 \times 10^{12}\,\mathrm{W/cm^2}$. The dashed lines indicate the respective centre of the pronounced low energy yield maximum at the lowest light intensity.

at 318 nm is generated by frequency doubling the output of a conventional Nd:YAG laser pumped pulsed dye laser operating at 636 nm (pulsewidth ≈ 5 nsec). With this setup we generate VUV radiation with a bandwidth of 1.7 cm^{-1} and $\approx 10^9$ photons per pulse, which can be tuned over the complete $B - X(3,0)$ absorbtion band.

Two IR light sources were applied to drive high intensity multiphoton transitions from the B-state, a Q-switched Nd:YAG and an amplified Kerr-lens mode locked Ti:Sapphire laser system. The Nd:YAG laser delivered pulses with ≈ 8 nsec pulsewidth at 1064 nm. The pulses of the Ti:Sapphire laser at 790 nm had a width of ≈ 0.7 psec. The light beams of both IR lasers were focused to reach intensities adjustable between $2 \times 10^{10} - 2 \times 10^{12}\,\mathrm{W/cm^2}$ in the interaction region with the hydrogen gas where they cross the VUV preparation light beam at right angles.

The interaction point of the laser radiation with the hydrogen molecules is located in an ultrahigh vacuum chamber which can be evacuated to a base pressure of 10^{-10} Torr. During runs hydrogen is introduced with a precision leak valve up to 10^{-5} Torr. The kinetic energy distribution of H^+ ions created in the multiphoton excitation process of B-state hydrogen molecules is measured by a time of flight spectrometer. The energy resolution of this device is presently limited by the thermal kinetic energy distribution in the room temperature gas sample we used. Further details of this experimental setup can be found in Ludwig et al. (1997) and Rottke et al. (1997).

RESULTS AND DISCUSSION

Using both infrared sources we measured the kinetic energy distribution of H^+ ions created by multiphoton absorption with selected $B\,^1\Sigma_u^+(v=3, J)$ ro-vibrational levels as initial state. Representative photoion spectra are shown in figure 1. 1a was taken with the long Nd:YAG laser IR pulses with $J=1$ as initial rotational state. We prepared this state by absorption of one VUV photon on the R(0) line of the $B - X(3,0)$

224

band. From bottom to top the IR light intensities 2.5×10^{10}, 4×10^{10}, 6×10^{10}, and $7 \times 10^{11}\,\mathrm{W/cm^2}$ were applied. In a similar way 1b was taken with the short Ti:Sapphire laser pulses at 4×10^{11}, 7×10^{11}, and $2 \times 10^{12}\,\mathrm{W/cm^2}$. The initial state in this case has rotational quantum number $J = 2$ and is prepared by excitation on the R(1) line of the $B - X(3,0)$ band. All spectra were taken with the polarization of the VUV and IR light beams pointing along the direction of detection of the H^+ charged dissociation products. For display purposes a varying offset has been added to the different spectra. In every spectrum the H^+ yield reaches zero at 2 eV kinetic energy. The dashed lines in the figure indicate the respective centre of the main H^+ yield maximum at the lowest light intensity applied. It appears at 265 meV in figure 1a and at 290 meV in figure 1b. As can be seen the position of this maximum shifts slightly to increasing kinetic energies in figure 1a whereas it remains fixed in figure 1b. Independent of the IR light source the shape of the kinetic energy distributions does not change significantly with increasing light intensity. With Ti:Sapphire laser excitation the main H^+ yield maximum shows a high energy shoulder which extends up to ≈ 1.25 eV. The height of this shoulder relative to the height of the main yield maximum decreases with increasing light intensity from $\approx 4 : 10$ to $\approx 2.5 : 10$.

Besides the difference in wavelength the decisive difference between the Nd:YAG and Ti:Sapphire laser pulses is their width compared to the width of the VUV preparation light pulses. The Nd:YAG IR and the VUV pulses are of similar width. Since they interact simultaneously with the H_2 gas sample B-state molecules are continuously supplied over the whole duration of the IR pulse. This means that there are certainly molecules in the B-state which interact with the IR light at the highest intensity reached in each pulse. Since the Ti:Sapphire pulse is much shorter than the VUV pulse no molecules are supplied to the B-state during application of this IR pulse. This may lead to a depopulation of the B-state before the IR pulse maximum is reached. The only low sensitivity of the kinetic energy distribution on an increase in light intensity seems to indicate that the main dissociation processes of H_2 we observed proceed at a moderate light intensity in this case.

The widths of the main low energy yield maxima in figure 1a and b are determined by the room temperature thermal kinetic energy distribution in the H_2 gas sample. Ionic products contributing to these peaks therefore originate in an H_2 dissociation process which gives rise to products with sharply defined kinetic energy in the H_2 centre of mass frame . These maxima of the yield show the present limit in the energy resolution of the experiment. Therefore the nature of the high energy shoulder of the peaks in figure 1b cannot unambiguously be determined. Either the corresponding dissociation mechanism creates products with a continuous kinetic energy distribution in the H_2 centre of mass frame, or a distribution of discrete product energies in the CM frame, broadened by the H_2 thermal kinetic energy distribution, gives rise to the observed unstructured shoulder in the laboratory frame of reference.

To discuss the dissociation mechanisms which can give rise to the observed H^+ kinetic energy distributions we show in figure 2 an H_2 potential energy diagram. It includes only potentials of electronic states which are expected to be important to this experiment. The energy in the molecule after absorption of Nd:YAG and Ti:Sapphire IR photons is given by the full and dashed lines, respectively, on the energy axis. The common initial vibrational state is $B\,^1\Sigma_u^+(v = 3)$. There are several possible pathways for formation of the H^+ ions we observe:

$$H_2 \xrightarrow{n\hbar\omega} H(1s) + H(nl) \xrightarrow{m\hbar\omega} H(1s) + H^+ + e \tag{1}$$

$$H_2 \xrightarrow{n\hbar\omega} H_2^+ + e \xrightarrow{m\hbar\omega} H(1s) + H^+ + e \tag{2}$$

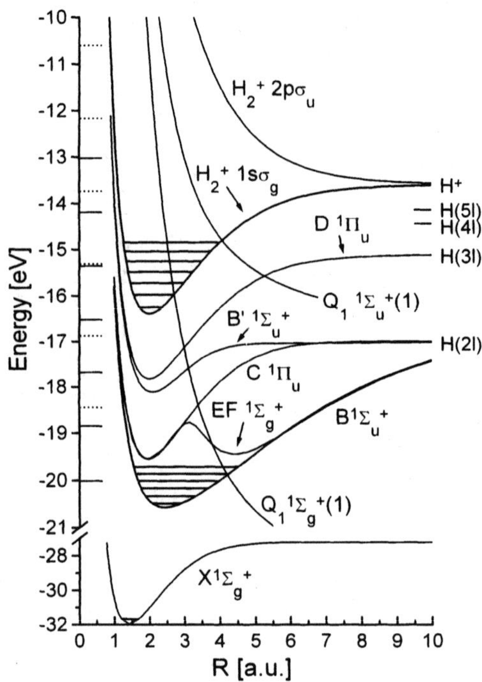

Figure 2. H_2 potential energy diagram for electronic states which are involved in the excitation mechanism. Vibrational levels are added to the H_2 ground and B-state and to the H_2^+ ground state. On the left hand axis the energy in the molecule after absorption of IR photons starting at $B\ ^1\Sigma_u^+(v = 3)$ is indicated (full lines: 1.165 eV Nd:YAG photons; dashed lines: 1.57 eV Ti:Sapphire photons). On the right hand side dissociation thresholds for $H(1s) + H(nl)$ product channels are given. The potentials are taken from Sharp (1971) and Guberman (1983).

$$H_2 \xrightarrow{n\hbar\omega} H(1s) + H^+ + e \tag{3}$$

Neutral dissociation after n-photon absorption can form one ground and one excited state H atom. Subsequently the excited state atom is ionized by absorption of further photons (process 1). This mechanism gives rise to narrow lines in a photoion spectrum which correspond to the different dissociation thresholds. The second possibility is multiphoton ionization (MPI) of H_2 and then photodissociation of H_2^+ after absorption of further photons (process 2). Or direct dissociative photoionization may result in H^+ ions (process 3). Processes 2 and 3 are expected to produce broad distributions of the H^+ kinetic energy. In process 2 MPI of H_2 will result in a distribution over many $H_2^+\ X\ ^2\Sigma_g^+$ vibrational states since the $H_2\ B\ ^1\Sigma_u^+$ and $H_2^+\ X\ ^2\Sigma_g^+$ potentials are very different in shape (figure 2). Through photodissociation of H_2^+ this distribution then appears in the H^+ kinetic energy spectrum in the form of closely spaced lines which we presently cannot resolve.

Processes 2 and 3 can therefore certainly be excluded as the origin for the narrow centre of mass H^+ kinetic energy distribution observed by excitation with the Nd:YAG pulses in figure 1a. The lowest order pathway which forms ions with the measured kinetic energy ($\approx$ 265 meV) via neutral dissociation is decay of H_2 after absorption of three infrared photons with formation of the products $H(1s) + H(2l)$ (process 1, figure 2). The angular momentum l of the excited hydrogen product atom may be 0 or 1. The

H(2l) atoms are then photoionized by absorption of three further infrared photons. The kinetic energy of the electrons formed this way is 96 meV. Alltogether six IR photons have to be absorbed for this minimum order process. It can be excluded that absorption of a VUV photon photoionizes the H(2l) atom. The VUV light intensity and H(2l) ionization cross sections are too small. The kinetic energy of the ions expected from this decay mechanism is 240 meV which is in good agreement with the measured 265 meV at the lowest light intensity. Dissociation of the molecule in this case is certainly strongly influenced by the lowest lying *gerade* doubly excited electronic state $Q_1\,^1\Sigma_g^+(1)$ which can be reached by three photon absorption (Rottke et al., 1997). The diabatic potential of this state which is repulsive is shown in figure 2.

The observed shift of the line centre in the spectra in figure 1a is in accordance with the proposed dissociation pathway. It may have two origins. Either the initial $B\,^1\Sigma_u^+(v = 3, J = 1)$ state or the final H(2l) product states may become AC-Stark shifted in the IR light pulse. Both shifts modify the dissociation threshold of the molecule. A line profile measurement of the rotational lines in the $B - X(3, 0)$ absorption band in the intense Nd:YAG light pulses (Ludwig et al., 1997a) showed that the shift of the B-state rotational levels is too small to be responsible for the shift observed in the photoion kinetic energy distributions (figure 1a). The AC-Stark shift of the H(2l) product states in the infrared laser field can be calculated in the intensity range of interest here. The calculation gives a lowering of the H(2l) energy levels with increasing IR light intensity. This gives rise to an increase in the kinetic energy of the dissociation products with increasing intensity of the 1064 nm light, in agreement with the observation (figure 1). Also the order of magnitude of the calculated AC-Stark shift agrees well with the observed shift of the peak position in the photoion spectra (Rottke et al. 1997).

Similar to the 1064 nm results the narrow ion yield peak (in the H_2 CM system) found at 290 meV in figure 1b is expected to have its origin in neutral dissociation of H_2 with subsequent photoionization of the hydrogen atom formed in an excited state. Based on an unperturbed initial and final dissociation product states it is not possible to identify a dissociation pathway which could be responsible for the observed kinetic energy of this structure. As already pointed out above the stability of the overall shape of the spectrum with increasing light intensity may indicate that the observed dissociation processes all proceed at moderate intensity (4×10^{11} W/cm^2) independent of the actual applied pulse peak intensity. Therefore already at moderate light intensity a substantial AC-Stark shift of the states involved has to be assumed.

There are only two reasonable neutral dissociation pathways which may form ions with 290 meV kinetic energy: four photon dissociation of $B\,^1\Sigma_u^+(v = 3, J = 2)$ with the formation of an H atom in the (4l) or in the (5l) state manifold with following one photon ionization of the excited atom. For unperturbed states H$^+$ ions with a kinetic energy of 355 meV are expected for the H(4l) product channel and 200 meV ions for H(5l). These kinetic energies differ from the measured one by at least 65 meV. The difference can only be accounted for by assuming an AC-Stark shift of the states involved. If mainly the final product state is shifted an increase of the dissociation threshold with increasing light intensity is expected. In this case the 290 meV ion yield maximum should correspond to dissociation with formation of H(1s) + H(4l). An AC-Stark shift of the $B\,^1\Sigma_u^+(v = 3, J = 2)$ initial rotational state will certainly increase the binding energy of this state. If this is the main shift H(1s) + H(5l) is probably the product channel responsible for the 290 meV ions.

Since intensity dependent AC-Stark shifts are involved and the observed 290 meV ions have a narrow intensity independent kinetic energy distribution in the H_2 CM frame one has to assume that dissociation proceeds at a well defined intensity in the light pulse. Two mechanisms may be imagined to be responsible for this behaviour. Photodissociation of $B\,^1\Sigma_u^+(v = 3, J = 2)$ may become saturated within a narrow intensity interval, or photodissociation proceeds preferably via resonance with a bound state of H_2 which is induced by the AC-Stark effect.

As figure 2 shows, the lowest lying *ungerade* doubly excited state $Q_1\,^1\Sigma_u^+(1)$ may be involved in four photon dissociation with formation of H(4l) or H(5l) products. The Franck-Condon factor for transitions to this state is certainly favourable. Dissociation can lead to population of atomic states with principal quantum number n $\geq$ 3 if the avoided curve crossings of $Q_1\,^1\Sigma_u^+(1)$ with the molecular Rydberg states are mainly traversed adiabatically.

The dissociation mechanisms described are lowest order pathways for formation of H^+ via neutral dissociation. Four photon dissociation can also populate atomic levels with principal quantum number n $\geq$ 6. One photon ionization of these products forms ions with kinetic energy smaller than 100 meV. In this energy range an increasing amount of H_2^+ contributes to the signal. Since we cannot discriminate between H_2^+ and H^+ we omitted this energy range in figure 1.

The high energy shoulder of the 290 meV peak in figure 2b may have contributions from two neutral dissociation processes. Four photon dissociation forming H(3l) creates ions with 685 meV kinetic energy. A second process needing the absorption of six 790 nm photons (three photon dissociation into H(1s) + H(2l) with three photon ionization of H(2l)) forms ions with 845 meV energy. For both pathways we assumed unperturbed molecular and atomic states to derive the ion kinetic energies. These are the only reasonable neutral dissociation mechanisms which can contribute to this shoulder. A further contribution may be dissociative photoionization after absorption of at least five IR photons (process (3)). This dissociation mechanism gives rise to a continuous distribution of the kinetic energy of the ions formed. The experimental energy resolution does not allow to distinguish between these different possibilities.

The experiment shows that the wavelength of the IR radiation influences the dissociation mechanisms of H_2 in a light intensity range intemediate between perturbative and non-perturbative. In this range the AC-Stark effect becomes important and may influence dissociation by inducing resonances into the H_2 excitation pathway or by shifting final product states.

REFERENCES

Guberman, S.L., 1983, The doubly excited autoionizing states of H_2, *J. Chem. Phys.* 78:1404

Ludwig, J., Rottke, H., and Sandner, W., 1997, Dissociation of H_2^+ and D_2^+ in an intense laser field, *Phys. Rev. A* 56:2168

Ludwig, J., Rottke, H., and Sandner, W., 1997a, The line profiles of single rotational lines of the H_2 $B - X(3,0)$ band in an intense infrared light field, *Z. Phys. D* 42:57

Rottke, H., Ludwig, J., and Sandner, W., 1996, H_2 and D_2 in intense sub-picosecond laser pulses: Photoelectron spectroscopy at 1053 and 527 nm, *Phys. Rev. A* 54:2224

Rottke, H., Ludwig, J., and Sandner, W., 1997, H_2 $B\,^1\Sigma_u^+$ photodissociation in an intense infrared light field, *J. Phys. B: At. Mol. Opt. Phys.* 30:2835

Shao, Y.L., Fraser, D.J., Hutchinson, M.H.R., Marangos, J.P., and Tisch, J.W.G, 1996, Ionization and dissociation of H_2 with intense and ultra-short laser pulses at 780 nm and 390 nm, *J. Phys. B: At. Mol. Opt. Phys.* 29:5421

Sharp, T.E., 1971, Potential-energy curves for molecular hydrogen and its ions, *At. Data* 2:119

Walsh, T.D.G., Ilkov, F.A., and Chin, S.L., 1997, The dynamical behaviour of H_2 and D_2 in a strong, femtosecond, titanium:sapphire laser field, *J. Phys. B: At. Mol. Opt. Phys.* 30:2167

Yang, B., Saeed, M., and DiMauro, L.F., 1991, High-resolution multiphoton ionization and dissociation of H_2 and D_2 molecules in intense laser fields *Phys. Rev. A* 44:R1458

Zavriyev, A., Bucksbaum, P.H., Muller, H.G., and Schumacher, D.W., 1990, Ionization and dissociation of H_2 in intense laser fields at 1.064 μm, 532 nm, and 355 nm, *Phys. Rev. A* 42:5500

ELECTRON DYNAMICS IN THE STRONG FIELD LIMIT OF PHOTOIONIZATION

B Sheehy,[1] B Walker,[1] R Lafon,[1] M Widmer,[1] L F DiMauro,[1] P Agostini,[2] and K C Kulander,[3]

[1]Brookhaven National Laboratory, Department of Chemistry, Upton, NY 11973
[2]Service des Photons, Atomes et Molécules, Centre d'Etudes de Saclay, 91191 Gif Sur Yvette, France
[3]TAMP Group, Lawrence Livermore National Laboratory, Livermore, CA 94551

High precision photoelectron energy and angular distributions in helium and neon atoms for a broad intensity range reflect the change in the continuum dynamics that occurs as the ionization process evolves into the pure tunneling regime. Elastic rescattering of the laser-driven free electron from its parent ion core leaves a distinct signature on the spectra, providing a direct quantitative test of the various theories of strong field multiphoton ionization. We show that it takes a relatively complete semi-classical rescattering model to accurately reproduce the observed photoelectron distributions. However, the calculated inelastic rescattering rate fails to reproduce the measured nonsequential double ionization yields.

INTRODUCTION

Strong field photoelectron spectra have attracted considerable attention over the past decade and a half, but a comprehensive understanding of the underlying dynamics which produce these spectra and related phenomena has solidified with the development of high repetition-rate, short pulse lasers which can span the entire intensity range of importance[1]. In 1965, Keldysh[2] showed that at infrared and visible wavelengths the dynamics of strong field atomic ionization undergoes a change in character as the laser intensity increases. In weak fields electrons are promoted into the continuum by the simultaneous absorption of enough photons to increase their energy above the ionization potential. This is called multiphoton ionization (MPI). However, as the laser intensity increases, a completely separate mode of escape becomes possible. At large distances from the nucleus the electrostatic attraction of

Applications of High Field and Short Wavelength Sources
Edited by DiMauro *et al.*, Plenum Press, New York, 1998

the ion core can be overwhelmed by the laser's instantaneous electric field, producing a barrier through which a valence electron can tunnel. In this regime a quasi-static tunneling picture becomes appropriate: the laser field varies so slowly compared to the response time of the electron that the ionization rate becomes simply the cycle-average of the instantaneous dc-tunneling rate. Tunneling becomes important when the ratio of the frequency of the applied field to the tunneling rate becomes smaller than unity. This ratio, known as the Keldysh or adiabaticity parameter, γ, is given by $[I_p/(2U_p)]^{1/2}$, where I_p is the binding energy of the electron and $U_p = 2\pi\alpha I/\omega^2$ is the ponderomotive energy (in atomic units) of a free electron in the laser field of frequency ω and intensity I, where α is the fine structure constant.

The majority of experimental studies on neutral atoms exposed to intense, short-pulse laser fields have been carried out in the MPI regime ($\gamma \geq 1$). A few experiments[3,4] have extended into the tunneling regime, but these measurements have been limited to observations of total ionization rates or electron energy distributions over a small dynamic range. In this paper, we report upon the first systematic experimental investigation in the strong field tunneling limit. Our study, by virtue of the enhanced dynamic range accessible with kilohertz laser technology, follows the change in the ion yield curves, electron spectrum and angular distributions as the ionization evolves from predominantly MPI to pure tunneling. The study examines both helium and neon atoms which have been established to tunnel ionize near the saturation intensity[5]. We find that the electron distributions in the tunneling regime are very different from any previous reports obtained in the MPI or mixed regimes[6,7,8]. We achieve a quantitative description of the electron spectra using a rescattering picture[8,9] which mimics the time evolution of a tunnel-ionized continuum wave packet in the *combined* fields of the laser and ion core. However, when the same model is applied to the e–2e inelastic rescattering process, it fails, both quantitatively and qualitatively, to describe the measured nonsequential production of doubly charged helium and neon ions. Since rescattering events are known to be important in other short-pulse, strong-field emission phenomena (e.g., harmonic generation) this investigation better defines the underlying dynamics of these processes.

QUASICLASSICAL MODEL IN THE TUNNELING LIMIT

The motion of the tunneling wave packet is controlled mostly by its interaction with the laser field, since it rapidly moves beyond the range where the ion core potential is effective. Its evolution can be reasonably approximated using the classical equations of motion for a laser–driven electron. In a simple quasiclassical (SQC) model[3], the bound electron becomes free at a particular phase, ωt_o, of the field, with zero kinetic energy, then undergoes oscillatory motion at the laser frequency, ω. The time dependent electron velocity is given in atomic units by $v(t) = (\mathcal{E}/\omega) \times (\cos\omega t - \cos\omega t_o)$, and the kinetic energy by $T = 2U_p \times (\cos\omega t - \cos\omega t_o)$, where $\mathcal{E} = (8\pi\alpha I)^{1/2}$ is the amplitude of the laser's electric field. The first term represents the electron's quiver motion: the time-averaged velocity is zero, while the average energy is U_p. The second, or drift, term represents the energy which would be detected in a short-pulse experiment. In the absence of further interactions with the ion core (rescattering) the maximum drift energy an electron can have is $2U_p$.

The rescattering picture[8,9] goes beyond the SQC model by recognizing that roughly half the electrons will recross the plane of the nucleus when the field has changed sign. We extend that model here to quantitatively model the photoelectron energy spectra (PES) in terms of elastic rescattering, and examine the model's potential for modeling multiple ionization as an inelastic rescattering process.

Only the first return to the core is considered. Due to dispersion of the tunneling electron wavepacket, the probability of rescattering on subsequent returns is negligible. We divide the optical cycle into a large number of equal time intervals. In each interval a trajectory is launched at the outer turning point of the suppressed effective potential with zero velocity. It is propagated in the combined fields of the laser and the helium (neon) ion core until either it escapes or returns to cross the plane of the nucleus. Those which escape contribute to the spectrum below $2U_p$ according to their drift velocities, as in the SQC model. The returning trajectories are assumed to be guiding a freely spreading gaussian wave packet whose width is given by $\alpha(\tau) = (\alpha(0)^2 + [2\tau/\alpha(0)]^2)^{1/2}$, where $\alpha(0)$ is the initial width and τ is the propagation time between initiation and return. Choosing $\alpha(0) = 4.0a_0$, gives a return width consistent with our numerical studies[5]. We calculate the differential elastic scattering cross section for this wave packet using[11]

$$\sigma(\theta) = \left| \frac{1}{2ik} \sum_{\ell=1}^{\ell_{max}} a_\ell (2\ell + 1) e^{2i(\eta_\ell + \delta_\ell)} P_\ell(cos\theta) \right|^2. \tag{1}$$

Here δ_ℓ is the Coulomb phase shift (for a charge of 1) and η_ℓ is the additional phase shift resulting from the short range part of the $He^+(Ne^+)$ potential. These phase shifts are obtained from numerical integration of the scattering equations for electron–$He^+(Ne^+)$ over the necessary range of energies and angular momenta. The partial wave amplitudes, a_ℓ are determined from the distribution of impact parameters in the returning wave packet ($\ell = mv_{ret}b$).

Equation (1) gives the field–free differential cross section. The laser field will distort this distribution. The transverse component of the outgoing velocity is conserved, but the velocity along the polarization direction has both a drift component and a quiver velocity which depend on the phase of the laser field at the return time. Drift velocities corresponding to PE energies as high as $10U_p$ can be produced if the trajectory is scattered by $\sim 180^\circ$. The energy and angular distributions from the wave packet in each time interval are weighted by the instantaneous tunneling rate. In these calculations we have actually used a scaled dc-tunneling rate which, when cycle averaged, gives the ADK rate[12]. This accounts for the initial state not being purely hydrogenic. The total angle-resolved electron distribution for a given laser intensity is obtained by summing the contributions from all time intervals. The e–2e inelastic process, which leads to the production of double ionization, is calculated using a modified Lotz cross section[13] which accounts for both excitation and ionization. Spatial and temporal averaging is performed for comparison to the experimental measurements.

TUNNELING PHOTOELECTRON DISTRIBUTIONS

In the experiments presented here a 120 fs, 1 kHz repetition rate, titanium sapphire laser operating at 0.78 μm was focused by f/4 optics into an ultrahigh vacuum chamber containing helium or neon. Under these conditions, the system is capable of producing a maximum intensity of 20 PW/cm^2 with typical pulse-to-pulse fluctuations $< 1.7\%$ for 10^6 laser shots. Electrons or ions are analyzed using a 30 cm time-of-flight spectrometer. The laser polarization is $> 99\%$ over the entire intensity range, and is rotated relative to the detector axis to study angular distributions.

Total PES are shown in Fig. 1 for three different intensities in both absolute and ponderomotive energy units. A single experimental run measures the PES for one polar angle, and these distributions are constructed by integrating the angle-resolved PES (ARPES) over the polar angle and assuming azimuthal symmetry around the polarization axis. The inset in Fig. 1a shows typical angular distributions at several energies.

The preponderance of photoelectron energies below $2U_p$ is evident, consistent with the simple quasi-classical (SQC) model[3]. The angular distributions of these electrons are strongly aligned along the laser polarization direction, becoming narrower towards the $2U_p$ limit. However, a striking change occurs above $2U_p$: there the ADs are significantly broader (with a weak narrowing as the energy increases towards $10U_p$), indicative of rescattering. In fact, this PES seems to be a superposition of two components: a "normal" narrow distribution that falls off rapidly with increasing energy between 0 and $2U_p$ and a much broader but weaker, almost flat energy distribution that extends out to high energies before abruptly truncating at $8 - 10U_p$. As the intensity increases, moving further into the tunneling regime, the fraction of electrons in this high-energy portion falls as $I^{-2.5}$. These results differ dramatically from all previous experimental reports[1]. For example, the PES for inert gas atoms[6,7] clearly show angle-dependent structures, as well as an abundance of electrons with energies $> 2U_p$. These differences reflect the pure tunneling nature in these experiments and provides a unique opportunity for quantitatively testing the rescattering picture.

The quantitative comparisons of the measured and calculated total (spatially and temporally averaged) PES are also shown in Fig. 1. The dashed-dotted line in Fig. 1(b) represents the SQC model. The obvious failure of the model in predicting the high energy portion of the spectrum indicates the necessity of including rescattering. The dashed lines are the calculated PES using the He^+ potential and give excellent agreement with the experimental measurement over the entire energy range for both "pure" tunneling cases. The poorer agreement seen in Fig. 1(c) signifies the transition into the mixed regime where the MPI contribution is becoming significant. Figure 1(b) also shows a PES calculated using a pure hydrogenic potential (dotted line). Coulomb scattering significantly underestimates the high energy plateau because the backscattering which produces the high energy electrons is most strongly affected by the more attractive short range part of the real potential.

Spectra were also obtained for the ionization of neon by $0.78\ \mu m$ radiation. The ion and electron spectra show an evolution consistent with "pure" tunnel ionization near the saturation intensity. Results for both helium and neon are compiled in Fig. 2 along with calculated curves. The plot shows the ratio of electrons with energies $> 2U_p$ over those with energy $< 2U_p$ as a function of intensity. The open and filled stars represent the values derived from the experiment for helium and neon respectively. The three theoretical curves are calculated using Coulomb (dashed), He^+ (solid), and Ne^+ (dotted) potentials, and include averaging over the laser intensity distribution. The helium data agrees well with the calculated curve (solid line) for the two highest intensities but deviates near the lowest intensity, due to the multiphoton contribution which is not included in our calculation. The neon ratio point (filled star) shows that at the same intensity as helium, neon is approximately ten times more efficient at producing high energy electrons. This result is quite consistent with the observed ten fold increase in the high harmonic emission[15] for neon over helium. This difference is reflected in the calculated curves and again demonstrates the importance of the short range part of the potential in accurately reproducing the measured results at these intensities. Further evidence is seen in the Coulomb curve, which underestimates the amount of elastic rescattering. However, as the intensity decreases the differences

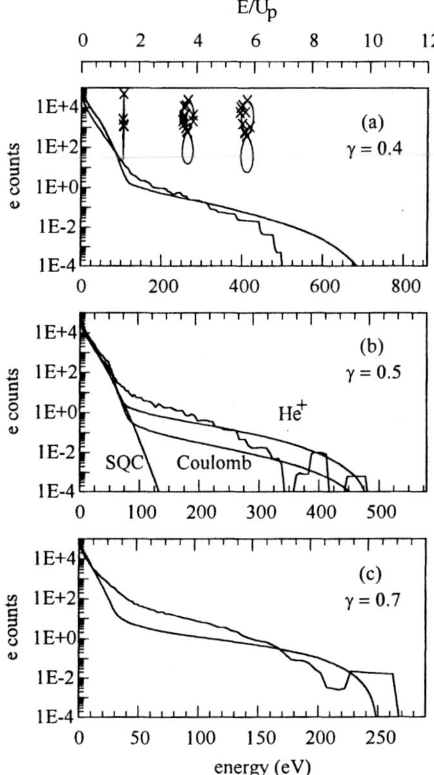

1. Total helium PE energy distribution for 0.78 μm excitation at (a) 12, (b) 8 and (c) 4×10^{14} W/cm^2. The experimental and calculated distributions using the complete semi-classical theory presented here correspond to the solid and dashed lines, respectively. In panel (b) the dotted line results from pure Coulomb rescattering and the dashed-dotted is without rescattering (SQC model). The Keldysh parameter, γ, is indicated for each intensity.

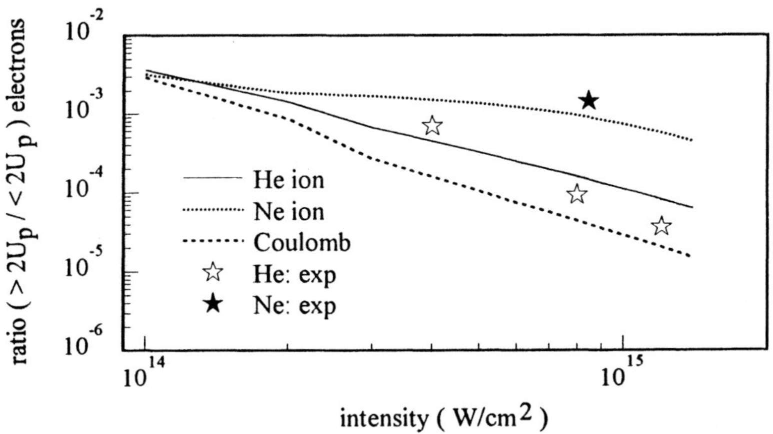

2. Compiled experimental and calculated ratio of total number of electrons with energies $> 2U_p$ over those $< 2U_p$.

233

between all three curves vanish and in fact approach the Coulomb scattering result. Since the electron's return energy falls linearly with intensity, collisions with the core are softer, consequently diminishing the influence of the short range physics. Of course, experimental access to this regime is impossible due to the overwhelming contribution of MPI ($\gamma > 1$).

DOUBLE IONIZATION AND THE RESCATTERING MODEL

Ion yields as a function of intensity were also collected for both helium and neon. Walker *et al* [5] have exploited the kilohertz laser technology to extend the range of these measurements by over five orders of magnitude over previous measurements and demonstrate the connection between the nonsequential (NS) double ionization rate and tunneling ionization of the neutral. A sensitive measure of the nonsequential dynamics is provided by plotting the intensity dependence of the X^{2+}/X^+ ratio (X = He or Ne) for both helium (solid circles) and neon (open circles) for 0.78 μm excitation, as shown in Fig. 3. To ensure accuracy, the two ions are concurrently collected at a fixed intensity and averaged for at least 10^6 laser shots. The plot shows that the measured NS yield is similar for both atoms, achieving a value of 0.0020[3] for helium and 0.0018[4] for neon at their saturation intensity. Below saturation, the ratio of each decreases by approximately a factor of 10 over the measured intensity range, although the total rates are changing by seven orders of magnitude. Furthermore, analysis shows that both the helium and neon ratio scale with the tunneling fraction of the ionization, again providing an important clue to the NS dynamics.

The complete quasiclassical calculation, described in Section 2, can be used to predict the double-to-single ionization ratio produced from e–2e inelastic rescattering. Figure 3 shows the results for both helium (solid line) and neon (dashed line). These results are calculated using the same initial conditions and core potentials used to calculate the photoelectron spectra of Figs. 1 and 2. A modified "field-free" e–2e Lotz

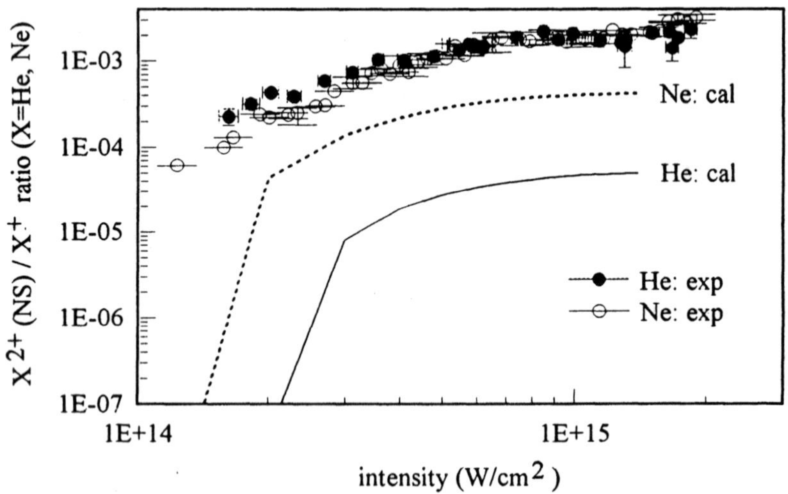

3. Compiled experimental and calculated ratio of nonsequential double ionization to single ionization for helium and neon for 0.78 μm excitation.

cross-section[13] is used to account for double ionization contributions from both core excitation and direct ionization. It has been shown[14] that the use of field-free cross sections is a reasonable approximation since the slowly varying electric field from the laser has a very small effect on the inelastic scattering processes. Clearly, the e–2e rescattering severely underestimates the absolute measured ratio, as well as the shape. The ratio of the experimental to calculated value at saturation is 47 for helium and 5 for neon. The lack of agreement is a clear indication that more than inelastic rescattering is involved in the physics of the nonsequential ionization. Reasonable assumptions on wave packet spreading and cross sections yield accurate predictions for the electron distributions seen in Fig. 1 and support the estimates used in our model.

The disagreement between the rescattering predictions and the experiment goes beyond underestimating the absolute value. As seen in Fig. 3, the experimental data shows a smooth decrease in the ratio with decreasing intensity, whereas the calculations show a sharp and abrupt cutoff. The origin of the cutoff is clear: as the intensity is lowered, the electron's return energy decreases to the point that it can no longer free the second electron. We see no evidence of this cutoff, even when we repeat the the experiment at $\lambda = 0.39\mu m$, where all of the NS ionization occurs at intensities below where the cutoff should occur, at a rate (0.0013 time the single ionization rate) very close to that at $0.78\mu m$. Additionally, it is difficult to rationalize in a rescattering picture why the double ionization ratios would be the same for helium and neon, considering the difference in the e–2e cross sections. The good agreement found between the complete quasiclassical calculation and the experimental electron distributions demonstrated the important distinction produced by the atom's short-range potential. Obviously the calculated curves in Fig. 3 reflect the difference in the ionization cross sections, while the experiment does not.

CONCLUSION

All of the evidence suggests that in the strong-field tunneling limit, the rescattering model captures, even quantitatively, the essential physics leading to the production of high energy electrons and harmonics. However, e–2e rescattering fails, even qualitatively, to reproduce most of the salient features of the experiment. We propose that strong-field double ionization is dominated by some other mechanism involving simultaneous two-electron ejection either through a shake-off[16] or threshold mechanism. While the underlying mechanism for this process remains unclear, various theoretical two-electron[17,18] treatments have begun. The results we have presented better define the underlying dynamics of these processes and provide more stringent limits on any future models.

ACKNOWLEDGMENTS

This research was carried out in part at Brookhaven National Laboratory under contract No. DE-AC02-76CH00016 with the U.S. Department of Energy and supported by its Division of Chemical Sciences, Office of Basic Energy Sciences, and in part under the auspices of the U. S. Department of Energy at the Lawrence Livermore National Laboratory under contract No. W-7405-ENG-48. L. F. D. and P. A. acknowledge travel support from NATO under Contract No. SA.5-2-05(RG910678).

REFERENCES

1. For a recent review, see DiMauro L F and Agostini P 1995 *Advances in Atomic, Molecular, and Optical Physics 35* Bederson B and Walther H, Eds. (San Diego: Academic Press)
2. Keldysh L V 1965 *Sov. Phys. JETP* **20** 1307–1311
3. Corkum P B, Burnett N H and Brunel F 1989 *Phys. Rev. Lett.* **62** 1259–1262
4. Mohideen U, Sher M H, Tom H W K, Aumiller G D, Wood II O R, Freeman R R, Bokor J and Bucksbaum P H 1993 *Phys. Rev. Lett.* **71** 509–512
5. Walker B, Sheehy B, DiMauro L F, Agostini P, Schafer K J and Kulander K C 1994 *Phys. Rev. Lett.* **73** 1227–1230
6. Yang B, Schafer K J, Walker B, Kulander K C, Agostini P and DiMauro L F 1993 *Phys. Rev. Lett.* **71** 3770–3773
7. Paulus G G, Becker W, Nicklich W and Walther H 1994 *J. Phys. B: At. Mol. Phys.* **27** L703–L708
8. Schafer K J, Yang B, DiMauro L F and Kulander K C 1993 *Phys. Rev. Lett.* **70** 1599–1602
9. Corkum P B 1993 *Phys. Rev. Lett.* **71** 1994–1997
10. Lewenstein M, Kulander K C, Schafer K J and Bucksbaum P H 1995 *Phys. Rev.* **A51** 1495–1507
11. Schiff L I 1968 *Quantum Mechanics* (New York: McGraw Hill)
12. Ammosov M V, Delone N B and Krainov V P 1986 *Sov. Phys. JETP* **64** 1191–1997
13. Lotz W 1968 *Z. Phys.* **216** 241–247
14. Kulander K C, Cooper J and Schafer K J 1995 *Phys. Rev.* **A51** 561–568
15. L'Huillier A and Balcou P 1993 *Phys. Rev. Lett.* **70** 774–777
16. Fittinghoff D, Bolton P R, Chang B and Kulander K C 1992 *Phys. Rev. Lett.* **69** 2642–2645
17. Becker A and Faisal F H M 1996 *J. Phys. B: At. Mol. Phys.* **29** L197–L202
18. Watson J B, Sanpera A, Lappas D G, Knight P L and Burnett K 1996 *Proceedings of ICOMP VII* (Bristol: IOP)

INTENSE LASER INTERACTIONS:
HIGHER ORDER RESONANCES AND HOT ELECTRONS

L. D. Van Woerkom, S. Evans, P. Hansch, and M. A. Walker

Department of Physics
The Ohio State University
Columbus, Ohio 43210-1106

INTRODUCTION

The kinetic energy spectrum of electrons produced by multiphoton and above threshold ionization of atoms has been studied in detail for nearly two decades. Much has been learned about the mechanisms behind the gross and fine structures present in these spectra over a wide range of experimental conditions. Theoretical models ranging from very simple and semiclassical to sophisticated quantum treatments have been developed to explain a great deal of the phenomena observed in experiments.

Unanswered questions do remain, however, and interesting physics is still emerging. A division is often made between portions of the spectrum which lie below $2U_p$ [1] in energy and those which lie above. We present two experimental results of interest. Our results fall into each of these categories, and could provide a link between the phenomena behind each region. Specifically, we have made detailed measurements of the transition to higher photon order resonances below $2U_p$ and narrow resonant structures above $2U_p$.

EXPERIMENT

The spectra presented were obtained using a Positive Light, Inc. Ti:Sapphire laser system operating at a 1 kHz repetition rate. The 120 fs, 800 nm output pulses have an energy of 950 μJ per pulse and are plane-polarized. Spectra at different intensities have been recorded using Intensity-Selective-Scanning (ISS). This technique has been described elsewhere [2]. Briefly, a 500 μm pinhole is placed at the front of the TOF tube and only electrons traveling on the line of sight from the ionization volume to the detector are measured. This technique yields data from a one-dimensional radial intensity distribution, thereby reducing the problems associated with spatial averaging over the entire laser focal volume. By scanning a pinhole across the Gaussian ionization volume, specific peak intensities can be selected with high precision.

BELOW $2U_p$: HIGH PHOTON ORDER RESONANCES

The low energy portion of the photoelectron spectrum has been studied extensively and is well understood for low intensity multiphoton ionization and very high intensity tunneling ionization. The gross structure in tunneling experiments can be explained by the semiclassical "simpleman's model" [3-5], and the fine structure seen under short pulse

Applications of High Field and Short Wavelength Sources
Edited by DiMauro *et al.*, Plenum Press, New York, 1998

conditions at low intensity is explained in terms of intensity-tuned crossings of ac-Stark shifted Rydberg states and the multiphoton-dressed ground state. For sufficiently high intensities, the Rydberg states should in principle be tuned through resonance with several harmonics of the ground state, but the intermediate nonperturbative intensity range in which this should occur has not been studied in much detail. We have observed low energy spectra in xenon, krypton and argon which clearly exhibit resonances with increasingly higher photon orders as the intensity is increased. In the case of argon, three photon orders are successively brought into resonance within the intensity range of the experiment.

While resonance with more than one photon order been seen in other experiments [6, 7], the detailed appearance and evolution of the photoelectron spectrum over a range of photon orders (not ATI orders) has never been studied. Figure 1 shows the typical transient resonance situation applied to xenon. For very low intensity, eight photons (at 800 nm or 1.55 eV) lie slightly above the lower spin-orbit ionization limit. As the laser intensity increases, the ionization limit and weakly bound Rydberg states are shifted to higher energy by an amount equal to U_p [8]. Specifically, the p- and f-Rydberg series are brought into eight photon resonance with the ground state of xenon (other states are as well but are not populated because of parity selection rules). The ionization rate is enhanced and photoelectrons are emitted with kinetic energy equal to the photon energy minus the Rydberg state binding energy. Absorption of additional photons in the continuum causes this fundamental structure to be repeated at multiples of the photon energy, producing several orders of ATI peaks. Further increase of the laser intensity brings the Rydberg states into nine photon resonance, where the allowed transitions are to d- and g-states.

The various Rydberg state resonances produce electrons having different kinetic energies, which account for the substructure within each ATI order. Given the resonant nature of this process, photoelectron peaks corresponding to specific transient resonances appear in the kinetic energy spectra and they do not shift with absolute laser intensity. Therefore, the appearance or absence of certain resonances is a good indicator of the peak laser intensity. However, as the intensity increases, other effects may play a role in determining the photoelectron yield. For example, the large number of photons involved (>8) and the high intensity ac-Stark shifts of bound levels could create coupling between states, leading to non-ponderomotive energy shifts. Calculations have shown this to be the case although experimental demonstration remains weak [9-11]. Thus, it is not obvious that the simple resonance model proposed by Freeman [8] is valid at higher intensities. We have observed that at higher intensities the process remains simple in xenon.

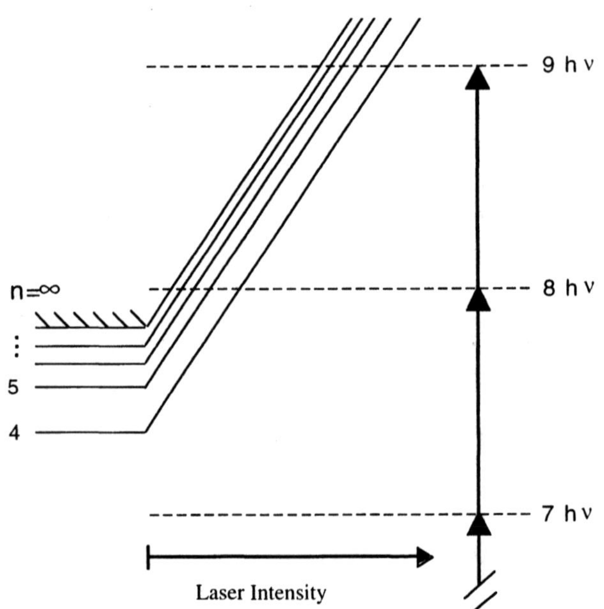

Figure 1. Schematic illustration of transient resonance process in xenon.

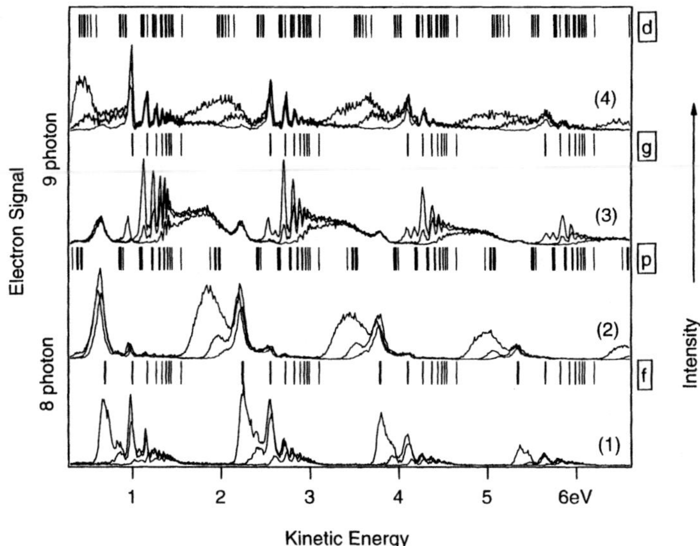

Figure 2. High resolution kinetic energy spectrum of xenon showing features due to successive resonances. The data are taken at intensities of (1)1.9, (2) 2.4, (3) 4.0, (4) 4.7×10^{13} W/cm^2. The vertical bars indicate expected kinetic energies for electrons produced by neutral Rydberg resonances. The boxed letters next to each set of bars indicate the orbital angular momentum of each Rydberg series.

Figure 2 shows the first four ATI orders of the photoelectron kinetic energy spectra for xenon over an intensity range of 1.9×10^{13} - 4.7×10^{13} W/cm^2. The structure is similar for the other ATI orders, although order-to-order differences in relative peak amplitudes are already apparent at 6-10 eV. Three different intensities are shown for each sub-figure. Each grouping of three curves (1-4) is normalized to a common maximum amplitude for better comparison of relative effects. The unprocessed vertical scale increases by a factor of 80 in going from Fig. 2-(1) to Fig. 2-(4). The ionization potential for leaving neutral Xe in the $^2P_{3/2}$ single ion state is 12.127 eV. As mentioned previously, the first parity-allowed multiphoton resonance for the f-series Rydberg states occurs with 8-photons.

The emergence of the 8-photon 4f and 5f-resonances is clearly seen as the intensity increases from Fig. 2-(1) to 2-(4). Above each grouping we have plotted the ATI replicated energy levels as indicated on the right hand side for the eight (f,p) and nine (g,d) photon resonances. As the intensity reaches the threshold for observing the 4f-state in Fig. 2-(1), the 4f-resonance clearly becomes the dominant spectral feature. In Fig. 2-(2), a broad feature appears below the 4f-state for intensities above 2.4×10^{13} W/cm^2. Due to fine structure, the 7p-state splits into a cluster of 6 levels with J-values ranging from 0 to 3 over an energy range of 0.11 eV. Furthermore, the broadening of this peak toward lower energy possibly indicates nonresonant contributions to the ionization.

Figs. 2-(3) and 2-(4) display the sudden emergence of the higher lying Rydberg g-and d-states once the ac-Stark shift brings them into 9-photon resonance. While the 8-photon 4f-state contribution remains fairly constant in Figs. 2-(1)-(4), the effects of the 9-photon resonances at higher intensities increase dramatically and dominate the spectrum. The onset of these 9-photon resonances occurs at about 3.3×10^{13} W/cm^2, and a similar evolution as with the 8-photon resonances begins again. The spectrum shown in Fig. 2-(3) at 4×10^{13} W/cm^2 follows the g-state evolution. Due to the high angular momenta and small quantum defects, the f- and g-states appear at the same kinetic energy positions. Fig. 2-(4) shows the appearance of the 5d manifold.

These resonant features have two important properties. First, since the intensity of the laser is responsible for bringing the atom into resonance with the field, the resonances occur *very abruptly at very specific intensities.* Second, since the ionization limit shifts by the same amount as the Rydberg states, ionization out of these states produces electrons *whose kinetic energies do not change* with changing laser intensity. These properties are characteristic of any spectral features produced by a resonant process.

ABOVE 2U_p: HOT ELECTRON PRODUCTION

Much of the recent interest in ATI has involved the unusual features observed in the high energy portion of the photoelectron spectrum of helium and the other rare gases. The first such observation was the discovery [12] by Paulus et al. of a so-called "plateau" region in these spectra, in which the peak amplitudes do not monotonically decrease as in the lower energy region, but instead show a marked change in slope when seen on semilog plots. This change occurs at kinetic energies around 2U_p, which is notable in itself since the "simpleman's theory" predicts a cutoff in the spectrum at this energy. The very existence of electrons with these energies requires modification of the theory for its explanation, and the detailed behavior of the spectra remains largely unexplained.

Since the discovery of the plateau, there has been considerable study of this phenomenon. Recent experiments by DiMauro *et al..* [13, 14] and Paulus *et al.* [12]; Paulus, 1994 #389; Paulus, 1994 #390; Paulus, 1995 #1672 have revealed the production of 'hot' electrons with kinetic energies as great as 11U_p. Rings have been observed [15] in the angular distributions of these electrons, and explanations for this as well as the gross features of the plateau have been given in terms of a semiclassical "rescattering model" [13, 16]. In this model the motion of tunnel-ionized electrons in the field and their interaction with the ion core accounts for their high energy and directionality. This model is unable to explain any fine structure in the spectrum of these "hot" electrons. Numerical solution of the Time-Dependent Schrodinger Equation (TDSE) using the Single-Active Electron (SAE) approximation [17] has produced spectra whose features qualitatively match those seen in helium and other rare gases. These models yield little insight however into the actual mechanism for hot electron production, and do not include possible multi-electron effects.

Our experiments on hot electron production in ATI have uncovered additional interesting results. First, the overall envelope of the photoelectron energy distribution shows strong modulation at high kinetic energy. Second, each ATI order is comprised of sub-structure that depends sensitively on the laser intensity and does *not* shift in energy as the intensity changes. Finally, the hot electron region provides a testing ground for observing multielectron and highly excited state effects due to ionization occurring in a region of the continuum filled with autoionizing levels. We present photoelectron kinetic energy spectra showing prominent features at high kinetic energies that cannot be explained by the simpleman/rescattering theory, traditional Rydberg state resonances or other current theoretical treatments.

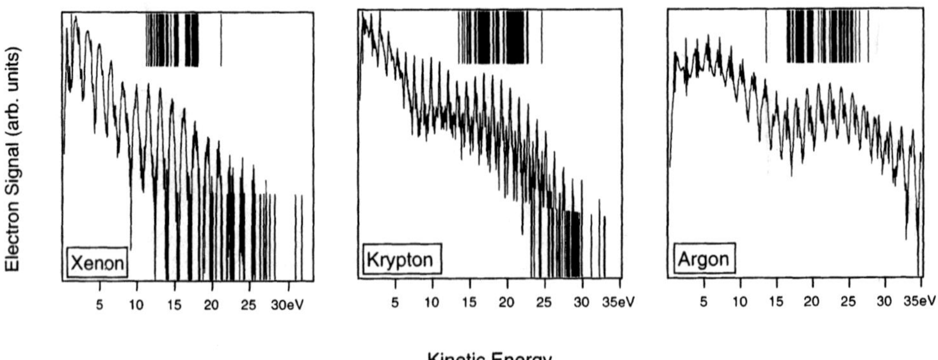

Figure 3. High resolution photoelectron spectra of xenon, krypton, and argon showing plateau features and excited levels of the each ion. The intensities are 5, 5.4, and 8.3x10^{13}W/cm^2 from left to right.

Figure 3 shows typical photoelectron energy spectra for xenon, krypton and argon from 0 to 35 eV. The data are plotted on a semilog-scale to clearly show the hot electron features. The high signal-to-noise ratio in data such as this has been made possible by advances in kilohertz repetition rate ultrashort pulse lasers. The intensity is different for each curve with approximate U_p = 3, 3.2, 5 eV for xenon, krypton and argon, respectively. All three exhibit the characteristic falloff at $2U_p$ which comprises over 90% of the photoelectrons produced. Beyond this point a change in slope is observed. As the laser intensity is further increased more modulation appears at higher kinetic energy.

Also shown on each curve in Figure 3 are the excited ion energy levels. Converging on each of these levels are many neutral autoionizing states. It is interesting to note that the change in slope for all three gases occurs in the same energy region as the excited states of the ion. The large density of autoionizing states in this region could give rise to a resonant enhancement in the production of photoelectrons with kinetic energies greater than $2U_p$.

Transient resonances account unambiguously for the low energy photoelectron structure found in experiments using linearly polarized light. For 800 nm light, Rydberg state resonances yielding kinetic energies up to the 2Up point can be clearly identified. In order to understand the sub-structure in the electron spectra for the hot electrons it is revealing to connect the low kinetic energy resonant structure to the high kinetic energy structure as the intensity is varied. This puts rather strict requirements on the experiment. The kinetic energy must be measured with a resolution better than 40 meV from .5 to 40 eV and be known absolutely to better than 40-60 meV. By using very fast counting electronics (EG&G Ortec 9308 pTA with picosecond discriminator) we are able to achieve sufficient resolution to follow the evolution of the photoelectron spectra. The absolute energy calibration is more difficult but is accomplished through a self-consistent iterative process.

As previously mentioned, the goal is to study the correlation between the low and high KE structures in the photoelectron spectrum. Figures 4 and 5 show argon spectra for low and high intensity regions, respectively. Each figure contains two plots for each intensity with the solid curve showing low kinetic energy (bottom axis) and the dotted curve showing high kinetic energy (top axis). The dotted curves have been shifted down by 11 photons to observe correlations with the low energy peaks.

Figure 4 shows photoelectron spectra of argon over the range of intensities from 5.9 - 6.6 $\times 10^{13}$ W/cm^2. These exhibit the emergence of the so called triplet structure first mentioned by Hertlein [18], who indicated that the middle peak of the triplet (shown in the figure by the vertical lines) mimicked the appearance of the 12 photon 4f-state resonance in neutral argon. We confirm with our high resolution studies that in fact the high kinetic energy structures follow the neutral Rydberg resonances reasonably well. There are, however, unexplained structures that appear in the positions near the 5p and 4d levels just near 2 eV (and at intervals of 1.55 eV above this near 3.55, 5.1, 6.65, and 8.2 eV). Argon has an accidental resonance with 1.55 eV photons between the 4p and 4d levels. Thus, features resembling d-states are seen for 12 photon resonances which are not allowed by direct absorption of 12 photons

Figure 5 shows a sequence of higher laser intensities (7.9-8.3x10^{13} W/cm^2) around the 13 photon g-state resonances. Again, the high KE features follow the evolution of the low KE bound Rydberg resonances. As mentioned for Fig. 4, the large dominant feature is most probably due to the p-d coupled states. Otherwise, the g- and f- states are identifiable for the allowed 11, 12, and 13 photon resonances. By following the development of the spectra as a function of kinetic energy and laser intensity, we can identify resonant peaks in the hot electrons with confidence. We have performed similar experiments in krypton and xenon, with similar results.

The match with Rydberg states is not perfect, however. Near photon energies of 1.55 eV there exists a resonant condition between the bound 4p and 4d levels. This coupling gives rise to the broad peak below the 4f resonance position. Although a resonant coupling should not give rise to a.c. Stark shifts that are ponderomotive, the location of all spectral features is consistent with a purely ponderomotive shift. Calculations are needed to study the case of a bound-bound resonance for the transient resonance model.

While the sub-structure appears to be due to neutral Rydberg state resonances induced by the ac-Stark shift, and the modulation of the spectral envelope coincides with access to excited ionic levels, many questions remain. First, the origin of the modulation in the hot electron spectrum remains unclear. Recent calculations suggest that an electron-electron correlation effect could be responsible [19]. However, the electron energies produced via the

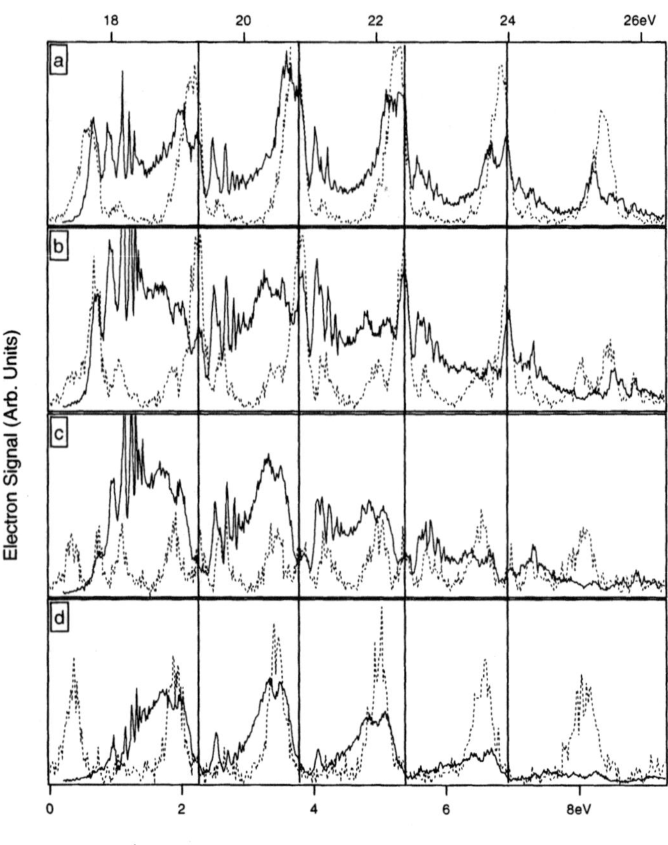

Figure 4. High resolution photoelectron spectra of argon in the intensity range 5.9-6.6x10^{13}W/cm^2 showing triplet structure described in the text. Solid curves correspond to the lower kinetic energy range shown at the bottom of the figure. Dotted curves correspond to the high kinetic energy range shown at the top, and have been shifted to lower energy by 11 photons to show appearance and position of similar features.

correlation model are identical to the energies produced using a multiphoton core excitation to autoionizing states that decay rapidly. Thus, electron spectra alone are not capable of distinguishing among the models. Experiments are underway in our laboratories to measure the angular distributions of the photoelectrons with high energy resolution for comparison with theory.

The apparent lack of multielectron and autoionizing effects is quite surprising in this energy range since all of the structure appears to be due to resonances related to the 11, 12, or 13 photon excitations through bound Rydberg states. This indicates that for high intensity photoionization into a structured continuum the fundamental process remains a single electron process. Most autoionizing states have lifetimes longer than the laser pulse duration of 100 femtoseconds. Thus, the signature of autoionization would be electron production over a range of kinetic energies not equal to the ATI ladder of bound Rydberg states since ionization would take place after the pulse passed and ac-Stark shifts subsided. Furthermore, it is not clear how autoionizing levels will shift

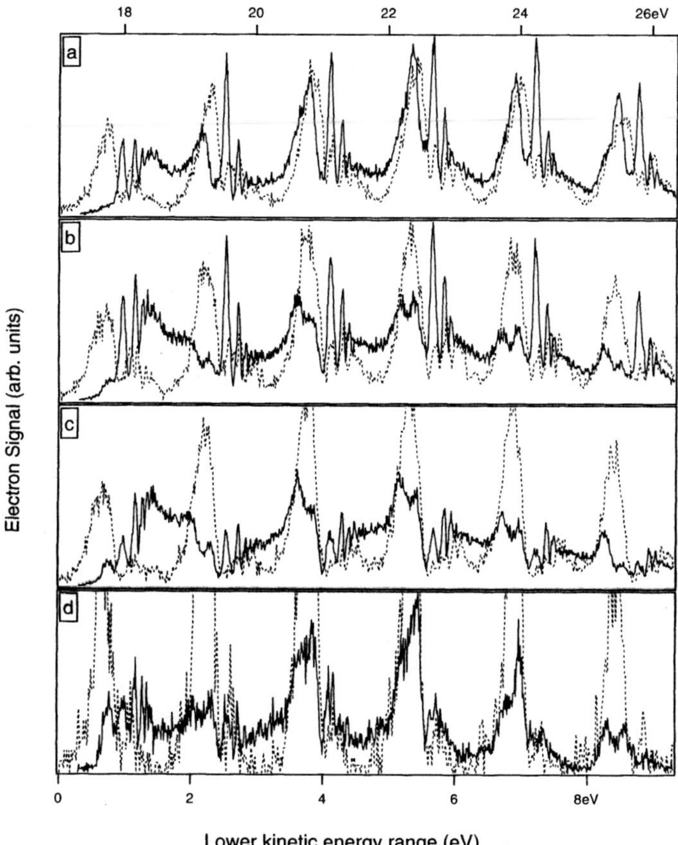

Figure 5. High resolution photoelectron spectra of argon in the intensity range $7.9\text{-}6.8.3\times10^{13}\text{W/cm}^2$. Solid curves correspond to the lower kinetic energy range shown at the bottom of the figure. Solid and dashed curves follow the same prescription used in Figure 4.

in the laser field since two electrons will be excited. If the second electron is excited to an orbital that is Rydberg like (with the wavefunction peaking far away from the core) then the autoionizing state will shift by $2U_p$ relative to the neutral ground state. Most likely the shift will be intermediate falling between U_p and $2U_p$.

CONCLUSION

In conclusion we have studied the photoelectron kinetic energy spectra of noble gases under high intensity, ultrashort pulse irradiation as a function of both kinetic energy and laser intensity. It is crucial to maintain high energy resolution in order to follow the emergence of resonant features in the hot electron portion of the spectra. A surprising amount of structure and modulation is found in the hot electron tail that can be directly related to neutral Rydberg resonances. Questions still remain as to the origin of the modulation and the details of the sub-structure due to bound-bound couplings.

243

REFERENCES

[1] The ponderomotive energy U_p is defined in atomic units as $I/4\omega^2$, where I and ω are the laser intensity and frequency, respectively. For 800 nm photons, $U_p = 5.98$ eV at 10^{14} W/cm^2

[2] P. Hansch and L. D. Van Woerkom, *Opt. Lett.* **21**, 1286 (1996).

[3] H. B. van Linden van den Heuvell and H. G. Muller, Eds., *Studies in Modern Optics No 8, Multiphoton Processes* (Cambridge University Press, Cambridge, 1988).

[4] T. F. Gallagher, *Phys. Rev. Lett.* **61**, 2304 (1988).

[5] P. B. Corkum, *Phys. Rev. Lett.* **71**, 1994 (1993).

[6] H. Helm and M. J. Dyer, *Phys. Rev. A* **49**, 2726 (1994).

[7] M. D. Perry , A. Szöke and K. C. Kulander, *Phys. Rev. Lett.* **63**, 1058 (1989).

[8] R. R. Freeman, *et al.*, *Phys. Rev. Lett.* **59**, 1092 (1987).

[9] M. Dörr and R. Shakeshaft, *Phys. Rev. A* **38**, 543 (1988).

[10] M. Dörr , R. M. Potvliege and R. Shakeshaft, *JOSA B* **7**, 433 (1990).

[11] G. N. Gibson , R. R. Freeman and T. J. McIlrath, *Phys. Rev. Lett.* **69**, 1904 (1992).

[12] G. G. Paulus, *et al.*, *Phys. Rev. Lett.* **72**, 2851 (1994).

[13] B. Walker , B. Sheehy , K. C. Kulander and L. F. DiMauro, *Phys. Rev. Lett.* **77**, 5031 (1996).

[14] B. Walker, *et al.*, *Phys. Rev. Lett.* **73**, 1227 (1994).

[15] B. Yang, *et al.*, *Phys. Rev. Lett.* **71**, 3770 (1993).

[16] G. G. Paulus , W. Becker , W. Nicklich and H. Walther, *J. Phys. B* **27**, L703 (1994).

[17] K. J. Schafer , B. Yang , L. F. DiMauro and K. C. Kulander, *Phys. Rev. Lett.* **70**, 1599 (1993).

[18] M. P. Hertlein , P. H. Bucksbaum and H. G. Muller, *J. Phys. B* **30**, L197 (1997).

[19] P. H. Bucksbaum , A. Sanpera and M. Lewenstein, *J. Phys. B* **30**, L843 (1997).

INFLUENCE OF RELATIVISTIC STRUCTURE AND RETARDATION IN TWO-PHOTON TRANSITIONS IN HYDROGENIC SYSTEMS

C. Szymanowski[1], V. Véniard, R. Taïeb, and A. Maquet

Laboratoire de Chimie Physique – Matière et Rayonnement ,
11 rue Pierre et Marie Curie , Université P. et M. Curie Paris VI ,
F - 75 231 Paris Cedex 05 , France
[1]Electronic address: szymanow@ccr.jussieu.fr

The experimental development of powerful coherent short wavelength sources, both from X-ray laser devices or from higher-order harmonics, enables now a wide range of possible applications[1]. A particular promising branch might be the performance of multi-photon bound-bound absorption experiments in inner-shells of heavy atoms or in highly-charged ions. Considerably more information could be obtained than in traditional (single-photon) X-ray absorption spectroscopy and related techniques. For example, two-photon ionization of Argon in the XUV frequency range has already been reported[2]. In this regime the usual non-relativistic dipole treatment of the atom-field interaction loses its validity. By considering the two-photon transitions from the $|1s\rangle$ ground state to the $|3d\rangle$ exited state of hydrogenic systems, meaning $|1^2S_{1/2}\rangle \rightarrow |3^2D_{3/2}\rangle; |3^2D_{5/2}\rangle$, both for linear and circular polarization, we discuss here, as a test case, in Dirac's relativistic theory the influence of the relativistic fine structure and the retardation of the field.

Regarding the currently achievable short-wavelength intensities, a perturbative approach is certainly valid. Therefore our treatment relies on the computation of the second-order S-matrix amplitude for the absorption of two identical photons. The infinite sums over the complete set of states are performed exactly using a discrete, Sturmian-like, expansion of the Dirac Coulomb Green's function[3]. Similar forms have been successfully used in calculations of γ-ray scattering amplitudes and of the dynamical polarizability of the ground state[4]. Our recently-obtained expression exploits the close similarities between the *second-order* Dirac equation and the non-relativistic Schrödinger equation in order to give a compact explicit expression for the *first-order* Dirac Coulomb Green's function which has proven to be very useful in actual two-photon calculations[5].

In order to determine the S-matrix element, the following twofold three dimensional integral over the spatial variables has to be evaluated:

$$\int d^3r_1 \int d^3r_2\, \psi_f^\dagger\, \tilde{V}\, G_E^{(1)}(\vec{r}_1,\vec{r}_2)\, \beta\, \tilde{V}\, \psi_{1S}\,, \tag{1}$$

where $\tilde{V} = \vec{\alpha}\cdot\vec{A}$ for both linear and circular polarization and r_i with $i \in \{1,2\}$ denote three dimensional space vectors. $\vec{\alpha}$ and β are Dirac matrices. In the case of the absorption of two identical photons with frequency ω the argument of the Green's function is, as a function of nuclear charge, $E = \epsilon_{1S} + \omega$ with $\omega = (\epsilon_f - \epsilon_{1S})/2$, where ϵ denotes the energies of unperturbed hydrogenic states. The retarded Dirac Coulomb Green's function $G_E^{(1)}$ obeys

$$\left[c\vec{\alpha}\vec{p} + \beta mc^2 - \frac{Z}{r_1} - E \right] G_E^{(1)}(\vec{r}_1,\vec{r}_2) = -\beta\, \delta(\vec{r}_1 - \vec{r}_2)\,. \tag{2}$$

In terms of Sturmian functions for the second order equation, the following expression for the Green's function of first order is obtained:

$$G_E^{(1)}(\vec{r}_1,\vec{r}_2) = \sum_{jlmnq} \begin{pmatrix} c_1 & c_2 \\ c_3 & c_4 \end{pmatrix}\,, \tag{3}$$

where the four coefficients c_i are

$$\begin{aligned}
c_1 &= \frac{Z^2/c^2}{4q\lambda}\Big[\Big(-\frac{E(\kappa+q\lambda)}{\eta(c^2-E)} - 1 + \frac{n+\zeta-q\lambda}{\nu r_1}\Big)\langle r_1|S_n^r\rangle \\
&\quad -\sqrt{(n+1)(n+2\zeta)}\,\frac{1}{\nu r_1}\,\langle r_1|S_{n+1}^r\rangle\Big]\frac{\langle S_n^r|r_2\rangle}{n-\eta+\zeta}\,\Omega_{jlm}(\hat{e}_1)\Omega_{jlm}^\dagger(\hat{e}_2)\,; \\[4pt]
c_2 &= \frac{Z/c}{(\kappa+q\lambda)}c_1\,(-i\vec{\sigma}\hat{e}_2)\,; \\[4pt]
c_3 &= \frac{Z^2/c^2}{4q\lambda}\frac{\kappa+q\lambda}{Z/c}\Big[\Big(-\frac{E(\kappa-q\lambda)}{\eta(c^2+E)} - 1 + \frac{n+\zeta-q\lambda}{\nu r_1}\Big)\langle r_1|S_n^r\rangle \\
&\quad -\sqrt{(n+1)(n+2\zeta)}\,\frac{1}{\nu r_1}\,\langle r_1|S_{n+1}^r\rangle\Big]\frac{\langle S_n^r|r_2\rangle}{n-\eta+\zeta}\,(-i\vec{\sigma}\hat{e}_1)\,\Omega_{jlm}(\hat{e}_1)\,\Omega_{jlm}^\dagger(\hat{e}_2)\,; \\[4pt]
c_4 &= \frac{Z/c}{(\kappa+q\lambda)}\,c_3\,(-i\vec{\sigma}\hat{e}_2)\,.
\end{aligned} \tag{4}$$

Here the following abbreviations are used: Ω_{jlm} denote the spherical spinor harmonics, κ is the eigenvalue of the spin-orbit operator, $\lambda = \sqrt{\kappa^2 - Z^2/c^2}$. The index $q = \pm 1$, and $\eta = (EZ/c)(c^4 - E^2)^{-1/2}$. The generalized Sturmian functions for the radial components of the second-order Dirac equation are

$$\langle \vec{r}|S_n^r\rangle \equiv \sqrt{\frac{n!(2\nu)^2}{\Gamma(n+2\zeta)Z}}\,(2\nu r)^{\zeta-1}\,\exp(-\nu r)\,L_n^{2\zeta-1}(2\nu r)\,, \tag{5}$$

where $\zeta \equiv \lambda + 1/2 + q/2$, $\nu \equiv (\sqrt{c^4 - E^2})/c$ and L_b^a denote generalized associated Laguerre polynomials.

The competing influences of retardation and relativistic effects, as compared to the simpler non-relativistic dipole approximation, have been a subject of active discussion. Notably, it was realized in single-photon ionization that the non-relativistic dipole, angle-integrated, cross section is accurate enough far beyond its expected range of validity[6]. A commonly accepted argument to account for this property of one-photon bound-free transition amplitudes, starting from an initial s-state, is due to partial cancellations between the contributions from retardation and relativistic structure to the cross sections. Concerning two-photon bound-bound processes, the state of the discussion is much less advanced, focussing on Rayleigh $|1^2S_{1/2}\rangle \rightarrow |1^2S_{1/2}\rangle$ scattering amplitudes. In this case, the presence of the Green's function in the second-order amplitude modifies the interplay between the contributions of retardation and relativistic effects and the conclusions drawn for the single-photon ionization cannot be transposed without care[7]. We have investigated this point in the case of the two-photon absorption from the ground state towards states in the L and M shell for linear polarization[5]. Here we concentrate on a comparison of the relativistic dipole and relativistic retarded results for linear and circular polarization for the transition towards $|3^2D_{3/2}\rangle$ and $|3^2D_{5/2}\rangle$.

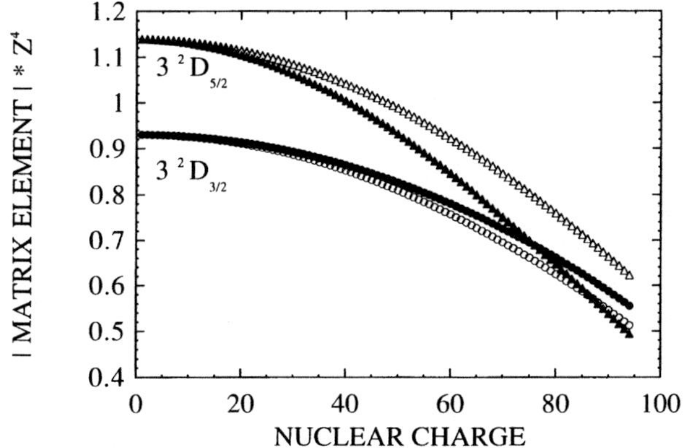

Figure 1. Linear Polarization. Modulus of the matrix elements M for the transition $|1^2S_{1/2}\rangle \rightarrow |3^2D_{3/2}\rangle$ (circles) respectively $|1^2S_{1/2}\rangle \rightarrow |3^2D_{5/2}\rangle$ (triangles) in atomic units (a.u.) multiplied by a factor of Z^4 and converged to one part in 10^6 as a function of nuclear charge Z. The open symbols denote the relativistic dipole (RD), and the full symbols the relativistic retarded (RR) results.

Consider Figure 1 for the case of linear polarization. Here, the general non-relativistic Z^4 dependence has been removed. For $|3^2D_{3/2}\rangle$ the relativistic dipole value (open circles) is $|M_{\mathrm{RD}}|(Z = 1) \simeq 0.9301\,\mathrm{a.u.}$, while for $|3^2D_{5/2}\rangle$ (open triangles) it is $|M_{\mathrm{RD}}|(Z = 1) \simeq 1.1392$ a.u.. This ratio of about $\sqrt{3/2}$ corresponds to the ratio $6 : 4$ of the number of accessible final m_j-substates in the excitation. With increasing Z the magnitude of the matrix element is lowered and due to fine structure splitting this $\sqrt{3/2}$-ratio does not hold any more. The differences amount to $\approx 0.5\%$ for $Z = 58$. Concerning retardation — full symbols for both states —, its influence on the $|3^2D_{3/2}\rangle$ transition is to increase the matrix element, while for the $|3^2D_{5/2}\rangle$ excitation the matrix element decreases. Several numerical values are compiled in Table 1.

Table 1. Linearly polarized light. Matrix elements M for the transitions $|1^2S_{1/2}\rangle \rightarrow |3^2D_{3/2}\rangle$ and $|1^2S_{1/2}\rangle \rightarrow |3^2D_{5/2}\rangle$ in atomic units converged to one part in 10^6 for several selected ions multiplied by Z^4 as a function of nuclear charge Z. The abbreviations are: relativistic dipole (RD), and relativistic retarded (RR).

	$3^2D_{3/2}$		$3^2D_{5/2}$	
Z	RD	RR	RD	RR
1	−0.930131	−0.930138	−1.139171	−1.139145
32	−0.880348	−0.887708	−1.076463	−1.051504
63	−0.739096	−0.764449	−0.900116	−0.818274
94	−0.512770	−0.555414	−0.622718	−0.494017

In Figure 2 the case of circular polarization is shown. Again, the general non-relativistic Z^4 dependence has been removed. For $|3^2D_{3/2}\rangle$ the relativistic dipole value (open circles) is $|M_{\mathrm{RD}}|(Z = 1) \simeq 0.805517$ a.u. , while for $|3^2D_{5/2}\rangle$ (open triangles) it is $|M_{\mathrm{RD}}|(Z = 1) \simeq 0.986551$ a.u. . Again, the ratio of about $\sqrt{3/2}$ corresponds to the ratio $3 : 2$ of the number of accessible final m_j-substates in the excitation. In this case, both the influence of the relativistic structure and of the retardation (full symbols) are to lower the matrix element with increasing Z for both fine structure components. Several numerical values are compiled in Table 2.

Table 2. Matrix elements M for the transitions $|1^2S_{1/2}\rangle \rightarrow |3^2D_{3/2}\rangle$ and $|1^2S_{1/2}\rangle \rightarrow |3^2D_{5/2}\rangle$ in atomic units converged to one part in 10^6 for several selected ions multiplied by Z^4 as a function of nuclear charge Z. The abbreviations are: relativistic dipole (RD), and relativistic retarded (RR). Circularly polarized light.

	$3^2D_{3/2}$		$3^2D_{5/2}$	
Z	RD	RR	RD	RR
1	−0.805517	−0.805515	−0.986551	−0.986535
32	−0.762404	−0.760028	−0.932244	−0.916846
63	−0.640076	−0.632148	−0.779523	−0.729226
94	−0.444072	−0.430938	−0.539290	−0.459802

In brief, the results of our original calculations show that, for two-photon bound-bound transitions, relativistic contributions are indeed significant, starting even before $Z = 20$. It appears also that no significant cancellations take place when adding up the contributions of retardation and relativity because we find that the contribution due to relativistic structure is dominating. Furthermore, the overall influence of relativity is to decrease the magnitude of the transition amplitudes as compared to the usual non-relativistic dipole treatments. Such an influence is opposite to the one observed in the case of single-photon bound-free transitions, where relativistic effects tend to increase the cross sections[4]. The magnitude and the sign of the contribution from retardation, however, depends crucially on the accesible final m_J sublevels as the analysis for different light polarization clearly shows.

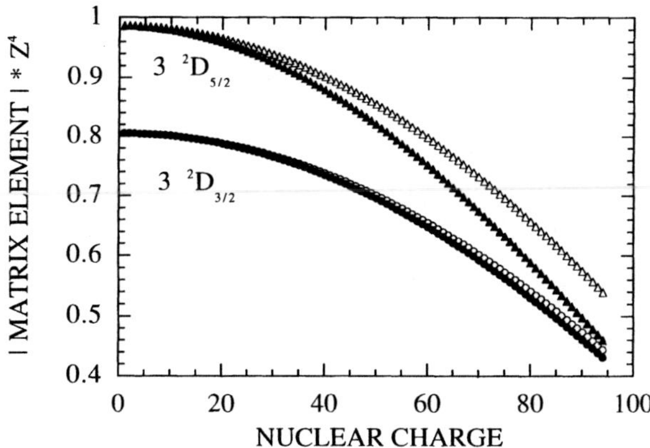

Figure 2. Circular Polarization. Modulus of the matrix elements M for the transition $|1^2S_{1/2}\rangle \to |3^2D_{3/2}\rangle$ (circles) respectively $|1^2S_{1/2}\rangle \to |3^2D_{5/2}\rangle$ (triangles) in atomic units (a.u.) multiplied by a factor of Z^4 and converged to one part in 10^6 as a function of nuclear charge Z. The open symbols denote the relativistic dipole (RD), and the full symbols the relativistic retarded (RR) results.

CS acknowledges a fellowship HSP II / AUFE from the German Academic Exchange Service (DAAD) financed by the German Ministry of Education, Science, Research and Technology. This research is supported as well in part by the European Union under the HCM contract ERB CHRX CT 940 470. The Laboratoire de Chimie Physique – Matière et Rayonnement is Unité de Recherche Associée au CNRS (URA 176).

1. see, for instance, "High-field interactions and short-wavelength generation", ed. by H. Milchberg and R. Freeman, special issue of *J. Opt. Soc. Am.* **B13**, 51:222 (1996); **B13**, 314:468 (1996).

2. D. Xenakis, O. Faucher, D. Charalambidis, and C. Fotakis, "Observation of two-XUV-photon ionization using harmonic generation from a short, intense laser pulse", *J. Phys. B* **29**, L457:L463 (1996).

3. C. Szymanowski, V. Véniard, R. Taïeb, and A. Maquet, "Two-photon transitions in hydrogenic atoms via a Sturmian basis for the Dirac Coulomb Green's function", *Europhys. Lett.* **37**, 391:396 (1997).

4. S. A. Zapryagaev and N. L. Manakov, "Primenenie funkcii grina uravneniya diraka k issledovaniyu relyativistskich i korrelyacionnych effektov v mnogozaryadnych ionach", *Izv. Akad. Nauk SSSR* **45**, 2336:2353 (1981); S. A. Zapryagaev, N. L. Manakov, and V. G. Pal'chikov. "Teoriya mnogozaryaryadnych ionov c odnim i dvumya elektronami", Energroatomizdat, Moscow (1985).

5. C. Szymanowski, V. Véniard, R. Taïeb, and A. Maquet, "Relativistic calculation of two-photon bound-bound transitions amplitudes in hydrogenic atoms", *Phys. Rev. A* **56**, 700:711 (1997).

6. A. Ron, I. B. Goldberg, J. Stein, S. T. Manson, R. H. Pratt and R. Y. Yin, "Relativistic, retardation and multipole effects in photoionization cross sections: Z, n, and l dependence", *Phys. Rev. A* **50**, 1312:1320 (1994).

7. A. Costescu, P. M. Bergstrom, Jr. , C. Dinu, and R. H. Pratt, "Retardation and multipole effects in Rayleigh scattering by hydrogelike ions at low and x-ray photon energies", *Phys. Rev. A* **50**, 1390:1398 (1994).

250

Laser Wakefield Acceleration Experiments

H. Kotaki[1], K. Nakajima[1,2], M. Kando[1,4], H. Ahn[1], T. Watanabe[3], T. Ueda[3],
M. Uesaka[3], H. Nakanishi[2], A. Ogata[2] and K. Tani[1]

[1]Japan Atomic Energy Research Institute (JAERI),
2-4, Shirakatashirane, Tokai, Ibaraki, Japan
[2]National Laboratory for High Energy Physics (KEK)
1-1, Oho, Tsukuba, Ibaraki, Japan
[3]Nuclear EngineeringResearch Laboratory (NERL),
The University of Tokyo,
2-22, Shirakatashirane, Tokai, Ibaraki, Japan
[4]Institute for ChemicalResearch, Kyoto, Japan, Kyoto University,
Uji, Kyoto, Japan

Laser-driven particle accelerators have been conceived over the past decade to be the next-generation particle accelerators, promising super-high field particle acceleration and a compact size compared with conventional accelerators [1]. Among a number of laser accelerator concepts, laser wakefield accelerators have great potential to produce ultra-high-field gradients of plasma waves excited by intense ultrashort laser pulses [2]. Recently wakefield excitation of the order of ~10GeV/m in a plasma has been directly confirmed by the use of a table-top-terawatt (T^3) laser [3].

Our project aims at achievinghigh energy particle acceleration to energies more than 1 GeV in a table-top scale owing to a channel-guided laser wakefield acceleration (LWFA) scheme by the use of 100 fs, 2 TW T^3 laser system.

We have demonstrated the self-channeling of ultrashort laser pulses with a relativistic intensity over a few cm [4]. We have achieved synchronization of a 10 ps electron beam with a 100 fs laser pulse within a few ps in order to accelerate injected electrons firmly due to wakefield induced by laser pulses in the rate of 10 Hz [5].

Fig. 1 shows the distribution of the energy gain of accelerated electrons at the Helium pressure of 20 Torr for the laser peak power of 0.5 TW. The energy gain of 100 MeV was obtained from this figure.

As a next step, we are planning to measure the frequency and the amplitude of the plasma wakefield by using interferometry technique [3].

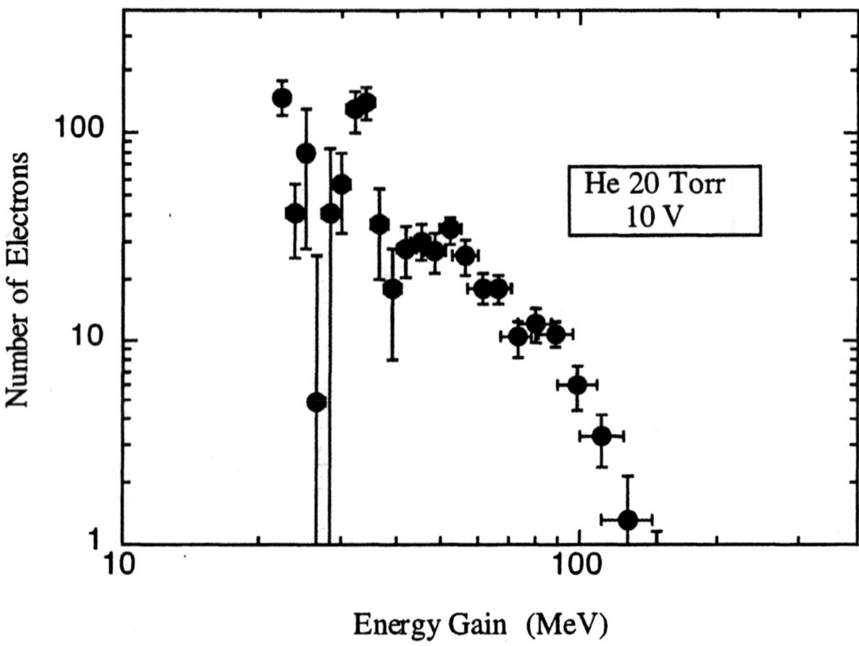

Fig. 1 Distribution of the energy gain of accelerated electrons at the Helium pressure of 20 Torr for the laser peak power of 0.5 TW.

References

1) T. Tajima and J. M. Dawson, Phys. Rev. Lett., **43**, 267 (1979)

2) K. Nakajima et al., Phys. Rev. Lett., **74**, 4428 (1995)

3) J. R. Marques et al., Phys. Rev. Lett., **76**, 3566 (1996) ; C. W. Siders et al., Phys. Rev. Lett., **76**, 3570 (1996) ; C. W. Siders et al., IEEE Trans. Plasma Sci., **24**, 301 (1996)

4) M. Kando et al., contributed to " '96 Advanced Accelerator Concepts", Lake Tahoe, California, USA, Oct. (1996)

5) M. Uesaka et al., contributed to " '96 Advanced Accelerator Concepts", Lake Tahoe, California, USA, Oct. (1996)

FEMTOSECOND HARMONIC LASER PHOTOEMISSION: PHYSICS AND CHEMISTRY

R. Haight
IBM T.J. Watson Research Center
PO Box 218
Yorktown Hts., N.Y. 10598

High harmonic generation, one of the main focal points of this conference, provides an exceptional experimental capability through its application in photoelectron spectroscopy. The high harmonics produced when an intense femtosecond laser pulse is focussed into a puff of rare gas represents, at present the only widely tunable laboratory source of radiation available. Additionally, with the added benefit of the femtosecond nature of the harmonic pulses, time resolved experiments can be carried out. Excite-probe experiments, in which electrons are excited into normally empty states of the system under investigation, and then probed with photoelctron spectrsocopy, provide the means to follow the dynamic evolution of those electrons with exceptional time resolution. A wide variety of studies of this nature have been described in a recent review[1].

Two recent experiments[2,3,4], the first involving the inorganic semiconductors $MoSe_2$ and WSe_2 and the second an organic "semiconducting" electron transporting material, tris (8-hydroxyquinoline) Al (Alq), highlight the wide application of this technique. To carry out these experiments, an amplified dye laser system operating at 610 nm was used to generate the high harmonics in Ar gas. The amplified dye laser system provided 0.5 mJ, 200-250 fs pulses at a repetition rate of 540 Hz. Fig. 1 shows a schematic of this system[5]. At the time of this writing, this system has been replaced by a more conventional and significantly more reliable kHz Ti:sapphire system.

The soft X-ray harmonic light for the photoelectron spectroscopy was generated by focussing the intense 610 nm pulses into a burst of Ar gas produced at the output of a pulsed valve situated within a vacuum beam line connected to the analysis chamber. Individual harmonics of interest were selected with a grazing incidence toroidal grating integral with the vacuum beam line. We worked with a variety of harmonics appropriate to the specific experiment being carried out. The beam line is connected to an ultrahigh vacuum analysis chamber which houses the sample mounted on a multiaxis manipulator, and a large collection angle (~1.1 sr.) time-of-flight electron energy analyzer. Overlap of the excite pulse consisting of either a 250 fs, 610 nm ($MoSe_2$ and WSe_2) or 70 ps, 355 nm (Alq) pulse of light, with the selected harmonic generates photoelectron spectra of the excited sample.

Applications of High Field and Short Wavelength Sources
Edited by DiMauro *et al.*, Plenum Press, New York, 1998

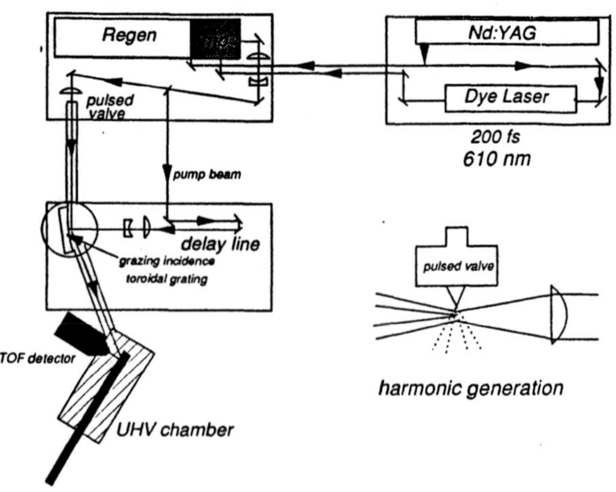

Fig. 1. Schematic of the femtosecond harmonic laser system, beam line and analysis chamber.

MoSe₂ and WSe₂

MoSe$_2$ and WSe$_2$ are members of a class of layered semiconductors called chalcogenides. These materials display interesting anistropies in their abilities to transport electronic charge[6]. Their strong absorption of visible radiation makes them attractive candidates for use in solar cells. In their crystalline form these materials form trilayers, consisting of a metal layer bounded above and below by chalcogen atoms. Van der Waals forces bind the tri-layers together to form a solid and this weak interaction allows for easy cleavage of the crystal to form clean surfaces. The nature of the bonding results in a band gap devoid of states and a surface which is nearly completely passive to chemisorption of foreign atoms.

Excite-probe photoemission experiments were carried out on these materials using a 250 fs 610 nm excitation pulse followed by a 22.4 eV (11th harmonic) probe pulse. Fig. 2 displays a representative photoelectron spectrum of photoexcited MoSe$_2$ collected at t=0, the nominal temporal overlap of the excite and probe pulses. States below 0 eV are normally occupied while those above, as evidenced by the small feature ~1 eV above the valence edge, are only observed when the system is photoexcited. A careful analysis gives a band gap of 1.1 eV which compares well with results from other experiments and the peak we observe is due to electrons occupying states at and above the conduction band minimum. The broad, asymetric peak indicates that at early times the excited electrons occupy states well above the conduction band edge. This conduction band signal was observed for harmonics ranging from 18.3 to 30.5 eV (9th-15th harmonics) although a resonant enhancement was observed for energies in the 22 to 26 eV range.

In order to more fully investigate the dynamics of these conduction electrons, the time dependence of the conduction band population was monitored as a function of the delay between the excite and probe pulses and is shown in Fig. 3. A rapid rise and extremely rapid decay is observed, particularly for MoSe$_2$. This rapid decay of the population, occuring mainly within the first 3 ps after excitation, cannot be fit with a model which incorporates simple diffusion of the electrons from the surface of the crystal.

To gain further insight into the electron dynamics, a detailed view of the excited conduction peak of WSe$_2$ is shown in Fig.4. At t=0 we observe a broad peak which has narrowed considerably by 1.33 ps. Importantly, note that the peak amplitueds at t=0 and 1.33 ps are virtually identical while nearly all of the decay in signal intensity is due to the loss of the high energy electrons. Further

254

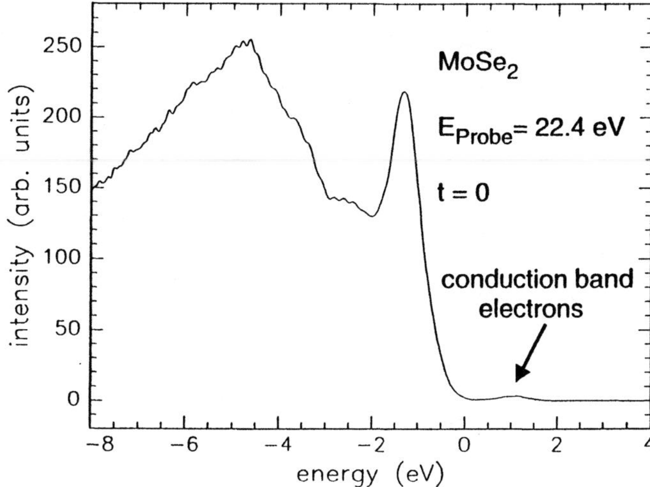

Fig. 2. Photoelectron spectrum of photoexcited MoSe$_2$ at t=0.

along in time we observe a continued narrowing of the signal while by 8 ps the peak amplitude has dropped as well. The bottom panel shows the difference between signals collected at t=0 and 1.33 ps showing nearly all of the change is due to loss of high energy electrons. This loss is observed to be bimodal in that at very early times the highest energy electrons transit rapidly into the interior of the crystal, while a much slower transport of electrons occurs at lower energies. This bimodal, energy dependent behavior can be understood in terms of the crystalline and energetic structure of the layered semiconductors. Transport within the plane of the layers is relatively free, while transport perpendicular to the planes is hindered by the energetic barrier which exists perpendicular to the planes. We also note that the wavefunctions near the conduction minimum are composed primarily of Mo-d_{xy} and Se-p_x character where x and y are in the plane of the layers[7]. At slightly higher energies, Mo-$d_z{}^2$ dominates, resulting in greater overlap of the wavefunctions of the layers. Therefore transport rates are highly energy dependent leading to the fast loss of carriers we observe at high energies and early times. As those carriers are lost, we observe at later times only those electrons which have occupied the lower energy states and which, as a result, diffuse into the interior much more slowly.

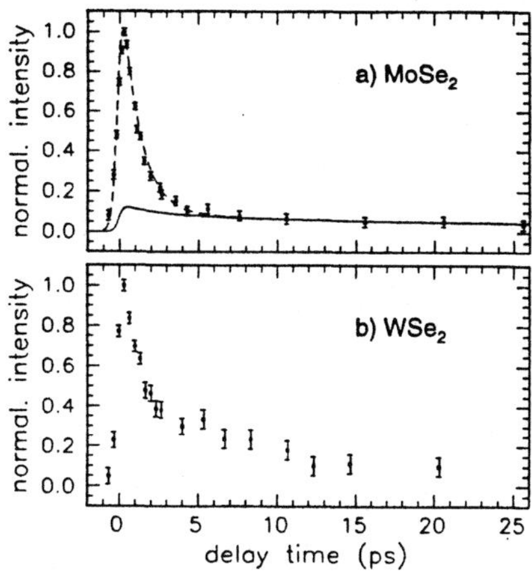

Fig.3 Delay curves for MoSe$_2$ and WSe$_2$

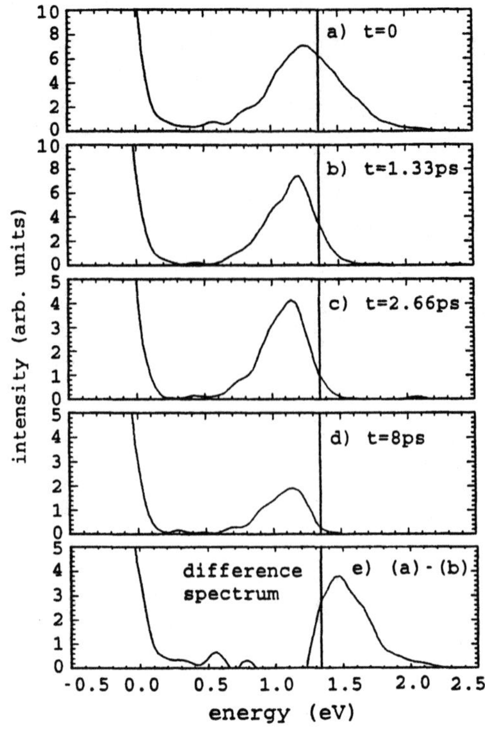

Fig. 4. Photoexcited conduction band peaks as a function of delay between excite and probe.

Tris (8-hydroxyquinoline) Al (Alq)

As a second example of the utility of harmonic laser photoemission, we trained our sights on a relatively new class of electron transporting materials which have found use in ultrathin electroluminescent structures known as organic LEDs (OLEDs) [8]. The basic structure of an OLED consists of an anode, typically a high work function material covered by a hole transporting layer. The hole transporting material is used to inject holes into an electroluminescent electron transporting layer called tris (8-hydroxyquinoline) Al (Alq) which finally covered with a low work function cathode material. Recombination of the electrons and holes occurs within the Alq layer giving rise to bright light emission in the green part of the spectrum. The entire OLED is only several thousand angstroms in thickness and is fabricated by sequentially evaporating the constituent organic materials. The ease of fabrication makes OLED an attractive alternative to crystalline inorganic semiconductor diodes fabricated from GaAs.

A number of important physical properties of Alq are presently under study and little is known about the surfaces and interfaces formed between Alq and a host of metals and semiconductors that are candidate electrode materials. In order to examine these issues experiments were carried out to investigate the formation of cathode/Alq interfaces. In such experiments a layer of Alq w first evaporated onto a conducting substrate. Subsequently, other materials were evaporated ont the Alq and photoemission spectra were collected in order to assess the changes in chemistry an structure at the interface.

Fig.5 shows a series of spectra collected when Ga was deposited onto a thin layer of Alq[3]. The bottom spectrum was collected from clean Alq and shows the distinct molecular structure of the

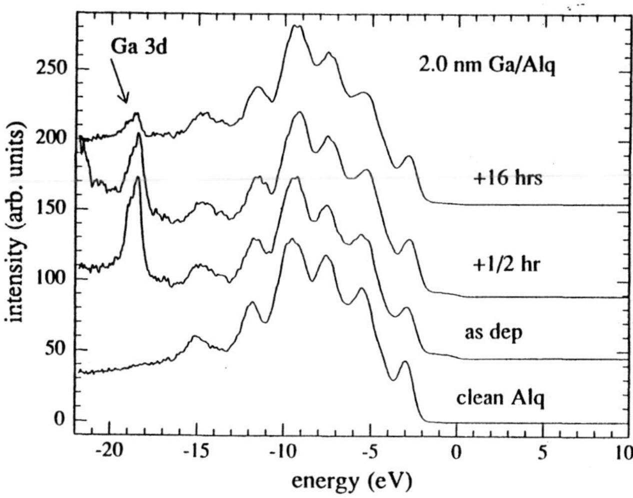

Fig. 5. Photoelectron spectra collected from clean and Ga covered Alq for photon energy of 34.6 eV.

organic. After deposition of a small amount of Ga a number of features are observed in the photoemission spectrum. We note a new feature at a binding energy of -18 eV which derives from the Ga 3d atomic core level emission. With the wide range of harmonics available it is possible to tune the wavelength of the light to energies sufficiently high to access atomic core levels. In addition we note that the presence of the Ga has little affect on the molecular levels of the Alq. With higher amounts of Ga we note the presence of a small foot of states above the highest occupied molecular orbital (HOMO) of the Alq. These observations show that the Ga interacts very weakly with the Alq and forms islands which display a metallic nature giving rise to the foot of states up to the Fermi level of the system. Quite importantly we note that after waiting for ½ hour, the Ga signal begins to diminish with significant loss at 16 hrs. This observation indicates that the Ga is diffusing into the Alq. Diffusion of materials into Alq was also observed for ZnS deposited at room temperature. In those experiments, the Zn 3d core level was monitored as a function of time and observed to decrease in a manner similar to the Ga. We concluded that materials which only weakly interacted with the Alq diffused into the interior of this material. Such diffusion has important and wide ranging implications regarding the operation of an OLED. If cathode materials are deposited onto the Alq which do not strongly interact with the organic, diffusion may occur which can result in electrical shorts thereby rendering the OLED inoperable. In order to address this issue, Ca was chosen as a candidate diffusion barrier material. Ca is a low work function material that might be expected to strongly interact with the Alq. To investigate the strength of this interaction, a series of valence spectra were collected with increasing Ca deposition (see Fig. 6). With only 0.2 nm Ca, a strong change in the molecular structure of the Alq was observed. Above ~ 1 nm, metallic Ca began to form as evidenced by the appearance of states up to the Fermi level as well as a peak at -4.3 eV associated with a Ca plasmon indicating the existence of Ca metal. Ca also proved to be an effective diffusion barrier with no loss of intensity of the Ga after deposition on top of 1 nm of Ca for up to several days after deposition.

It was also of interest to determine if it was possible to photoexcite electrons into the lowest unoccupied molecular orbital (LUMO) of the Alq[4]. Since the band gap of Alq[9] is reported to be 2.9 eV, we used 70 ps, 355 nm (3.5 eV) pulses of light generated from the frequency doubled output of our Nd:YAlG regenerative amplifier to photoexcited the Alq. Fig. 7 shows the unexcited and photoexcited photoelectron spectra collected from a clean Alq sample using 22.4

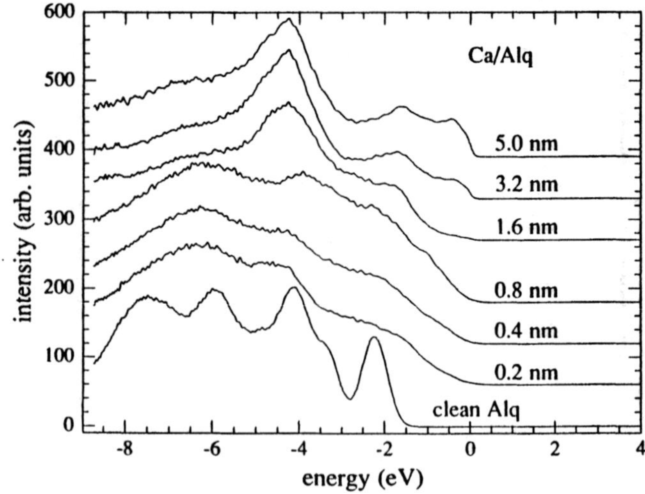

Fig. 6. Photoelectron spectra of Alq valence states for increasing coverages of Ca. Photon energy=22.4 eV.

eV (11th harmonic) light. We observe a small but clear signal from LUMO electrons when the excite and probe pulses are overlapped in time.

The time dependence of the LUMO signal was determined and is shown in Fig. 8 out to 400 ps. Since the fluorescence lifetime of the Alq has been measured to be 12 ns[9], the decay we observe cannot be due to bulk electron-hole radiative recombination. Electron mobilities of 10^{-4} cm^2 /Vs cannot, within a simple diffusion equation model, account for the loss of carriers within the

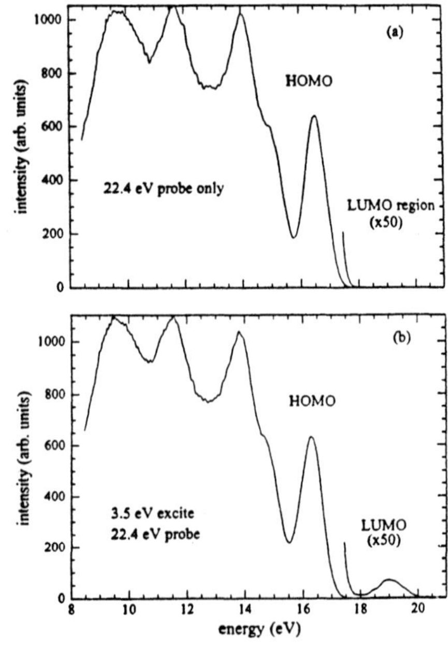

Fig. 7. Photoelectron spectra of unexcited and photoexcited Alq. Photon energy=22.4 eV.

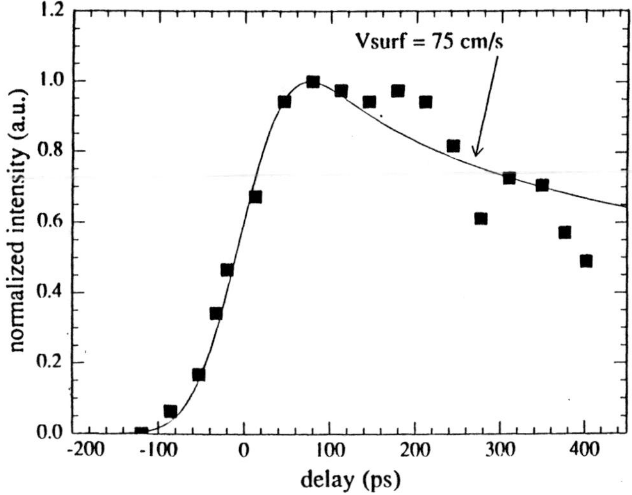

Fig. 8. Time dependence of Alq LUMO state.

LUMO on the observed time scale. By including a surface recombination velocity of 75 cm/s an acceptable fit was obtained. A surface recombination velocity is a phenomenological term which refers to a flux of electrons and holes which reach the surface and recombine due to the two-dimensional nature of the surface as well as the existence of defects which give rise to non-radiative electron-hole recombination. For a comparison a heavily damaged Si surface can have surface recombination velocities as high as 10^6 cm/s while for an etched high quality surface the velocity drops to ~100 cm/s. Close inspection of the region between the HOMO and the LUMO reveals a non-zero density of states which suggests that such a non-radiative mechanism may be operating at the surface of Alq.

CONCLUSIONS

In this brief paper, dynamics involving electrons as well as atoms have been described. Photoelectron spectroscopy carried out with the harmonics of intense femtosecond lasers has proven to be a powerful tool for studying dynamics on conventional as well as "exotic" materials. As laser technology improves and our understanding of the harmonic generation process increases new experiments utilizing harmonics will provide even greater insight into the properties and behavior of condensed matter.

Acknowledgements

I wish to express my gratitude to Armin Rettenberger from the University of Konstanz, and Matthias Probst from the University of Stuttgart, for their important contributions to this work. Both Rettenberger and Probst were primarily responsible for the collection and analysis of the data presented in this paper.

REFERENCES

1. R. Haight, Surf. Sci. Rep., 21, 275 (1995)

2. A.R. Rettenberger, P. Leiderer, M. Probst, R. Haight, Phys. Rev. B, 56, 12092 (1997).

3. M. Probst, R. Haight, App. Phys. Lett. 70, 1420 (1997).

4. M. Probst, R. Haight, App. Phys. Lett., 71, 202 (1997).

5. R. Haight, D.R. Peale, Rev. Sci. Instrum. 65, 1853 (1994).

6. R.C. Firas, Ph.E. Schmid, in "Optical and Electrical Properties of Layered Semiconductors", D. Reidel Pub. Co (1976).

7. R. Coehoorn, C. Haas, R.A. Groot, Phys. Rev. B 35, 6203 (1987).

8. C.W. Tang, S.A. van Slyke Appl. Phys. Lett. 51, 913 (1987).

9. I Sokolik, R. Priestly, A.D. Walser, R. Dorsinville, C.W. Tang Appl. Phys. Lett 69, 4168 (1987).

TWO-PHOTON IONIZATION AND THREE-PHOTON ABOVE-THRESHOLD IONIZATION OF ARGON

A. Bouhal[1], G. Hamoniaux[1], A. Mysyrowicz[1], A. Antonetti[1],
P. Breger[2], P. Agostini[2],
R. C. Constantinescu[3], H. G. Muller[3]
L. F. DiMauro[4]
[1]Laboratoire d'Optique Appliquée, ENSTA-Ecole-Polytechnique,
91125 Palaiseau, France
[2]Service des Photons, Atomes et Molécules
DSM-DRECAM CEA Saclay, 91191 Gif-sur-Yvette France
[3]FOM-Institute for Atomic and Molecular Physics Kruislaan 407, 1098 SJ
Amsterdam, The Netherlands
[4]Brookhaven National Laboratory, Chemistry Department
Upton 11973 NY

INTRODUCTION

Studies of nonlinear laser-matter interaction have been so far limited to wavelengths from the near Ultraviolet to Infrared, because of the low brightness of currently available sources outside this range. However nonlinear processes in the VUV/Soft Xray domain would initiate multiphoton innershell spectroscopy, XUV nonlinear optics and applications of such processes to metrology (for instance, autocorrelation measurements of ultrashort XUV pulses).

The probability of multiphoton transitions decreases rapidly with the number of photons involved. A typical two-photon bound-free transition has a rate of the order of .1 ps^{-1} at 10^{12} W.cm^{-2} and scales as the square of the intensity. Such an intensity is therefore required to saturate a two-photon transition with a 100 fs pulse. This is difficult to achieve with the present state-of-the-art techniques for producing intense XUV pulses. To our knowledge, only two cases of such transitions have been reported so far. The first one is a two-photon ionization of argon by the third harmonic of a KrF laser[1]. Since the photon energy (15 eV) is just below the ionization energy (15.75 eV) for argon, the transition is quasi-resonant. The second case is a two-photon ionization of helium by the 9th harmonic of a Ti:S laser[2]. In the latter case an autocorrelation measurement of the harmonic pulse has been reported. In the present work, we report on two-photon ionization of argon at 133 nm (9.3 eV) from the third harmonic of a frequency doubled Ti:S laser, and a three-photon above-threshold ionization involving two 9.3 eV photons and one 3.1 eV photon.

Applications of High Field and Short Wavelength Sources
Edited by DiMauro *et al.*, Plenum Press, New York, 1998

SETUP

The output of a CPA Ti:Sapphire laser (800nm, 150fs, 30 mJ, 10 Hz) is frequency-doubled and focused to a 50 micron diameter spot by a 1 m focal length lens into a pulsed xenon jet whose backing pressure is 0.5 bar. The unconverted 800 nm radiation is absorbed in a BG38 color filter. The nearly flat-top spatial mode of the 400 nm light is converted into an annular profile by a circular disk blocking the center of the beam and an iris limiting the outer diameter. The disk is imaged on a pinhole (located at 90 cm downstream from the jet) by the lens. The pinhole stops the 400 nm beam while transmitting the harmonics produced in the jet. The efficiency of this simple filtering device is not perfect though: the forward scattering in the gas, the beam refraction caused by the free-electron density resulting from ionizing the jet and the diffraction of the disk cause leakage of fundamental radiation into the spectrometer. The amount of residual 400 nm light in the beam after the pinhole is estimated to be of the same order as that of a typical harmonic. The amount of 800 nm is completely negligible due to the combined transmission of the BG38 filter and the disk-pinhole assembly. This gives a comparable efficiency to other type of filtering schemes but with the advantage that the

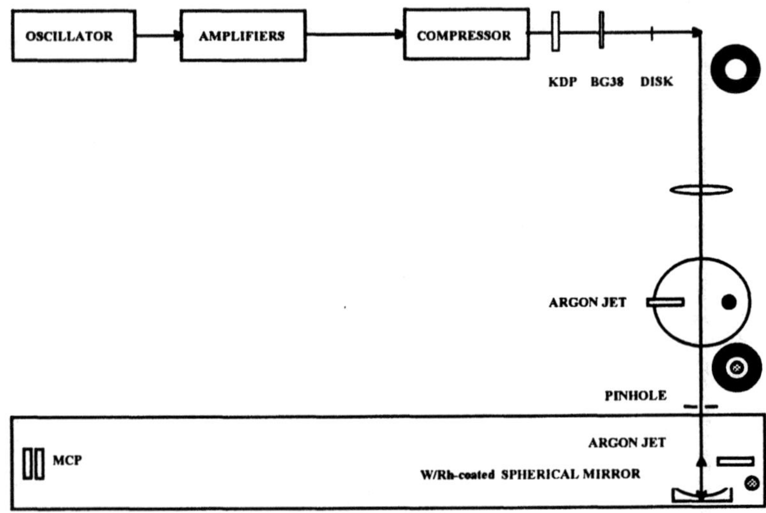

Fig. 1: Experimental setup. The beam cross-sections of the 400 nm (black) and harmonic (gray) light are shown at various positions.

harmonic pulse is not affected by the filtering process: in particular, no pulse broadening results from the use of a diffraction grating or similar dispersive devices.

In general, multilayer coated spherical mirrors provide both frequency selectivity and focusing. However in the present experiment, we use a broad-band coated mirror with a reflectivity of about 30% for the low-order (3rd-7th) harmonics (133-57 nm) (see Fig. 1). The focal length of the mirror is 35 mm and the focus position is externally adjustable.

The argon target gas is injected into interaction region of the spectrometer through a pulsed valve backed by a 0.5 bar pressure.

RESULTS

Figure 2 shows a typical electron energy spectrum of argon. Only two peaks are evident: a peak near 3 eV energy and a weaker peak at 6 eV. The first peak corresponds to a total

262

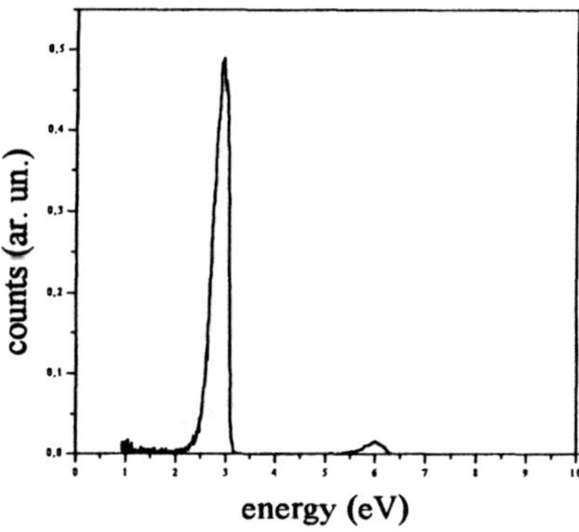

Fig. 2: Electron energy spectrum from ionization of argon by a combination of 133 nm, 80 nm, 57 nm and 400 nm photons (see discussion).

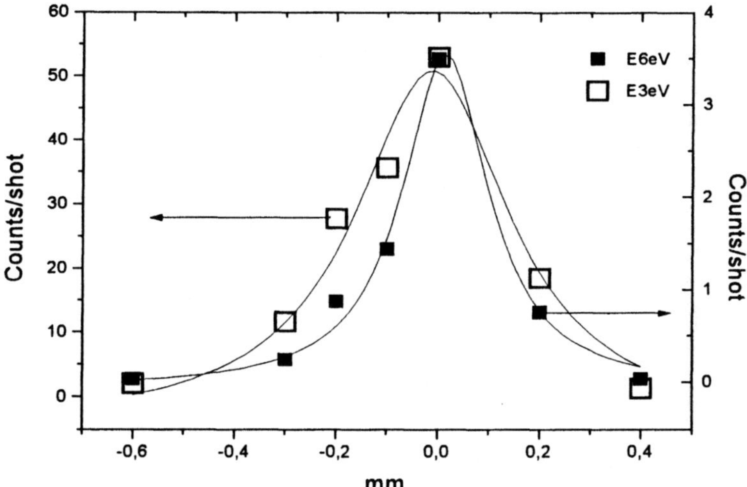

Fig. 3: Amplitudes of the two peaks in Fig. 2 versus the focus position inside the sensitivity zone of the spectrometer (squares). The solid lines are lorentzian fits to the data.

energy of six 400 nm photons. It can therefore be assigned either to the absorption of a fifth-harmonic photon (80 nm) and one 400 nm photon or two third-harmonic photons. The total energy of the second peak corresponds to the absorption of seven 400 nm photons. It can be interpreted in three ways: (1) one-photon ionization from the seventh-harmonic, (2) one fifth-harmonic plus two 400 nm photons and (3) two third-harmonic plus one 400 nm photons.

Additional experimental information is shown in Fig. 3 as a variation of the two peak amplitudes as a function of the mirror focus position inside the spectrometer. Both peaks are strongly position-dependent, i.e. on the intensity of the radiation producing them, whatever wavelengths are involved in the ionization process.

DISCUSSION

One difficulty in assigning the electron peaks is the lack of direct information on the respective intensities at 400, 133, 80 and 57 nm (neglecting the remaining 800 nm). For the 400 nm radiation, to weak to be measured directly, we rely on the estimate discussed above yielding an intensity of 5×10^{13} W cm^{-2}. The number of photons at the third harmonic (133 nm) deduced from known third order susceptibilities[3] at 1054 nm (6.46×10^{-26} m/V^2) should be of the order of 10^{11} per pulse, assuming a fundamental intensity of 5×10^{14} W cm^{-2} and an atomic density of 10^{16} cm^{-3} in the harmonic jet. This value is supported by direct measurements[4] which also provide the relative conversion efficiencies for harmonics 3, 5 and 7 at three fundamental wavelengths (1055, 616 and 308 nm). Interpolating these results to 400 nm yields a ratio of the intensity at 133 nm to that at 80 nm of 10^3. The intensity at 57 nm should not exceed 10^{-6} of that at 133 nm from the same data. Using a mirror reflectivity of 30% and a pulse duration determined by perturbative scaling $\tau/\sqrt{2}\sqrt{q}$ (where τ is the pulse duration at 800 nm and q is the harmonic order), we estimate the intensities at 400, 133, 80 and 57 nm as 2×10^{13}, 10^{13}, 10^{10} and 10^7 W cm^{-2}, respectively. It follows that the two-photon ionization rate at 133 nm is more than 100 times larger than the rate corresponding to the absorption of one 80 nm and one 400 nm photon, assuming (in the absence of known values) the two generalized cross-sections equal. The same argument rules out higher-order processes, involving more than one 400 nm photon. We conclude that the 3eV peak is mostly due to the two-photon ionization at 133 nm with a very weak contribution from the (80 + 400) nm transition.

The stronger dependence of the 6 eV peak on the focus position rules out a direct, one-photon ionization from the seventh harmonic, which would be linear in intensity and therefore independent upon focusing, consistent with the above calculation. The most likely mechanism for this peak is an ATI transition involving two-133 nm photon + one excess-photon at 400 nm.

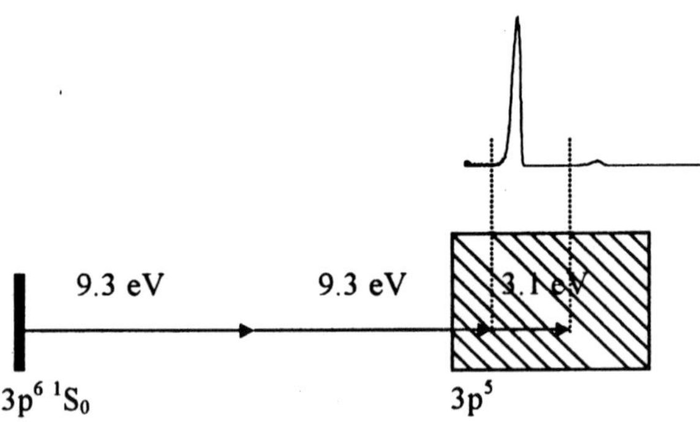

Fig. 4: Multiphoton transitions observed in argon.

Since the signal-dependence on the focal position reflects essentially the 300 μm size of the spectrometer sensitivity zone, the intensity dependence of the electron count cannot be deduced from the data in Fig. 3. Let us just stress again that the observed FWHM is consistent with a diffraction-limited focusing of the 133 nm radiation and a peak intensity at 133 nm of about 10^{13} W.cm^{-2}.

In summary, the electron energy spectra produced by focusing low-order harmonics of 400 nm radiation in argon are consistent with the two and three-photon ionization transitions illustrated in Fig. 4. This result is one of the first multiphoton transitions reported in the VUV range and the first one involving an ATI transition. The peak intensity at the higher harmonics is still too low, in the present state-of-the-art, to permit observation of nonlinear processes in the XUV domain. Both the conversion efficiencies and multilayer mirrors performance must be improved to obtain the intensity necessary to observe more interesting multiphoton innershell transitions. Nevertheless, it is already possible to perform autocorrelation measurements on large two-photon signals (like the 3 eV electron peak) as shown by Kobayashi e al[2].

REFERENCES

[1] Xenakis, O. Faucher, D. Charalambidis, and C. Fotakis, J. Phy. B, **29**, L457 (1996).

[2] Y. Kobayashi, T. Sekikawa, Y. Nabekawa, and S. Watanabe, post-deadline paper, Quantum Electronics and Laser Spectroscopy, Baltimore (1997).

[3] H. J. Lehmeier, W. Leupacher and A. Penzkofer, Opt. Comm, **57**, (1985).

[4] P. Balcou Thèse, Université Paris VI, 1993 Unpublished.

[5] P. Kruit and F. Read J. Phys. E **16**, 313 (1983).

[6] B. Carré, L. Lederof and P. Salières (Private Comm.).

ULTRA-FAST TIME-RESOLVED X-RAY DIFFRACTION DETECTED BY AN AVERAGING MODE STREAK CAMERA

Jörgen Larsson [1], Zenghu Chang[3], Ellen Judd[1], Phillip A. Heimann[2],
Aaron M. Lindenberg[1], Henry C. Kapteyn[3], Margaret M. Murnane[3],
Richard W. Lee[4], Anton Machachek[5], Justin S. Wark[5], Howard
A. Padmore[2], Roger W. Falcone[1]

[1] Department of Physics, 366 LeConte Hall, University of California
Berkeley, Berkeley CA 94720-7300, USA, E-mail:
jorgen@physics.berkeley.edu

[2] Advanced Light Source, Acceleration and Fusion Research Division,
Lawrence Berkeley National Laboratory, Berkeley, California 94720

[3] Center for Ultrafast Optical Science, The University of Michigan, Ann
Arbor, MI 48109-2099

[4] Lawrence Livermore National Laboratory, Livermore, USA

[5] Clarendon Laboratories, Oxford, UK

We have set-up an experiment in order to use ultra-fast time-dependant x-ray diffraction as a diagnostic to probe laser-induced phase transitions in solids. Such studies have previously been performed using laser probes in the infra-red, visible or ultraviolet spectral range. However, in order to directly probe structures in solids, and large molecules it is necessary to use x-ray radiation. In an ongoing project at the ALS synchrotron aimed at probing structural dynamics in solids using with a sub-picosecond time we have incorporated a streak camera with a temporal jitter smaller than 2 ps. We have used this streakcamera to observe a decrease in bragg reflectivity from single crystal InSb following laser irradiation.

We have established an experimental setup [1] (Figure 1) which utilize a Ti:Al_2O_3-based 100 fs, 1 kHz laser synchronized to a electron storage ring with jitter less than 10 ps. A

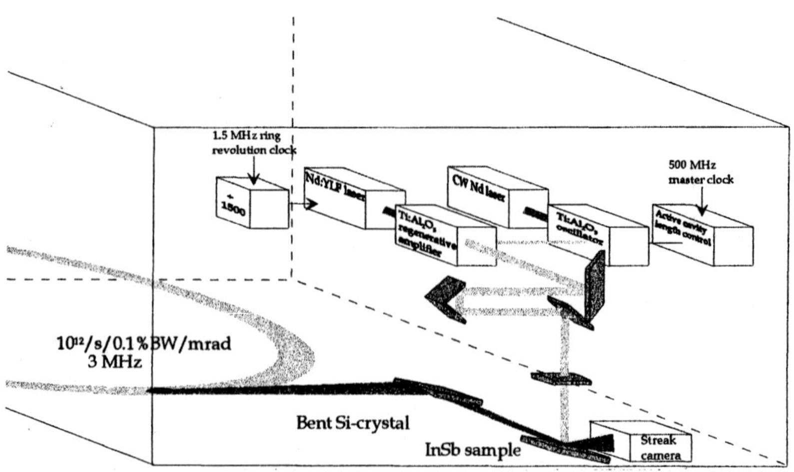

Figure 1. Experimental set-up

bending magnet beamline source emits a broad spectrum of x-ray radiation, extending to a useful photon energy of about 12 keV. The delay between the laser and the x-ray pulse can be varied using an optical delay line in the laser beam path. X-rays within a 0.4 mrad aperture are monochromatized and focused to a line using a cylindrically bent silicon wafer

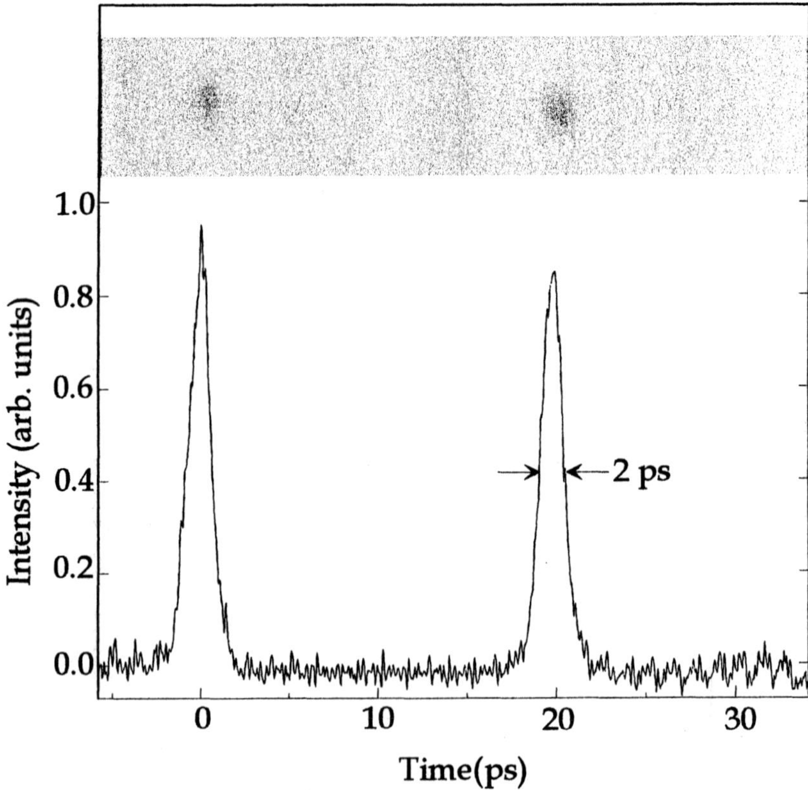

Figure 2. The temporal response of the ultrafast, photoconductive-switch triggered streak camera was measured to be 2 ps when averaging 5000 pulses.

cut in the (111) plane. The width of the focused line is typically 100 mm. In this study, the Bragg angle was chosen so that the x-ray photon energy was 5 keV. We introduced a 1 mm horizontal slit between t he Si wafer and the sample. This allowed us to uniformly irradiate, using the laser beam, a well-defined diffracting region of the sample. In our experiment we studied the disordering of laser-illuminated InSb. This well-studied material was used mainly from experimental considerations. Due to the fact that it is is a heavy element, the x-ray cross section is high leading to a wide rocking curve (strong signal) and high photo-absorption (x-ray penetration is matched with the laser penetration depth). Diffracted x-rays are detected by a streakcamera[1,2] operating in averaging mode[3]

The jitter of the camera was studied using UV pulses obtained by frequency mixing in KDP crystals. Since we did not obtain any single-shot data we can not give an accurate value for the trigger jitter, but specify the temporal multiple shot response to be 2 ps when averaging 5000 laser pulses. As a test of being able to look at picosecond time scale events in the x-ray spectral region we measured the pulse duration of the ALS operated at an electron energy of 1.9 GeV and a beam current of 30 mA in 2-bunch mode. We show the results of the measurements in Fig. 3. The data was obtained by averaging for 15 minutes at a 1kHz repetition rate. The obtained pulse duration of 70 ps is in good

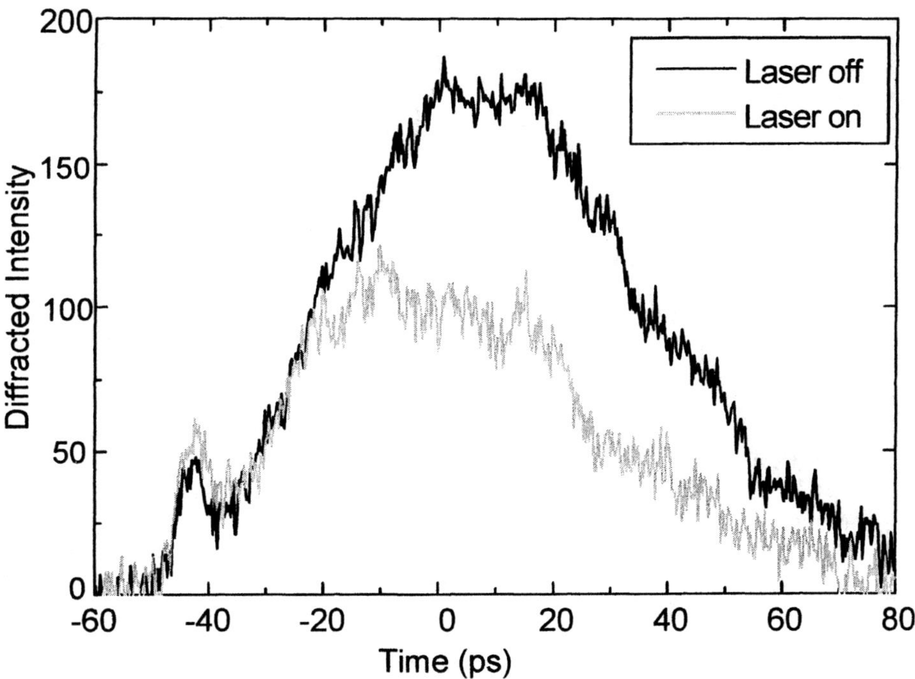

Figure 3. Time-resolved x-ray diffraction using a streakcamera. The laser is incident at t=-15 ps. This absolute timing has an error of 10 ps.

agreement with previous measurements of the ALS pulse duration using a fast photo diode and a streak camera sensitive to visible light[4].

From the lineouts in Figure 3, it can also be seen that when the laser irradiates the crystal, the intensity drops. From this relatively noisy data, it is not possible to set a time-scale for the disordering, but work on a high-flux beamline is in progress.

References

1. Z. Chang, A. Rundquist, J. Zhou, M. M Murnane, H. C. Kapteyn, X. Liu, B. Shan, J. Liu, L. Niu, M. Gong and X. Zhang: Appl. Phys. Lett. 69, 133 (1996)

2. Z. Chang, A. Rundquist, H. Wang H. C:. Kapteyn, M. Murnane, X. Liu, B. Shan, J. Liu, L.Niu, M. Gong, and X. Zhang, R. Lee, (To appear in *Proceedings of the 22nd International Congress on High Speed Photography and Photonics*, SPIE vol 2869, Santa Fe, NM, October 1996)

3. J. Larsson, Z. Chang, E. Judd, P.J. Schuck, R.W. Falcone, P.A. Heimann, H.A. Padmore, H. C. Kapteyn, P. Bucksbaum, M.M. Murnane, R. W. Lee, A. Machachek, J. S. Wark Opt Letters 22, 1012 (1997)

4. R. Keller, T. Renner and D. J. Mussiletti: Proceedings of the 7th Beam Instrumentation Workshop, Argonne, IL, May 6-7, 1996

RADIOLOGICAL APPLICATIONS OF HARD X-RAY

EMISSION FROM A LASER-PRODUCED PLASMA

M. Grätz[1], C. Tillman[1], A. Nykänen[1], L. Kiernan[1], C.-G. Wahlström[1],
S. Svanberg[1], K. Herrlin[2]

[1]Department of Physics, Lund Institute of Technology,
P.O. Box 118, S-221 00 Lund, Sweden
[2]Department of Radiology, University Hospital, S-221 85 Lund, Sweden

INTRODUCTION

Terawatt laser systems based on the chirped-pulse amplification technique readily deliver intense, ultra-short laser pulses, which can be focused down to intensities well above 10^{17} W·cm^{-2}. Focusing such intense pulses onto solid targets results in the generation of intense, energetic bursts of x-ray radiation[1]. Unique properties of this x-ray emission have lead to a strong interest in possible radiological applications which cannot be implemented with conventional x-ray tube technology.

Key-points for advanced applications are the very short x-ray pulse duration, the small source size and the spectral composition of the x-ray emission. These features allow for improved image quality and, in principle, a reduced patient exposure. Further, a laser-produced plasma source has the advantage of being relatively compact and less expensive than for example a synchrotron, which is another x-ray source that can be used for novel medical applications[2]. Apart from radiological applications discussed in this paper, x-ray radiation from laser-produced plasmas is also used in a large number of more research-orientated applications, e.g. ultrafast diffraction experiments and x-ray pump-and-probe schemes.

Most of the work on hard x-ray emission (> 10 keV) from laser-produced plasmas has been devoted to fundamental research in physics, and issues of importance for radiological applications have frequently not been taken into account. Only more recently, reports on application-relevant characteristics of laser-produced x-ray sources or novel imaging methods have been published[3-7].

In this paper, we summarize some of our earlier results on x-ray source characterization and present new results on novel imaging methods. The emphasis will be placed on results about scatter-reduced imaging. Finally, we give a comparison between conventional x-ray sources and a laser-produced x-ray source in the context of radiological applications.

Applications of High Field and Short Wavelength Sources
Edited by DiMauro *et al.*, Plenum Press, New York, 1998

X-RAY GENERATION AND CHARACTERIZATION

The Lund Laser Centre terawatt laser system[8] is used in the presented work. It typically delivers pulses of 110 fs duration, at 794 nm and 10 Hz repetition rate, with about 120 mJ of energy on target. In the target chamber, the laser pulses are focused with a parabolic mirror onto a rotating tantalum target[3]. The chamber is held at a backing pressure of about 10 hPa. The focused intensity on the target surface is of the order of 10^{17} W·cm^{-2}.

X-ray radiation from the laser-produced plasma is emitted in a broad spectral range, extending from below 1 keV up to several 100 keV or even MeV[1]. The spectrum consists of characteristic line radiation and a Bremsstrahlung-like continuum. We use single-photon analysis techniques[9] and crystal spectroscopy to study the x-ray spectrum[10]. In our time-resolved experiments, we use an x-ray streak camera (Kentech Instruments Ltd.) with a customized cathode (CsI, 5µm), optimized for hard x-ray detection[5]. The temporal resolution is limited to about 15 ps due to trigger jitter. From simulations based on hot electron diffusion, the hard x-ray pulse duration is predicted to be about 1 ps[11]. The x-ray source size is measured by pinhole imaging. The pinhole has a diameter of 5 µm in a platinum substrate (175 µm thick). The x-ray source size is measured to be about 35 µm (FWHM), at an average detected photon energy of about 40 keV (see Fig. 1). The source size does not change upon varying the laser pulse energy within a factor of 5, neither upon a change in prepulse-to-main-pulse contrast ratio from about 10^{-6} to 5×10^{-4}.

APPLICATIONS

Different radiological applications are feasible with a laser-produced x-ray source, and dosimetry becomes an important aspect. Therefore, we performed a comparative study on cell survival, where living cells were exposed to various equivalent doses of either laser-produced or conventionally produced x-rays[12]. The cell survival rate upon radiation from conventional sources was similar to the survival rates obtained from a laser-produced x-ray source. According to these measurements on purely cell survival no additional problems should arise when using laser-produced x-rays for imaging of living organisms.

Magnification Imaging and Short-Pulse Imaging

The small source size allows for sharp x-ray imaging of very small objects by magnification imaging, which is shown in Fig. 2 for the example of a star-test pattern. Line-patterns can be resolved down to a spatial frequency of about 23 lp/mm. This corresponds quite well to the source size measured with the pinhole. The ability to detect such small objects is of importance both in medical (e.g. microcalcifications in breast tissue) and technical imaging (e.g. detection of microcracks).

The most straight-forward use of the very short pulse duration is fast imaging with an

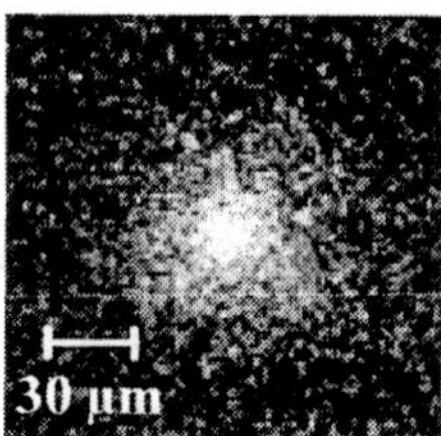

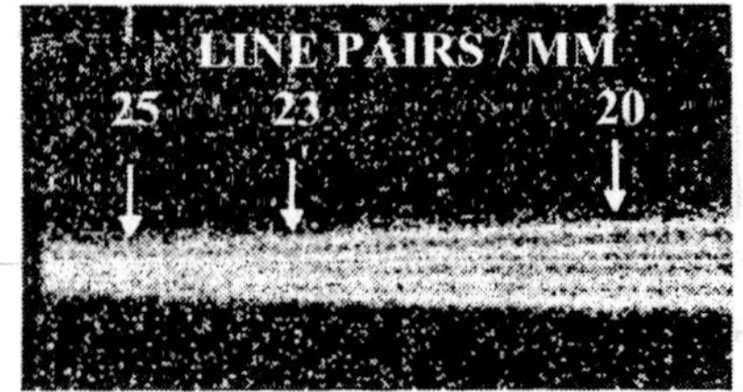

Figure 1: Pinhole image of the x-ray source. The FWHM is ~ 35 µm.

Figure 2: Image of a line pattern obtained with the x-ray source. The pattern can be resolved down to about 23 lp/mm.

exposure time of only a few picoseconds. The x-ray intensity is sufficiently high to obtain images of extended objects with a single x-ray pulse[3]. We imaged crater-formation processes induced by ns-pulse laser ablation, as an application example for such a technique[13]. The very short exposure time could be used for imaging of solid-state phase transitions or fast rupture processes. However, processes of radiological interest are quite rare on such short time scales.

Scatter Reduction by Time-Gated Imaging

Image deterioration due to scattered radiation has always been a major concern in medical radiology[14]. The potential of scatter-reduced imaging is not only an improved image quality (with unchanged dose exposure for the patient), but also a possible reduction in the dose exposure to the patient (while preserving the image quality).

The information content in an x-ray image corresponds to an absorption pattern within the imaged object and is produced by unscattered (=ballistic) photons. Scattered photons have lost most of their information, but contribute to the signal on the detector. Therefore, they produce a slowly varying background, which reduces the contrast of any structure in the image. The image information can be retrieved by contrast enhancement methods, though at the cost of an increased noise level. This can be compensated for by an increased exposure level, leading to an unwanted additional patient exposure.

Instead, spatial and temporal scatter-suppression methods can be used. The spatial suppression is realized by anti-scatter grids or air-gaps inbetween object and image detector, which suppress photons that are not following straight paths originating from the x-ray source[14]. Temporal filtering, which requires very short x-ray pulses, is based on the delay in arrival time between ballistic (unscattered) and scattered photons, due to their slightly longer path lengths. This concept of time-gated imaging, earlier applied in the visible domain[15,16], can also be used in x-ray imaging[5,6].

In this work, we present numerical simulations as well as experiments to investigate the theoretical limits and the practically achievable improvements through time-gated imaging. Special account is taken to the influence of the photon energy, the time resolution and the tissue composition.

Simulations. In order to distinguish between the influence of different parameters, we perform detailed simulations of photon scattering in tissues by means of a Monte Carlo photon migration code[5]. Data from this code give information on the temporal and spatial shape of an x-ray pulse after passage through a homogeneous, laterally infinite medium. The amount of scattered radiation can be significant, as illustrated in Fig. 3 for different thicknesses of adipose (i.e. fat) tissue. The scatter reduction by time-gating for different gate

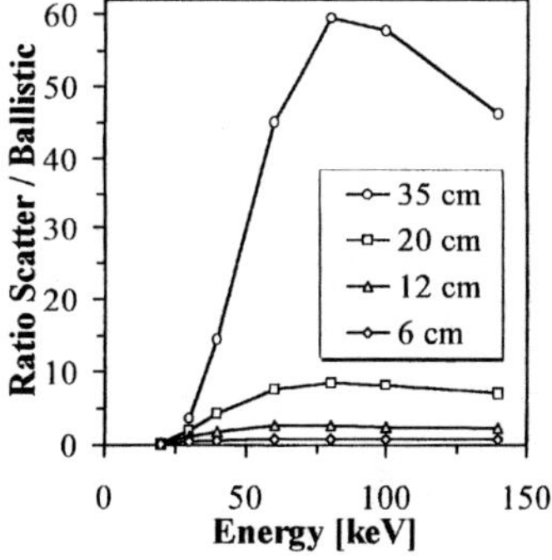

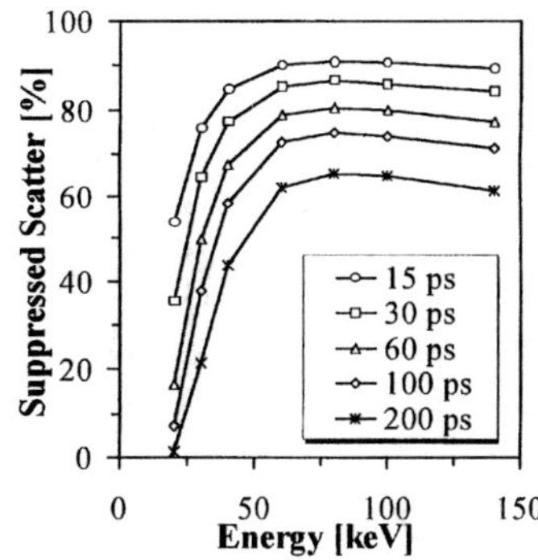

Figure 3: The total amount of scattered radiation in relation to the non-scattered radiation is shown (left). The percentage of scattered radiation that can be suppressed using different gate widths (right).

lengths is shown in the same figure. In general, scattering is most significant at high photon energies and for thick tissues. A shortening of the gate widths results in a better scatter suppression. There is a need for substantially short gate widths, especially at low photon energies. The simulations show, that most benefit could be gained from scatter-reduced imaging when imaging thick objects at relatively high energies (above ~ 40 keV). We observed similar behavior for several types of soft tissue.

Experiments. We chose time-resolved one-dimensional imaging as an appropriate tool to investigate the time dependence of the image contrast and to test the simulation results. Obviously, future applications will demand two-dimensional imaging. However, the one-dimensional information obtained in our experiments gives access to all relevant imaging properties, moreover with a much better time resolution than with available two-dimensional x-ray detectors.

A small lead object is imaged onto the photocathode of the x-ray streak camera, using a tantalum target. Time-integrated images (cathode images without time dispersion) are recorded at different thicknesses of a water phantom placed inbetween the object and the streak camera. The images obtained without phantom and with maximum phantom thickness are shown in Fig. 4. The contrast decrease is clearly visible and is attributed to the large amount of scattered radiation. The contrast (ratio of the dip-depth to the total intensity) is measured to be 0.94, 0.68, 0.35 and 0.16 for phantom thicknesses of 0, 3, 9 and 15 cm, respectively.

Time-resolved line images, corresponding to the time-integrated images in Fig. 4, are shown in Fig. 5. The bright horizontal line corresponds to the ballistic peak, whereas radiation detected at later times is due to temporally delayed scattered radiation. The gain in contrast at early times is obvious from these images. The scattered radiation is clearly visible when imaging through the water phantom. Compared to the time-integrated case, the contrast is increased by factors of 1.5, 2.8 and 6.1 for water phantoms of 3, 9 and 15 cm thickness, respectively.

Another series of experiments was carried out, resembling the different situations of mammography and thick-tissue imaging. These situations were modeled by using a tin target (at <E> ~ 25 keV) with a mammography test phantom (5 cm thick) and a tantalum target (at <E> ~ 60 keV) with a thick water phantom (15 cm thick).

The time-integrated and time-resolved images are shown in Fig. 6 and Fig. 7. In the mammography case, most of the scattered radiation arrives shortly after the ballistic peak,

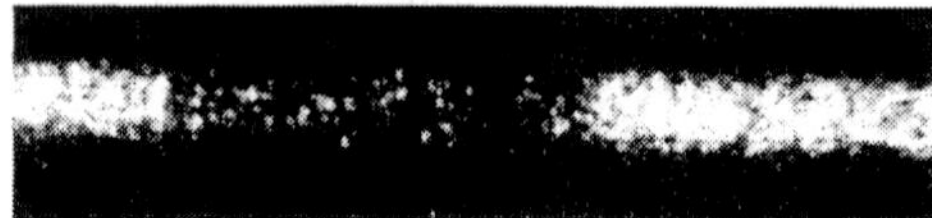

Figure 4: Time-integrated images of a small phantom. The shadow is clearly visible in the absence of a scatterer (left). When imaging through a scattering medium (15 cm of water), this shadow is hardly visible due to the large amount of scattered radiation (right).

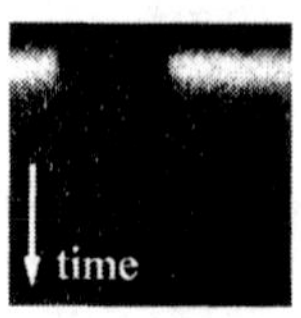

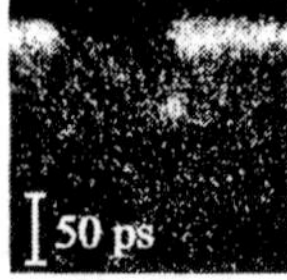

Figure 5: Time-dispersed images of a small phantom, corresponding to the images in Fig. 4. In the absence of a scatterer, only prompt radiation is visible (left). When imaging through the scattering medium, the shadow is clearly visible at early times, whereas the scattered radiation arrives delayed (right).

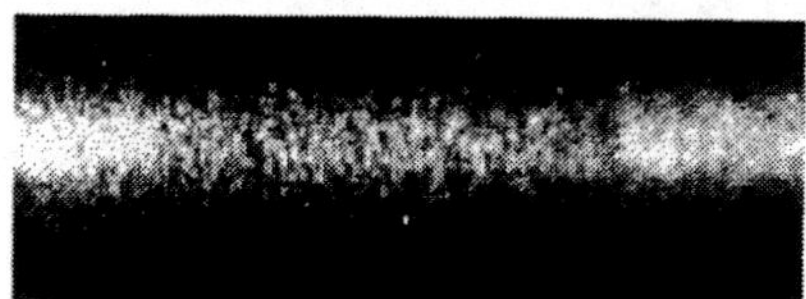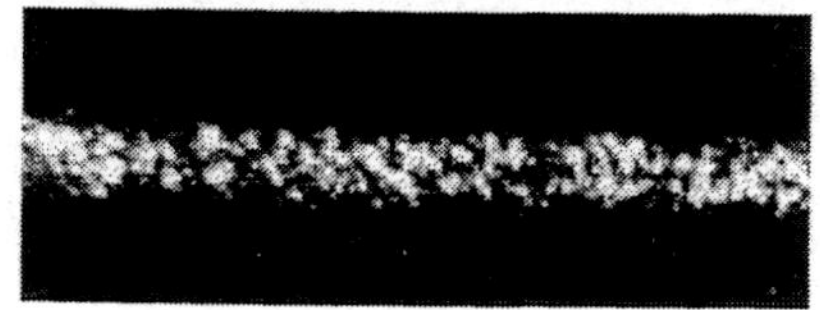

Figure 6: Time-integrated images of a phantom: An object within a mammography phantom (4.5 cm thick) is imaged with radiation from a tin target (left). A water phantom (15 cm) and a tantalum target is used for the image on the right side.

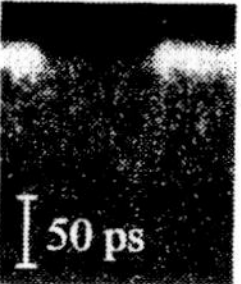

Figure 7: Time-dispersed images corresponding to the images in Fig. 6. In the case of the mammography phantom (left), the scattered radiation is less delayed and more concentrated shortly after the ballistic peak than for the water phantom (right), where the scattered radiation is more dispersed in time.

whereas the scatter is more delayed in the thick-tissue image. The contrast improvement factor is 1.5 and 6.1, respectively. These results correspond well to the results from the simulations.

COMPARISON WITH OTHER X-RAY SOURCES AND DISCUSSION

A comparison of source size and average x-ray intensity for laser-produced x-rays and different conventional x-ray sources is shown in Table 1. At the present stage, the *average* x-ray intensity is still much lower than for conventional x-ray devices. However, an intensity increase can be achieved by increasing the laser intensity, the repetition rate, or both.

The intrinsic advantage of time-gated imaging is that all ballistic photons exiting the patient reach the detector. This is in contrast to anti-scatter grid techniques, where some part of the radiation is attenuated by the collimating structure. This leads to a dose penalty. A comparison of achievable contrast improvements and corresponding dose penalty is shown in Table 2.

Table 1: Comparison of source size for a laser-produced x-ray source and conventional sources

source type	size	relative average intensity
laser-produced x-ray source	~ 10-30 μm	1
microfocus x-ray tube	~ 10 μm	~ 10
mammography x-ray tube	~ 40 μm	~ 100
standard x-ray tube	300 - 600 μm	~ 1000

Table 2: Comparison between time-gated imaging and anti-scatter grids.

	time-gated imaging		anti-scatter grids	
	contrast increase	dose penalty	contrast increase	dose penalty
mammography (~ 20 keV, 4.5 cm)	~ 1.6	*none*	1.3 - 1.5	2 - 3
body imaging (~ 60 keV, 15 cm)	~ 6	*none*	4 - 7	6 - 10

Therefore, at comparable detector sensitivities, time-gated imaging would be superior. However, at present times the limitations in x-ray sensitivity of time-gated image detectors result in an inferior performance compared to anti-scatter grids. Fortunately, the rapid technical development of detection techniques might soon overcome this problem.

ACKNOWLEDGEMENTS

Financial support from the Swedish Natural Sciences Research Council (NFR), the Swedish Medical Sciences Research Council (MFR) and the EU through its TMR program (No. FMRX-CT96-0080) is greatly acknowledged.

REFERENCES

1. J.D. Kmetec, C.L. Gordon III, J.J. Macklin, B.E. Lemoff, G.S. Brown and S.E. Harris, MeV x-ray generation with a femtosecond laser, *Phys. Rev. Lett.* 68:1527 (1992)
2. R. Lewis, Medical applications of synchrotron radiation x-rays, *Phys. Med. Biol.* 42:1213 (1997)
3. C. Tillman, A. Persson, C.-G. Wahlström, S. Svanberg and K. Herrlin, Imaging using hard x-rays from a laser-produced plasma, *Appl. Phys. B* 61:333 (1995)
4. K. Herrlin, C. Tillman, M. Grätz, C. Olsson, H. Pettersson, G. Svahn, C.-G. Wahlström and S. Svanberg, Contrast-enhanced radiography by differential absorption, using a laser-produced x-ray source, *Invest. Radiol.* 32: 306 (1997)
5. M. Grätz, A. Pifferi, C.-G. Wahlström and S. Svanberg, Time-gated imaging in radiology: Theoretical and experimental studies, *IEEE J. Select. Top. Quant. Elect.* 2:1041 (1996)
6. Z. Jiang, A. Ikhlef, J.C. Kieffer, A. Krol, D.A. Bassano, C.C. Chamberlain and S.C. Prasad, Laser-driven hard x-ray source for angiography, *see these proceedings*
7. C.L. Gordon III, G.Y. Yin, B.E. Lemoff, P.B. Bell and C.P.J. Barty, Time-gated imaging with an ultrashort-pulse laser-produced-plasma x-ray source, *Opt. Lett.* 20:1056 (1995)
8. S. Svanberg, J. Larsson, A. Persson and C.-G. Wahlström, Lund High-Power Laser Facility - Systems and first results", *Phys. Scr.* 49:187 (1994)
9. C. Tillman, S. Johansson, B. Erlandsson, M. Grätz, B. Hemdal, A. Almén, S. Mattson and S. Svanberg, High-resolution spectroscopy of laser-produced plasmas in the photon energy range above 10 keV, *Nucl. Instr. Meth. A* 394:387 (1997)
10. G. Hölzer, E. Förster, M. Grätz, C. Tillman and S. Svanberg, X-ray crystal spectroscopy of sub-picosecond laser-produced plasmas beyond 50 keV, *J. X-ray Scie. Techn.* 7:50 (1996)
11. M. Schnürer, R. Nolte, T. Schlegel, M.P. Kalachnikov, P.V. Nickles, P. Ambrosi and W. Sandner, On the distribution of hot electrons produced in short-pulse laser - plasma interaction, *J. Phys. B* 30:4653 (1997)
12. C. Tillman, G. Grafström, A.-C. Jonsson, I. Mercer, S. Svanberg, B.-A. Jönsson, S.-E. Strand and S. Mattson, Survival of V79-CH cells studied *in vitro* after extremely high absorbed dose irradiation by x-rays from a alser-produced plasma, *submitted to Radiology*
13. L.-J. Hardell, *High-Speed Radiography using Laser-Produced X-Rays: Characteristics and Applications*, Diploma Thesis, Lund Reports on Atomic Physics, LRAP-217, Lund (1997)
14. G.T. Barnes, Contrast and scatter in x-ray imaging, *Radiographics* 11:307 (1991)
15. L.-H. Wang, P.P. Ho, C. Liu, G. Zhang and R.R. Alfano, Ballistic 2D-imaging through scattering walls using an ultrafast optical Kerr gate, *Science* 253:769 (1991)
16. S. Andersson-Engels, R. Berg, O. Jarlman and S. Svanberg, Time-gated transillumination for medical diagnostics, *Opt. Lett.* 15:1179 (1990)

BIOLOGICAL X-RAY MICROSCOPY WITH A COMPACT LASER SYSTEM

Martin Richardson, Masataka Kado & David Torres
LPL, CREOL, University of Central Florida
4000 Central Florida Blvd, Orlando, FL 32816-2700
Tel # 407-823-6819, Fax # 407-823-3570
E-mail: mcr@creol.ucf.edu

Dr. Yoshimasa Yamamoto & Dr. Herman Friedman
University of South Florida, College of Medicine
12901 Bruce B. Downs Blvd, MCD Box 10, Tampa, FL 33612

Dr. Jayshree Rajyaguru & Dr. Michael Muszynski
Arnold Palmer Hospital for Children & Women
The Nemours Children's Clinic
85 West Miller Street, Suite 303, Orlando, FL 32806

ABSTRACT

We describe progress being made in an x-ray imaging technology that provides high-resolution single frame x-ray images of *in-vitro* specimens captured in a time sufficiently short that radiation damage mechanisms to the structure are not recorded. Several different biology and medical research groups find this type of microscopy particularly well-suited to the detailed analysis of sub-cellular features, and to the study of live organisms subjected to various forms of external stimuli. This technology utilizes bright x-ray sources produced by compact pulse laser systems. The incorporation of advanced x-ray optical and electron-optical systems will lead to the development of a compact, real-time x-ray microscope, having a broad range of applications.

INTRODUCTION

Research at synchrotron facilities has shown that x-ray microscopy fills a vital analytical gap between optical and electron techniques[1]. Optical microscopy offers the ability to observe live organisms, in their natural environment, with the limited spatial resolutions (confocal) of ~ 200 nm, and depths of field of ~ 300 nm. These depths of field allow for the visualization of whole cell structures and phenomena, but are limited by the spatial resolution in being unable to depict fine cell structure. Electron microscopy (SEM) offers high resolutions of ~ 1 nm, but small depths of field (~ 1 nm). The price for this high resolution is elaborate preparation of the specimen, that includes: drying, staining, sectioning, and metallic coating the specimen. This type of preparation eliminates the possibility of observing live specimens, in a natural environment. The small depth of field typical in electron techniques, eliminates the ability to view whole cell phenomena. X-ray microscopy, with resolutions of ~10 nm, and depths of field > 1 μm, bridges the gap nicely. X-rays offer high enough resolution to view the fine cell structure, without the morphology altering preparation involved in electron techniques. This method delivers, for the first time, the combination of live, whole cell, in vitro, observation with high resolution. In addition, the ability of x-rays to penetrate organic matter, makes available the imaging of internal cell structure. Today's biologist makes use of all these techniques, each with its own merits, to give them an overall or "global imaging" view of the micro-organic world. The development of x-ray optical techniques has lead to some spectacular advances in the deployment of state-of-the-art x-ray optical elements, particularly zone-plate lenses, in sophisticated x-ray microscopes on dedicated beamlines at major synchrotron facilities [2,3,4]. This intense activity has resulted in the development of several approaches to x-ray microscopy including scanning transmission [5], scanning phase contrast [6], and scanning dark field [7], and their use for high resolution imaging [8] of biological specimens in their natural and dried state, with significant progress being made towards intriguing options such as elemental and chemical mapping at the sub-cellular level [9,10]. There are however, some insurmountable drawbacks to this technological direction. X-ray microscopes attached to a beamline of a major national accelerator facility will never become a mobile, user-friendly device available at short notice for diagnostic analysis. It is also unlikely that this technical path will lead to the design of compact devices that become the seed from which corporate R&D takes the lead in developing this technology. Moreover, we have found that many biologists rely on analysis from several different microscopes, (confocal, fluorescence, electron, etc.) Easy access and proximity to other microscopes becomes an important logistical requirement of an x-ray microscope technology. Another technical limitation stems from the fact that with current synchrotron-based x-ray microscopies it takes many seconds of x-ray exposure to capture an image, during which time the damaging effects of the x-rays to the biological structure of the organism can be registered on the image of the specimen. Various measures have been devised to mitigate this effect, including the use of chemical fixation, specimen dehydration [11] or fast freezing [12]. With the exception of the latter, all these measures are precisely those limitations of electron microscopy for biology that x-ray microscopy was meant to avoid. A larger long-term penalty of long exposure times is the inability to study fast kinetics in biological structures, such as the response of an organism to some specific physical or chemical stimulus or the interaction of two or more separate biological organisms.

The drawbacks associated with the use of synchrotron x-ray sources, has led to a search for alternate sources of x-rays. Laser-produced-plasmas can emit short pulses, (< 10 ns), of tunable, bright, (> 4×10^8 W), of x-rays. Furthermore, with the development of high power, tabletop lasers, these systems are modular, and have the potential for the development of a stand-alone device. In this paper, we describe progress in the development of this type of x-ray microscopy, and its application to the biology and medical sciences. In particular we give some examples of recent data obtained from a dedicated micro-radiography facility, and indicate some of the practical applications of this type of x-ray microscopy. Secondly, we outline a development pathway for a laser-plasma x-ray source based microscopy that produces images in real time. This system depends on the use of novel, high-resolution x-ray optical components and a special x-ray opto-electronic image tube. We believe that this is the optimum architecture for a stand-alone mobile microscopic system.

A LASER PLASMA X-RAY MICROSCOPY FACILITY FOR BIOLOGISTS

We have deployed a laser plasma x-ray microscope facility that is both user-friendly and useful to the biological community, in order to demonstrate that x-ray microscopy, based on compact modular systems is a viable alternative to those based on large synchrotrons. Our strategy was to build a reliable facility based on projection micro-radiography. This approach, perhaps the oldest used for x-ray microscopy [13, 14] provides the high spatial resolution (~ 10 nm) required for small samples with a relatively compact and flexible system. Its principal disadvantages are that it employs 'line of sight' or projection imaging which suffers loss of resolution with thick (i.e. > 100 nm) specimens. Secondly, since the image is recorded with unity magnification, the resolution is set by the grain-size or pixel size of the recording media. In practice this means resorting to high resolution photo-resists as recording media, these having a grain size of ~ 5 nm (PMMA), since x-ray film and active array detectors have effective pixel sizes many times larger than this. Several methods have been devised to read the image registered within the resist. These all require chemical development of the resist, a somewhat variable procedure, and the subsequent analysis of the three-dimensional imprint of the image in the resist. Several methods can be used for the latter, including the use of a SEM, a TEM [15], and atomic force microscopy [16]. To date, the use of an AFM is the least ambiguous approach, and is relatively easy to implement. Nonetheless, each image requires several hours of specialized resist processing, drying and analysis. Notwithstanding these limitations, this facility provides the biological community with a flexible, open, x-ray microscope that is receiving increasing interest in the biology community.

The principal elements of this facility are shown in Fig.1. The biological samples are loaded into a special, vacuum-sealed sample cell in their natural fluid. The sample cell has a thickness of ~10 μm. The samples are encapsulated between the PMMA photo-resist, which is deposited on a Si substrate on one side, and a thin (~100 nm) SiN window on the other side. The overall exposed area is a few square millimeters. The cell is exposed to one single burst of x-rays for a laser plasma source in a configuration shown schematically in Fig.1. The vacuum-sealed cell is located ~ 2 cm from the plasma source, which is a small ~100 μm diameter high density plasma created from an Y or Au metal target. The plasma is produced by a ~10 ns burst of focused 1064 nm laser radiation having a total energy of 6 –20 J, generated from a Nd;glass laser system comprising a seed-injected,

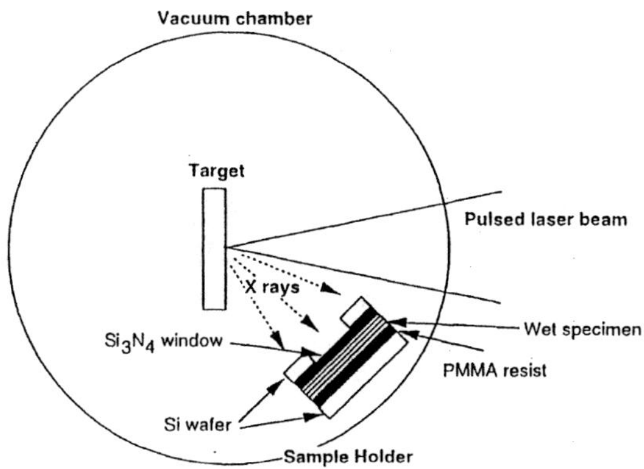

Fig.1 Schematic arrangement of laser plasma x-ray micro-radiography system

single-mode Q-switched oscillator followed by a passive four-pass amplifier incorporating an SBS phase conjugate mirror, and two additional linear amplifiers. The plasmas produced by this laser emit x-rays characteristic of a Planckian source with a temperature of ~ 100 eV, together with broad band N shell emission emanating from multiple-ion transitions in the 1-5 nm range. This is close to the so-called "water window" region, (2.3 – 4.5 nm) in which a strong contrast in x-ray absorption (close to a factor of ~10) exists between protein material containing C and N, and the typical hydrogenous fluids in which they are contained. For the current conditions the conversion efficiency of laser light into x-rays in this region is ~ 10%. Thus the number of x-ray photons incident on each square 10 nm pixel of the photo-resist is ~ 30 photons/ J of absorbed laser energy.

APPLICATIONS IN BIOLOGY

This facility has been used to examine a variety of biological and physical structures. Some of these features are illustrated below. The spatial resolution of this form of microscopy is demonstrated by the image shown in Fig.2. This shows an early image taken of in-vitro human chromosomes[17]. The chromosomes were stretched across a water surface, and "floated" onto the photoresist in such a way as to ensure a thin layer of chromosome in the projection direction. This minimized the thickness of material through which the x-rays had to pass, and maximized the spatial resolution. Careful analysis of the

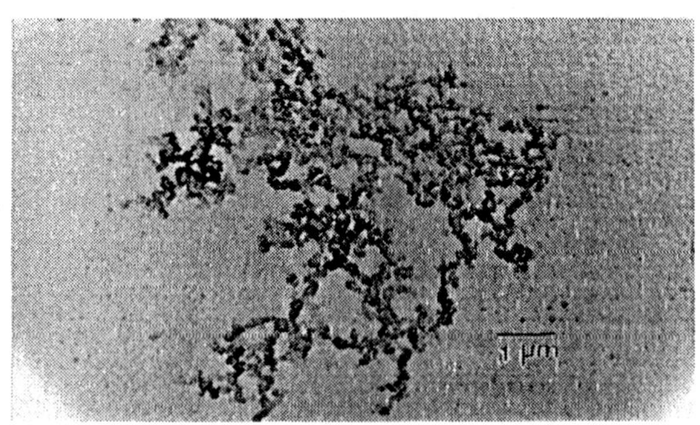

Fig.2. X-ray microradiograph of in-vitro human chromosome

images[18] could reveal individual nucleosomes with feature sizes in the region of ~ 10 nm. This is well beyond the resolution of an optical microscope. Moreover, such features, although resolvable with electron microscopy, could not have been observed in their natural state.

Another advantage of x-ray microscopy based on laser plasmas derives from its capability to capture an image with a single burst of x-rays from the source. This allows for an image to be registered in the resist before any biological, chemical or kinetic change can occur within the organism as a consequence of absorption of the x-rays. X-ray microscopy based on synchrotron sources cannot avoid this problem at present with room-temperature organisms in their natural state. Moreover, the use of short bursts of x-rays permits time-dependent studies to be made of transient phenomena in live organic specimens. Many of the users of the CREOL facility have expressed interest in, and the potential for time dependent studies with x-ray microscopy. An example of a recent study is shown in Fig.3.

This shows an x-ray micrograph of a cell of *Pseudomonas aeruginosa* before and after it had been exposed to the drug, Gentamicin. Considerable changes in shape, and in the structure of the outside structure of the cell is observable. Other groups are interested in observing kinetic changes in specimens, such as motion in live muscle tissue, and the effects of radiation on specimens.

Time resolved microscopy can also be used to examine the effect of one organism on another. As an example, Fig.4. shows an x-ray micrograph of the interaction between peritoneal macrophage cells from a mouse and *Candida albicanas*, a pathogenic yeast which causes candidiasis of immuno-compromised patients. Time-resolved x-ray microscopy is the only observational microscopy that has the resolution to observe the sub-cellular features on the nanometer scale in live specimens in this type of interaction.

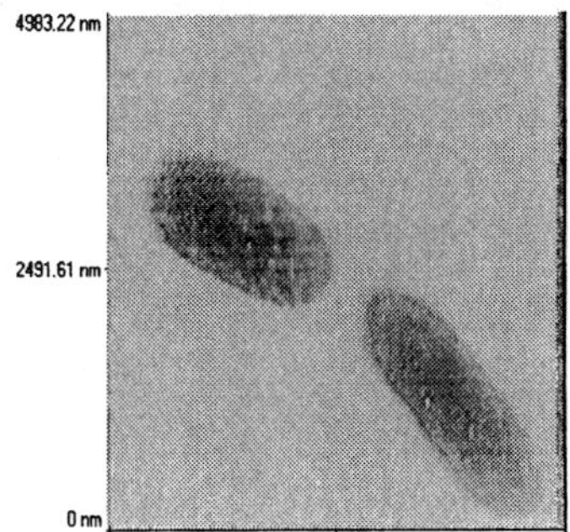
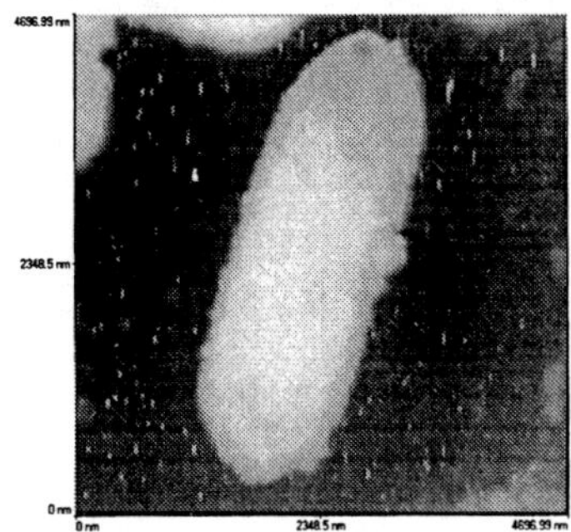

Fig.3. Timed x-ray micrographs of *Pseudomonas aeruginosa* under the effects of Gentamicin

Fig. 4. Interaction of macrophage cells with *Candida albicanas*

Other features of laser plasma based x-ray microscopy that are attractive for micro-biological studies include the possibility of elemental analysis, possible approaches to tomographic imaging, and the potential for short-pulse laser-dependent probe or micro-interaction studies.

DEVELOPMENT OF A REAL-TIME, STAND ALONE X-RAY IMAGING MICROSCOPE

In order to extend the capabilities of our current system, we are developing a new kind of x-ray microscope. This microscope will have three principal advantages over the

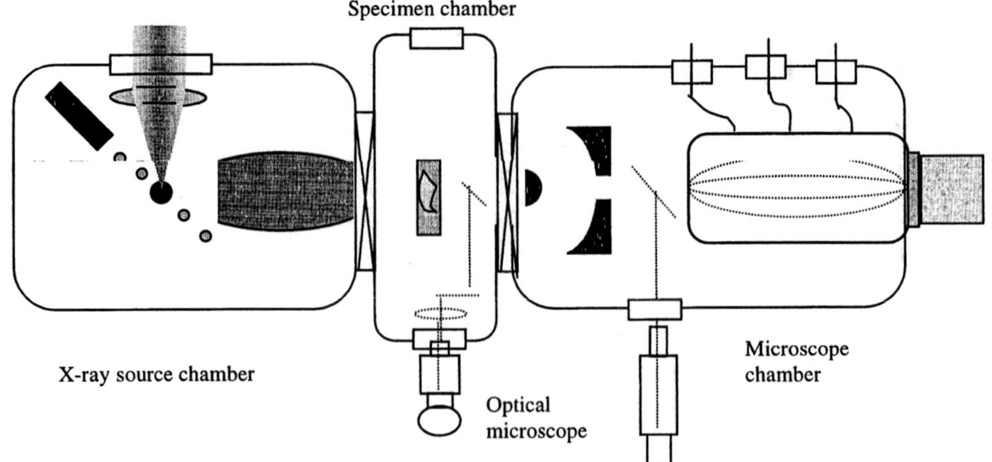

Fig. 5. Design features of a real-time compact x-ray microscope

existing system. Firstly it will provide real-time imaging, without the requirement of post-shot processing procedures and delays. Secondly, this system will incorporate x-ray imaging optics, eliminating the problems of line-of-sight imaging associated with the current radiographic approach, and allowing thicker specimens to be analyzed. Final the system architecture will be changed to facilitate a separate specimen chamber, amenable to changing the specimen environment, introducing external stimuli, and rapid specimen insertion for time resolved studies.

This microscope, depicted in Fig 5, will draw upon several different technologies that have not previously been linked together in this way. The x-ray light source will be a compact laser plasma light source of a type we have recently developed and patented [19,20], that eliminates the plasma debris, a potentially harmful attribute of all other pulsed x-ray sources. The optical condenser section of the microscope will incorporate a novel form of x-ray collimator [21, 22]. The specimen holder will be flexible in design, to allow adaptation of the microscope to dynamic biological studies and will include a co-axial optical microscope for precise specimen registration. The design of the imaging section of the microscope will depart from all previously used concepts for high-resolution x-ray imaging. As a first stage it will use a high numerical aperture, high-resolution, two element Schwarzschild optic [23, 24]. The image from this optic will be displayed with modest magnification onto the photocathode of a novel x-ray sensitive electron-optical image magnifier [25, 26, 27, 28, 29]. This precision electron tube will both magnify and intensify the electron image before converting it, via an phosphor converter to an image recorded by a high-resolution CCD that is the processed by conventional image processing techniques. The overall resolution will be set by the imaging x-ray optic, expected to be in the range of 30 nm. The time resolution of the instrument will, in its shortest mode of operation be in the range of ~ 100 ps, the shortest duration of the x-ray burst from a single laser pulse. We calculate that the instrument should detect single x-ray photons per pixel in the image plane. The overall image magnification will be 1000-10,000. The spectrum of the recording radiation will initially be set by the specifications of the imaging optic. However in a second phase of the development, we expect to change the x-ray optics of the instrument to allow x-rays over a broad range of energies to be recorded.

SUMMARY

We have demonstrated the use and potential of x-ray microscopy based on single shot laser-plasma x-ray sources. This approach to x-ray microscopy has some advantages over those using x-ray radiation from synchrotron sources. We have formed a diverse Users Group to exploit this technology. Micro-biologists, medical scientists and organic chemists are finding important applications of this type of microscopy. We are now in the process of developing the next generation of this type of microscope that will incorporate real-time imaging and quick, easy access for time-dependent analysis of live organisms. In the future we foresee that this system being amenable to commercial development into system that would provide biology and medical science, and other disciplines with a new diagnostic tool for microstructure analysis.

This work is supported by AFOSR (contracts F49620-94-1-0371 and F49620-93-1-0148), by NSF (contract ECS-9412008), and by the State of Florida.

REFERENCES

[1]Kirz, J., C. Jacobsen & M. Howell, Quart. <u>Rev. of Biophysics</u>, 28, 1, 33, (1995)

[2]Rarback, H., Kenney, J.M., Kirz, J., Howells, M.R., Chang, P., Coane, P.J., Feder, R., Houzego, P.J. Kern, D.P. & Sayre, D. In Schmahl & Rudolph, Springer Series in Opt. Sciences 43, Berlin: Springer-Verlag, (1984)

[3]Jacobsen, C., Williams, S., Anderson, E. Browne, M.T. Buckley, C.J. Kern, D. Kirz, J., Rivers, M. & Zhang, X., Optics Comm. 86, 351 (1991)

[4]Meyer-Ilse, W., Koike, M., Berguiristain, H.R., Maser, J. & Attwood, D.T. in Jacobsen, C. & Trebes, J., eda, SPIE Procs. 1741 (19 92)

[5]Buckley, C.J., Rarback, H., Alforque, R. Shu, D. Ade, H. Hellman, S. Iskander, N. Kirz, J. Lindaas, S. McNulty, I. Oversluizen, M., Tang, E. Attwood, D., DeGennaaro, R. Howells, M. jacobsen, C. Vladimirsky, Y. Rothman. S., Kern, D. Sayre, D., Rev. Scient. Instrum. 60, 2444 (1989)

[6]Morrison, G.R., in Benattar, R., ed., SPIE Procs. 1140 (1989)

[7]Chapman, H.N., Rev. Scient. Instrum. 66, 1332 (1994)

[8]Jacobsen, C., Anderson, E., Chapman, H., Kirz, J., Lindaas, S., Rivers, M. Wang, S., Williams, S., Wirick, S. & Zhang, X. in Aristov, V.V. & Erko, A.I., X-ray Microscopy, IV, Chernogolovka Moscow Region: Bogorodski Pechatnik (1994)

[9]Ade, H. Zhang, X., Cameron, S. Costello, C. Kirz, J. & Williams, S., Sci 258, 927 (1992)

[10]Buckley, C.J., Bone 13, 100 (1992)

[11]Haddad, W.S., I. McNulty, J. E. Trebes, E.H. Anderson, R.A. Levesque & L. Yang, <u>Science</u>, 228, 1213 (1994)

[12]Maser, J., C. Jacobsen, J. Kirz, A. Osanna, S. Spector & S. Wang, <u>Cell Vision,</u> 4, 2, 215 (1997)

[13]Sayre, D., J. Kirz, R. Feder, R. Kim, & E. Spiller, Science, 196, 1339 (1977)

[14]Cheng, P.C., X-ray Microscopy: Instrumentation and Biological Applications, ed. P.C. Cheng & G.J. Jan., publ. Springer, Berlin, p 65 (1977)

[15]Shinohara, K., S. Aoki, M. Yanagihara, A. Yashshita, Y. Iguchi & A. Tanaka, Photochem. Photobiol. 44, 401 (1986)

[16]Richardson, M.C., A. Vasiliev, S. Grantham, K. Gabel & M. Kado, SPIE Procs. 2015, 97 (1993)

[17]Richardson, M.C., K. Shinohara, K. Tanaka, Y. Kinjo, N. Ideda & M. Kado, SPIE Procs. 1741, 133 (1992)

[18]Kinjo, Y., K. Shinohara, A. Ito, H. Nakano, M. Watanabe, Y. Horike, Y. Kikuchi, M. Richardson & K. Tanaka, Journ. of Microscopy, 176 (1994)

[19]Torres, D., Jin, F., M. Richardson & C. DePriest, OSA 4, 75 (1996)

[20]Richardson, M., D. Torres, C. DePriest, F. Jin & G. Shimkaveg, Opt. Comm. (1997)

[21] Espry, S., D.B. Ohara, S. Scarborough & M.L. Price, Proc. SPIE, 2279, 110 (1994)

[22]Richardson, M. & D. Torres, SPIE Annual Meeting (1997)

[23] Underwood, J. & Knotright, J., Center for X-ray Optics, Lawrence Berleley Lab (1989)

[24] Shinohara, K., H. Nakano, Y. Kinjo & M. Watanabe, J. of Microscopy, 158, 335 (1990)

[25] Polack, F. & S. Lowenthal, Rev. Sci. Instr. 52, 207, (1981)

[26] Polack, F. S. Lowenthal, D. Phalippou & and D. Fournet, X-ray Microscopy, ed., D. Sayre, et al., publn., Springer Verlag, 220 (1988)

[27] Tonner, B., D. Dunham, T. Dronbay, J. Kikama, J. Denlinger, E. Rotenberg & A. Warwick, J. Electr. Spectr. 75, 309 (1995)

[28] Watts, S. Liang, Z.H. Levine & T.B. Lucatorto (submitted for publication)

[29] Grantham, S., E. Miesak, P. Reese & M. Richardson, SPIE Proc. 2273, 108 (1994)

HIGH HARMONICS AS A PROBE FOR FEMTOSECOND
LASER-PRODUCED PLASMAS

W. Theobald, R. Häßner, and R. Sauerbrey

Institut für Optik und Quantenelektronik
Friedrich-Schiller-Universität Jena
Max-Wien-Platz 1
D-07743 Jena, Germany

INTRODUCTION

The interaction of high intensity femtosecond laser pulses (10^{19} W/cm^2) with matter is a topic of current interest. Due to the short laser pulse duration, this leads to rapid ionization before a considerable expansion and it becomes possible to heat solids to mean electron energies of several keV at almost constant ion density. Plasmas with high electron densities and short scale lengths are created with relatively moderate laser energies. Much work has concentrated on the generation and characterization of ultrashort X-ray bursts,[1] investigations of the plasma dynamics[2-4] and on transport properties of dense plasmas through reflectivity measurements.[5-8] Production of matter in such a state allows the study of basic plasma physics in regimes relevant to astrophysics, atomic physics, or in strongly correlated plasmas.[9]

There are only few experimental methods to study the evolution of femtosecond laser produced plasmas and up to now only time-resolved x-ray spectroscopy could reveal substantial information.[3,4,10] We have demonstrated a new experimental method that uses high harmonics produced by intense laser radiation to probe the evolution of ultrashort laser-produced plasmas.[11] Differential absorption of the harmonics allows the measurement of plasma electron densities well above the critical density of the fundamental wavelength. It is shown that this pump-probe technique gives direct access to ultrahigh electron densities and electron temperatures in conjunction with a time resolution of a few hundred femtoseconds. For the measurement of the electron density it achieves a precision of about 10% at zero time delay.

In addition, we have studied the interaction of 110 femtosecond Ti:sapphire laser pulses with thin foil targets at intensities exceeding 10^{18} W/cm^2. High harmonics up to the 31st order were observed in transmission through an overdense plasma in the direction of the incident beam indicating a compression of the plasma by the ponderomotive force of the laser pulse.[12]

Applications of High Field and Short Wavelength Sources
Edited by DiMauro *et al.*, Plenum Press, New York, 1998

EXPERIMENTAL INVESTIGATIONS

The principle of our new method which probes the optical properties of a dense plasma is shown in Fig. 1 while a detailed description of the experimental setup can be found in ref. 11. The idea is to measure the transmittance of a plasma at two or more frequencies in such a way that some frequencies are below the plasma frequency while others are above. Light can only propagate in a plasma with an electron density less than the critical density. Since the electron densities in femtosecond laser produced plasmas from solid targets are on the order of 10^{23} cm^{-3} and the temporal evolution occurs on a subpicosecond time scale ultrashort XUV-pulses have to be used to transmit these plasmas. A convenient way to produce an XUV-spectrum of distinct frequencies is with high order harmonic generation.[13,14] Depending on the electron density, high-order harmonics are transmitted through the femtosecond laser-produced plasma while lower-order harmonics are absorbed or reflected. The electron density is then inferred from the measured ratio of the transmitted harmonic signal. By measuring the opacity of the plasma at various evolution stages the electron density and the mean electron energy can be mapped out in time.

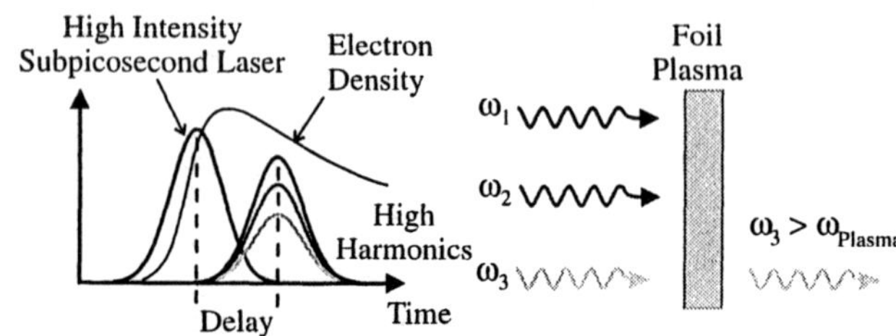

Figure 1. Principle of the method. A femtosecond laser pulse creates in a thin foil a dense plasma which is eventually irradiated by temporally correlated high harmonics after a certain delay time. Depending on the density higher order harmonics are better transmitted than lower order harmonics and from the measured ratio of the transmitted signal the electron density can be deduced.

A short-pulse hybrid dye-KrF laser system was used in the experiment providing 0.7 ps, 40 mJ laser pulses at a wavelength of 248.5 nm.[15] The laser beam was divided into two parts and one beam was focused to an intensity between 10^{14} W/cm^2 and 5×10^{14} W/cm^2 onto a thin polycarbonate foil target (Lexan 101, GE Plastics) in order to generate a dense plasma. After the foil burns through the maximum expected electron density is on the order of 10^{23} cm^{-3} while at later times hydrodynamic expansion and recombination eventually will lead to a decreasing electron density in time. The second temporally correlated laser pulse produced high harmonics in an inert gas jet down to a wavelength of 19.2 nm (13th harmonic of KrF-laser) acting as a probe beam. In our case the 5th and the 7th harmonic of the KrF-radiation at 49.7 nm and 35.5 nm respectively were intense enough to be used. A piezoelectric drive pulsed valve was used to release a gas jet with a particle density of about 3×10^{17} cm^{-3} into the vacuum. The plasma was now irradiated by the XUV-harmonics at a certain time delay which could be adjusted with a precision of about 60 fs by a delay line in one of the beam paths.

Single shot spectra of the harmonics transmitted through foil plasmas of various thickness (70 nm to 120 nm) were taken at delay times up to 12 ps in typically 0.5 ps time steps. Without a foil target [Fig. 2(a)], i.e., at zero electron density the 5th harmonic appears about 3 times more intense than the 7th harmonic when the harmonics are produced in neon gas. In contrast, when a dense plasma is present the 7th harmonic is stronger than the 5th at zero delay (pump and probe are overlapping in time), as can be seen in Fig. 2(b). When the probe precedes the plasma creating pulse only plasma self emission as a background at all delay times is observed. The intensity ratio is gradually changing with increasing time delay and reaches after several picoseconds the original ratio indicating that both harmonics are transmitted equally. At high electron densities the transmission of the higher frequency (7th) is much better than for the lower frequency (5th) while at long delay times and reduced density the original ratio is recovered.

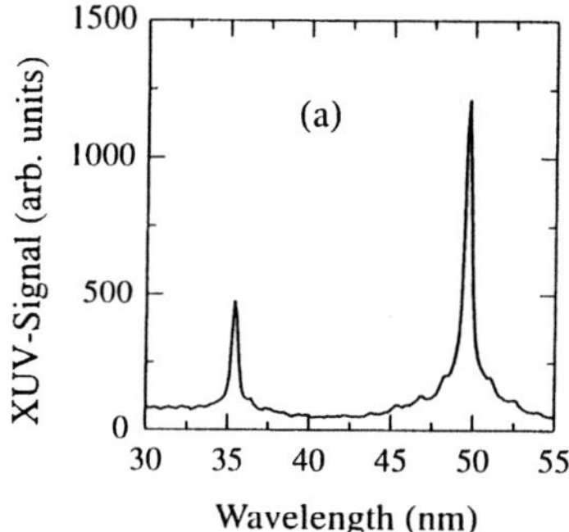
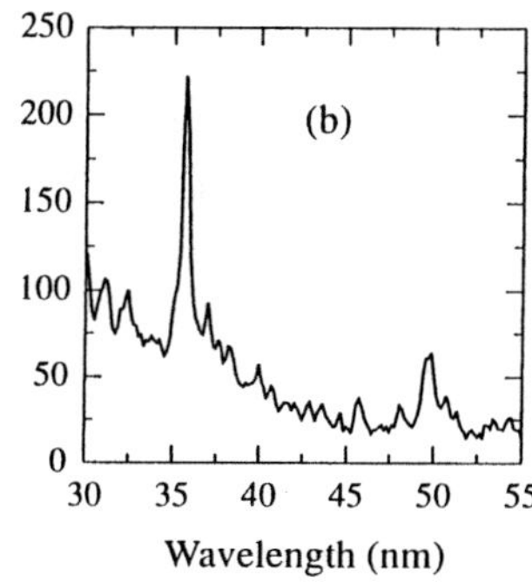

Figure 2. (a) shows part of the harmonic spectrum without a foil target while in (b) the spectrum is shown when a dense plasma is present.

Three different absorption mechanisms of the harmonics passing through the foil plasma have in principle to be considered, namely, bound-bound, bound-free, and free-free absorption. Absorption of the fifth harmonic at 49.7 nm due to bound-bound transitions can only arise from the $2p^2$-2p3d (λ=49.9 nm) transition in CIII and the 1s3p-1s5d (λ=49.7 nm) transition in CV. The seventh harmonic at 35.5 nm might be absorbed by the 3-5 transition (λ=35.6 nm) in CVI. It has to be taken into account that the ionization energy of ions in a dense plasma is lower than their single particle value which is attributed to the lowering of the continuum edge. Due to the large reduction of the continuum edge none of the states of the above mentioned transitions are bounded any more and all fall into the continuum which means that bound-bound absorption is not further considered. The contribution of bound-free absorption or photoionization is estimated to be about ten times smaller than free-free absorption, i.e., $\alpha_{ff}/\alpha_{bf}\geq 10$. Only collisional heating is therefore considered in the following analysis.

Fig. 3(a) shows the measured ratio 7th/5th (square symbols) as a function of time after transmission through a plasma which has been generated at an intensity of 2×10^{14} W/cm^2 in a 100 nm thick foil target. The initial ratio at zero delay is 9.2($\pm$3.3) and decreases with time to 1.2($\pm$0.13) at 12 ps. The curves shown in Fig. 3(a) are calculations of the temporal evolution of the ratio for various initial mean electron energies. The calculations assume that after the foil burns through and the laser pulse ceases, the plasma undergoes an adiabatic expansion due to the lack of thermal heat transport and negligible radiation cooling. To a good approximation the plasma expansion is uniform and planar at the two outer edges since the lateral plasma dimension (about 70 µm) is much larger than the thickness of the expanding plasma sheet. The plasma expansion is estimated to be smaller than 1 µm for the short time delay under consideration. The curve with an initial energy of 50 eV fits best to the measured values and we are now able to infer the electron density from the transmitted signals versus time as shown in Fig. 3(b). The deduced electron densities are plotted as solid circles while the curve represents the calculation for the density decay. The initial plasma parameters ($\Delta t = 0$) are $n_{e0} = (3.2^{+0.2}_{-0.3}) \times 10^{23}$ cm^{-3} and $T_{e0} \approx 50$ eV.

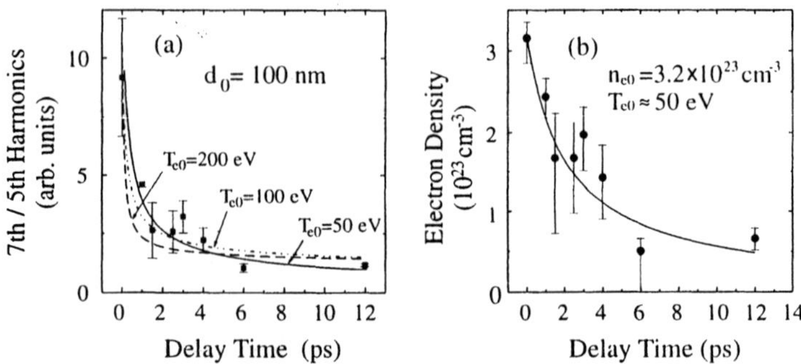

Figure 3. (a) The measured ratio 7th/5th (solid squares) is shown as a function of the delay time. The plasma was generated in a 100 nm thick plastic foil irradiated at an intensity of 2×10^{14} W/cm^2. The curves are calculations for the time dependence of the ratio assuming various initial electron energies between 50 eV and 200 eV. (b) The evolution of the electron density inferred from (a) (solid dots) is shown together with a calculation for an initial electron energy of 50 eV (solid curve).

The Drude model is used to derive the plasma refractive index and the Fresnel formulas are applied in order to describe the reflection of the harmonics. This is justified as long as the density gradient length is smaller than the absorption skin depth. Here, the skin depth[*] is on the order of the initial foil thickness, i.e., $1/\alpha \approx 30\text{-}100$ nm while the density scale length[2,6] is estimated to be about $\lambda/15 \approx 17$ nm. The assumption of a steep plasma gradient is therefore not critical for the quantitative results. The probe radiation impinges perpendicular to the plasma surface and therefore the fraction of the reflected radiation is given by $R = |(\eta-1)/(\eta+1)|^2$, where η is the complex refractive index of the plasma. The fraction T of the transmitted radiation through a homogeneous plasma sheet of thickness d with a box like electron density profile that is probed under normal incidence can be calculated by $T = (1-R)^2 \exp[-\alpha d]$ taking into account that part of the radiation is back-reflected at the rear density step. Refraction of the harmonic radiation in the plasma due to a lateral density gradient has to be considered only for radial scale lengths shorter than 10 µm. The part of the radiation which is in this case refracted out of the acceptance angle of the spectrograph never exceeds 20% of the total transmitted radiation. This was taken into account in the uncertainty of the determined electron density.

It should be emphasized that the time evolution of the ratio 7th/5th is only a function of the initial electron density and the initial mean electron energy. These two parameters are directly obtained from the measurement. The effect of recombination is considered in this simple model by estimating the evolution of the degree of ionization with the Saha equation. For delay times in excess of about 5 ps an additional decrease in electron density by at most a factor of 2 has been estimated.

---

[*] In the visible the skin depth can be well estimated by $\delta = c/\omega_p$ yielding $\delta \approx 10$ nm while for the XUV-radiation the calculation with $\alpha = 2(\omega/c)\text{Im}(\eta)$ results in larger values.

288

In a second experiment the plasma was generated in a 70 nm foil with a laser intensity of about 5×10^{14} W/cm^2 yielding the initial parameters of $n_{e0}=(4.3^{+0.2}_{-0.3})\times10^{23}$ cm^{-3} and $T_{e0}\approx150$ eV. Fig. 4(a) shows the measured and calculated ratio as a function of the delay time and Fig. 4(b) the extracted density decay with time. The faster decrease in the measured ratio indicates a higher initial mean electron energy to be consistent with the higher degree of ionization.

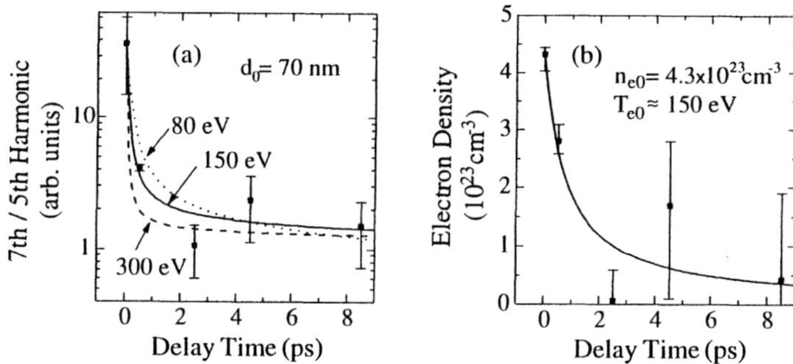

Figure 4. (a) The measured ratio 7th/5th (solid squares) is shown as a function of the delay time. The plasma was generated in a 70 nm thick foil at a laser intensity of 5×10^{14} W/cm^2. The curves are calculations for the time dependence of the ratio for various initial electron energies. (b) The evolution of the electron density inferred from (a) (solid squares) is shown together with a calculation for an initial electron energy of 150 eV (solid curve).

Experiments have been also performed using 110-fs, 2.4 TW Ti:sapphire laser pulses interacting directly with thin foil targets of 120 nm thickness at intensities exceeding 10^{18} W/cm^2. High harmonics were observed in transmission through an overdense plasma in the direction of the incident beam. A time integrating detector has been used in the experiment and in some of the spectra evidence of harmonic radiation is found, but it is hardly distinguishable from the strong plasma self-emission.[12] It is expected that the harmonic-generation efficiency is slightly enhanced in the case of oblique incidence and p-polarized light which has been found also in particle-in-cell simulations.[16] In order to suppress the plasma background the spectra measured at 45° angle of incidence with p-polarization were divided by the spectra obtained at 0°.

Fig. 5 shows the resulting spectrum which displays a significant enhancement of the spectral components at the positions of the odd and even harmonics. The positions of the odd and even harmonics which are calculated from the known laser wavelength of 790 nm are marked by solid and dashed bars respectively. Subtraction of the spectrum at 0° from that at 45° angle of incidence results in a spectrum (45°-0°) that exhibits the same features as those shown in Fig. 5. It is expected that the harmonic spectrum should have a cutoff to the long-wavelength range as predicted by one-dimensional particle-in-cell simulations.[16] In fact, in the wavelength range between 34 nm and 70 nm no harmonics were observed, but the first visible harmonic of 25th order appears just below 32 nm. Assuming the cutoff at 33 nm an electron density of $\sim1\times10^{24}$ cm^{-3} is inferred from the measured spectra. Harmonics above the 31st order are not distinguishable from the plasma background which is attributed to the strong decrease of the harmonic intensity toward shorter wavelength. The measured electron density is about twice as high as one expect it from a fully ionized foil plasma if no expansion is assumed indicating a density profile steepening by the large ponderomotive pressure in the high-intensity laser field.

At very high laser intensities the radiation pressure of a high-contrast laser pulse will balance the thermal plasma pressure and will push the critical surface into the solid target.[17-19] Spectroscopic measurements of the reflected laser radiation at various intensities revealed that during the laser interaction time the plasma motion is dominated by acceleration of the critical surface rather than by a constant expansion velocity and accelerations on the order of 10^{17} g -10^{18} g are reported.[20,21]

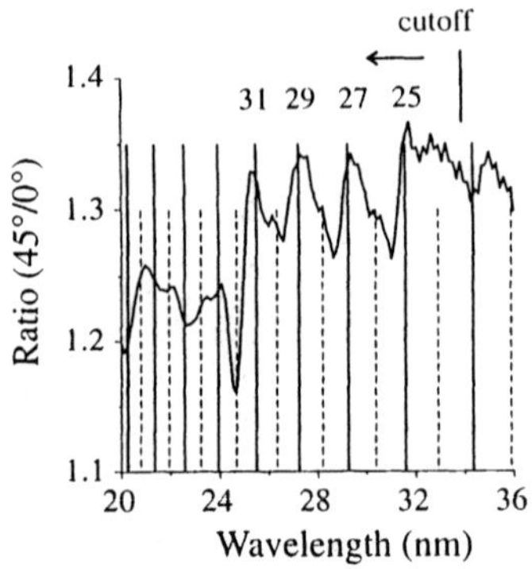

Figure 5.[12] High harmonics are observed in transmission through the foil plasma that was generated by the interaction of a 110-fs Ti:sapphire laser pulse with a 120 nm plastic foil at an intensity in excess of 10^{18} W/cm^2. From the cutoff of the spectrum an electron density of about 1×10^{24} cm^{-3} is inferred. Reproduced with kind permission of the Optical Society of America.

OPTICAL PROPERTIES OF DENSE PLASMAS

Although only the measured ratio of the harmonics need be used to determine the electron density the relation between the electron density and the refractive index is essential for the interpretation of the results. Much of the understanding of femtosecond-laser produced plasmas depends on the knowledge of the optical properties of dense plasmas in the visible and the X-ray regime.[22] The optical properties are determined by the complex dielectric function which is equal to the square of the refractive index η. Usually, for a laser-produced plasma the generalized Lorentz-Drude dielectric function is given by:

$$\varepsilon = \varepsilon_{10} + i\varepsilon_{20} - \frac{\omega_p^{\,2}}{\omega(\omega + i\nu_{ei})} \tag{1}$$

where $\omega_p=[n_e e^2/\varepsilon_0 m_e]^{1/2}$ is the local plasma frequency, ν_{ei} is the electron-ion collision frequency, and ε_0 is the permittivity of free space. The factors ε_{10} and ε_{20} describe the contribution of the bound electrons to the dielectric function which are $\varepsilon_{10}=1$, $\varepsilon_{20}=0$ for free electron metals or fully ionized plasmas.

Further theoretical investigations by several groups[23-25] have been performed refining the Drude model by taking screening effects and density dependent transport cross sections in Born and T-Matrix approximation into account and by considering the dipole polarizabil-

290

ity of the ions. The refractive index were calculated for a carbon plasma with a temperature of 150 eV and an average degree of ionization of 5.2. The calculations for the dielectric function including the dipole polarizability of the ions were performed with the Thomas-Fermi model to evaluate the charge density distribution inside an ion and the two fluid plasma model to account for the coupling between the electrical high frequency field and the plasma. No collisions were considered in this model. Fig. 6 shows the real (a) and imaginary (b) part of the refractive index for the 5th harmonic of the KrF wavelength as a function of the electron density. The solid curve represents the calculation for an ideal Drude plasma while the dotted curve and the squared symbols are the results of the Born and T-Matrix calculations, respectively, taking high density effects into account. The dashed curve displays the results of the calculation considering the ion dipole polarizability. Since no collisions were included the dashed curve has a singularity at the critical density. For electron densities below 4×10^{23} cm^{-3} Born, T-Matrix and ion polarizability show almost no deviation in the real part of the refractive index with respect to the Drude model, while in the imaginary part a discrepancy of up to a factor of two appears. At electron densities exceeding 5×10^{23} cm^{-3} one observes strong deviations in the real part showing that the Drude model is not applicable in this density range.

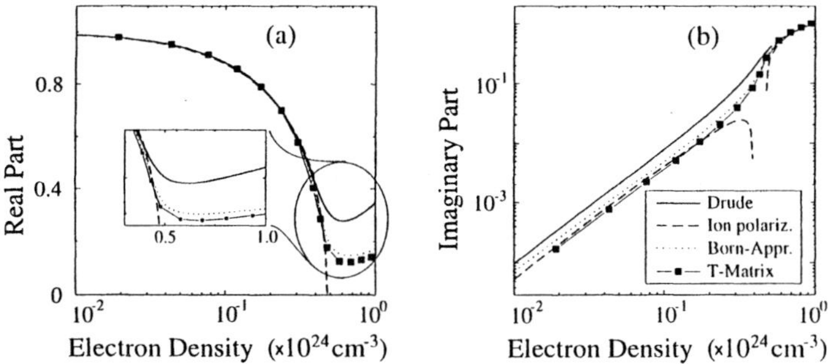

Figure 6. Real (a) and Imaginary part (b) of the refractive index of a dense carbon plasma as a function of electron density. Four different calculations were performed for an electron temperature of 150 eV and an average degree of ionization of 5.2. The solid curve represents the result of the simple Drude model while the dotted curve and the square symbols are the result of calculations including high density effects. The dashed curve has been obtained by calculating the refractive index considering the ion dipole polarizability of the ions.

CONCLUSION

A novel method has been presented to measure both the evolution of the plasma free electron density above 10^{23} cm^{-3} and the mean electron energy with subpicosecond time resolution. High-order harmonic generation provides coherent light pulses down to a wavelength below 10 nm corresponding to a critical density of 10^{25}cm^{-3} or about 1000 times solid density for a hydrogen plasma. Therefore, the method described here is well suited for applications in the field of inertial confinement fusion (ICF), for x-ray laser development, and in general for the investigation of high density laser-produced plasmas.

REFERENCES

1. M.M. Murnane, H.C. Kapteyn, and R.W. Falcone, High-density plasmas produced by ultrafast laser

pulses, *Phys. Rev. Lett.* 62:155 (1989).

2. H.M. Milchberg, R.R. Freeman, S.C. Davies, R.M. More, Resistivity of a simple metal from room temperature to 10^6 K, *Phys. Rev. Lett.* 61:2364 (1988).

3. J.C. Kieffer, Z. Jiang, A. Ikhlef, C.Y. Cote, and O. Peyrusse, Picosecond dynamics of a hot solid-density plasma, *J. Opt. Soc. Am. B* 13:132 (1996).

4. J. Workman, A. Maksimchuk, X. Liu, U. Ellenberger, J.S. Coe, C.Y. Chien, and D. Umstadter, Control of bright picosecond x-ray emission from intense subpicosecond laser-plasma interactions, *Phys. Rev. Lett.* 75:2324 (1995).

5. U. Teubner, J. Bergmann, B. van Wonterghem, F.P. Schäfer, R. Sauerbrey, Angle-dependent x-ray emission and resonance absorption in a laser-produced plasma generated by a high intensity ultrashort pulse, *Phys. Rev. Lett.* 70:794 (1993).

6. R. Fedosejevs, R. Ottman, R. Sigel, G. Kühnle, S. Szatmari, and F.P. Schäfer, Absorption of subpicosecond ultraviolet laser pulses in high-density plasma, *Appl. Phys. B* 50:79 (1990).

7. D.F. Price, R.M. More, R.S. Walling, G. Guethlein, R.L. Shepard, R.E. Stewart, and W.E. White, Absorption of ultrashort laser pulses by solid targets heated rapidly to temperatures 1-1000 eV, *Phys. Rev. Lett.* 75:252 (1995).

8. U. Teubner, I. Uschmann, P. Gibbon, D. Altenbernd, E. Förster, T. Feurer, W. Theobald, R. Sauerbrey, G Hirst, M.H. Key, J. Lister, D. Neely, Absorption and hot electron production by high intensity femtosecond uv-laser pulses in solid targets, *Phys. Rev. E* 54:4167 (1996).

9. W.D. Kraeft and M. Schlanges (eds.). *Quantum Statistics of Charged Particle Systems*, World Scientific, Singapore (1996).

10. F. Ráksi, K.R. Wilson, Z. Jiang, A. Ikhlef, C.Y. Côté, and J.-C. Kieffer, Ultrafast x-ray absorption probing of a chemical reaction, *J. Chem. Phys.* 104:6066 (1996).

11. W. Theobald, R. Häßner, C. Wülker, and R. Sauerbrey, Temporally resolved measurement of electron densities ($>10^{23}$ cm^{-3}) with high harmonics, *Phys. Rev. Lett.* 77:298 (1996).

12. R. Häßner, W. Theobald, S. Niedermeier, H. Schillinger, and R. Sauerbrey, High order harmonics from solid targets as a probe for high density plasmas, *Opt. Lett.* 22:1491 (1997).

13. A. L'Huillier and Ph. Balcou, High-order harmonic generation in rare gases with a 1-ps 1053-nm laser, *Phys. Rev. Lett.* 70:774 (1993).

14. J.J. Macklin, J.D. Kmetec, and C.L. Gordon III, High-order harmonic generation using femtosecond pulses, *Phys. Rev. Lett.* 70:766 (1993).

15. G. Almàsi, S. Szatmàri, and P. Simon, Optimized operation of short-pulse KrF amplifiers by off-axis amplification, *Opt. Commun.* 88:231 (1992).

16. R. Lichters and Meyer-ter-Vehn, in: *Proceedings of the 7th International Conference on Multiphoton Processes*, P. Lambropoulos, ed., Vol. 154 of Institute of Physics Conference Series, Institute of Physics, London (1997), p.221.

17. X. Liu and D. Umstadter, Competition between ponderomotive and thermal forces in short-scale-length laser plasmas, *Phys. Rev. Lett.* 69:1935 (1992).

18. M.P. Kalashnikov, P.V. Nickles, Th. Schlegel, M. Schnürer, F. Billhardt, I. Will, and W. Sandner, Dynamics of laser-plasma interaction at 10^{18} W/cm^2, *Phys. Rev. Lett.* 73:260 (1994).

19. M. Zepf, M. Castro-Colin, D. Chambers, S.G. Preston, J.S. Wark, J. Zhang, C.N. Danson, D. Neely, P. A. Norreys, A.E. Dangor, A. Dyson, P. Lee, A.P. Fews, P. Gibbon, S. Moustaizis, and M.H. Key, Measurements of the hole boring velocity from Doppler shifted harmonic emission from solid targets, *Phys. Plasmas* 3:3242 (1996).

20. R. Sauerbrey, Acceleration in femtosecond laser-produced plasmas, *Phys. Plasmas* 3:4712 (1996).

21. R. Häßner, W. Theobald, S. Niedermeier, K. Michelmann, H. Schillinger, T. Feurer, and R. Sauerbrey, Relativistic accelerations in laser-produced plasmas, to be published.

22. Y.T. Lee and R.M. More, An electron conductivity model for dense plasmas, *Phys. Fluids* 27:1273 (1984).

23. B.U. Felderhof, T. Blenski, B. Cichocki, Dielectric function of an electron-ion plasma in the optical and x-ray regime, *Physica A* 217:161 (1995).

24. Th. Bornath, D. Kremp, W.D. Kraeft, M. Schlanges, Kinetic equations for a nonideal quantum system, *Phys. Rev. E* 54:3274 (1996).

25. G. Röpke, C.V. Meister, K. Kollmorgen, and W.D. Kraeft, Long wave limit of the complex dielectric function for Coulomb systems with bound states, *Ann. Physik* 36:377 (1979).

292

LASER-ASSISTED MULTICOLOR PHOTOIONIZATION OF ATOMS WITH HIGHER HARMONICS

Valérie Véniard, Richard Taïeb and Alfred Maquet.

Laboratoire de Chimie Physique-Matière et Rayonnement
Université Pierre et Marie Curie
11, Rue Pierre et Marie Curie
75231 Paris Cedex O5 FRANCE.

1. INTRODUCTION

The recently observed generation, by samples of atomic rare gas, of high-order harmonics of the fundamental frequency of an infrared laser opens the possibility to have at one' s disposal a table-top source of coherent and pulsed VUV, (or even soft X-ray) radiation[1-3]. The main motivation of the present paper is to address the possibility to perform a new class of (N+1)-color (N≥1) photoionization experiments, which fully exploit the unique characteristics of high-order harmonic radiation. Among these characteristics is the fact that the harmonic emission spectrum is constituted of coherent UV radiation lines with equally spaced frequencies and comparable intensities, the spacing between two lines being twice the frequency ω_L of the infrared laser which has been used to generate the harmonics. The idea is to expose simultaneously an atomic target to the fields of a high order harmonic radiation with frequency ω_H and the fundamental of the laser ω_L. Then, if the harmonic order is high enough, i.e. $\omega_H > E_I$, where E_I is the ionization energy of the atom, Laser-Assisted-Single-Photon-Ionization (LASPI) can be observed. Experimental observations have already been reported[4,5].

One of the main outcome of our simulations is to evidence the prominent role of the harmonic phase. Indeed, the phase difference between the laser and its harmonic is one of the key parameters which govern the relative intensities of the photoelectron peaks in the 2-color spectra[6,7]. In multicolor spectra, involving several harmonics the phase difference between harmonics becomes also important. Furthermore, one of the objective of this paper is to show that conversely, a detailed analysis of multicolor photoionization spectra, involving high harmonics and the (attenuated) laser field can provide interesting informations on the phase differences between successive harmonics [8].

The discussion is supported by numerical simulations of photoelectron spectra obtained via the solution of the time-dependent Schrödinger equation for the (3-D) hydrogen atom in the simultaneous presence of the two fields. The time propagation of the solution was

Applications of High Field and Short Wavelength Sources
Edited by DiMauro *et al.*, Plenum Press, New York, 1998

ensured via a standard Peaceman-Rachford scheme coupled to an inverse iteration procedure. The integration technique is similar to the one previously derived by Kulander *et al* [9]. The photoelectron spectra have been determined with the help of a spectral analysis of the atomic wave function immediatly after the pulse. We have used a technique developed in ref. 10, which is based on the use of a window operator. Here, unless otherwise stated, we have chosen the laser frequency as $\omega_L=1.55$eV (Ti:Sapphire laser). The pulse is assumed to have a trapezoidal shape with a duration of eight laser cycles with one-cycle turn-on and turn-off. An additional assumption is that the duration of the harmonic pulse is the same as the one of the laser. More specific details regarding the technique of our computations can be found in ref 7.

2. TWO-COLOR PHOTOIONIZATION

Two possibilities are to be considered, depending on whether the UV frequency $\omega_H = (2N+1)\omega_L$ is smaller or larger than the atomic ionization frequency (here $E_I = 13.6$ eV).

In the first case, for the relatively low harmonic intensities and short pulses considered here, the ionization probability is vanishingly small in the absence of the laser. In the presence of the laser, additional IR photons can be absorbed and the atom can be ionized, the signature being a photoelectron spectrum much similar to the ones resulting from standard "Above Threshold Ionization", (ATI).

To illustrate this case, we have chosen the laser frequency $\omega_L = 1.77$ eV and the high frequency is its seventh harmonic, which is not far from resonance with the n=3 state. The intensities for the laser radiation and its harmonic are respectively $I_L=10^{12}$ W/cm^2 and $I_H=3.10^8$ W/cm^2. In Fig. 1, we have shown both the 2-color and the ATI spectra, the latter being as it would be obtained with the laser alone. A most remarkable feature of the laser-assisted process is that the corresponding photoelectron peaks can be several orders of magnitude more intense than the ones which would result from ATI. One observes many peaks in the

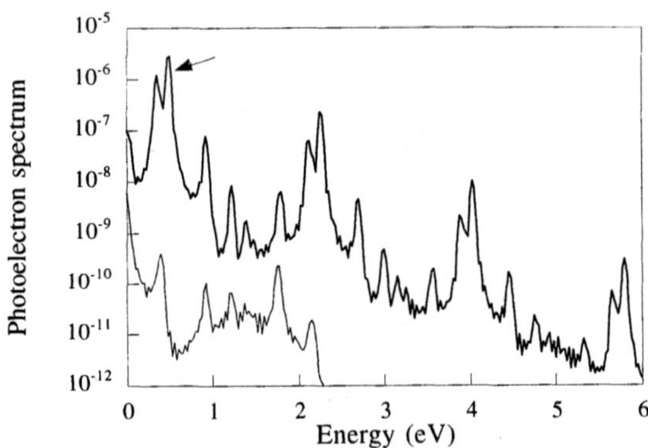

Figure 1. Photoelectron spectrum for a radiation pulse containing the frequency $\omega_L=1.77$ eV, $I_L=10^{12}$W/cm^2 and its seventh harmonic. We have shown also, for the sake of reference, the ATI spectrum one would obtain with the same laser pulse. Thick line: two-color spectrum with $I_H=3.10^8$W/cm^2; thin line: ATI spectrum ($I_H=0$). The main peak of the two-color process (simultaneous absorption of one UV and one IR photon) is indicated by the arrow.

spectrum, corresponding to resonances which can be unambiguously assigned to the n=3,4,5... states. These resonances, somewhat enhanced, are due to the trapezoidal shape and the short duration of the pulse (32 cycles in the present case) leading to a large frequency bandwidth. The peak indicated by the arrow corresponds to the simultaneous absorption of one UV and one IR photon. On the left, one can see the peak resulting from the resonance via the n=3 state. We have checked however, that these resonant structures are strongly dependent on the pulse shape.

In the second case, when one harmonic photon can ionize the atom, these structures disappear, as expected. Nevertheless, the presence of the laser field modifies the photoelectron spectrum with respect to that obtained with the UV radiation alone. For the sake of illustration, we present in Fig. 2 the influence of the laser intensity I_L on a typical two-color spectrum when the laser frequency is ω_L=1.55 eV, the high frequency field consisting of the 13th harmonic of the laser, i.e. ω_H=20.15 eV with a fixed intensity I_H = 3. 10^8 W/cm^2.

The curve (a) represents the single-photon ionization spectrum obtained with the harmonic radiation alone. LASPI begins to show up at a relatively moderate laser intensity, see Fig. 2b corresponding to I_L=5.10^{11} W/cm^2. Two satellite peaks are clearly visible on each side of the main peak, located at E_0. At such a low laser intensity, the two satellites can be unambiguously associated to two-photon transitions, implying the absorption of a high-frequency photon ω_H and the absorption (or emission) of a low frequency one ω_L[11].

At higher laser field intensities, the number of satellite peaks increases, as can be seen in Figs. 2c and 2d, corresponding to I_L=3. 10^{12} W/cm^2 and 8. 10^{12} W/cm^2 respectively. We are clearly in a nonperturbative regime as large numbers of photons are exchanged with the atomic system. However, for the short pulse considered here, the laser itself does not contribute to the ionization process. In these cases, the role of the laser is to redistribute the photoeletrons into continuum dressed states.

Interferences between Above-threshold-ionization (ATI) and LASPI are observed at higher laser field intensity, as shown in Figs. 2e and 2f, corresponding to I_L=1.75 10^{13}

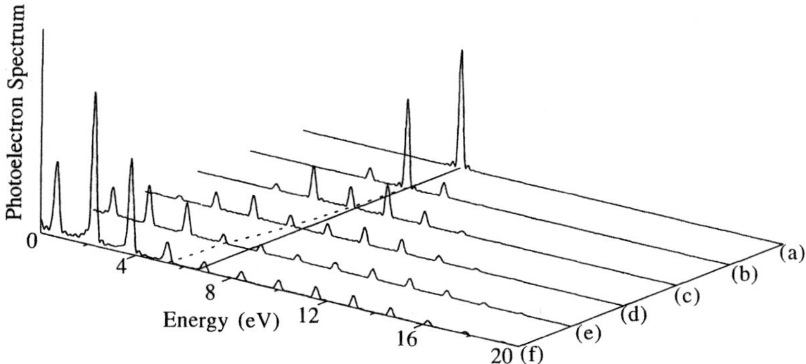

Figure 2. Effect of the laser intensity on the two-color photoelectron spectra, for radiation pulses containing the fundamental frequency of a Ti:Sapphire laser, ω_L = 1.55 eV and its 13th harmonic with a fixed intensity I_H = 3. 10^8 W/cm^2. (a) : I_L=0 (single-photon ionization); (b): I_L= 5. 10^{11} W/cm^2 ; (c): I_L=3. 10^{12} W/cm^2; (d): I_L= 8. 10^{12} W/cm^2 ; (e): I_L= 1.75 10^{13} W/cm^2 ; (f): I_L= 2. 10^{13} W/cm^2. Note that the peaks are shifted towards the lower energies due to the ponderomotive shift.

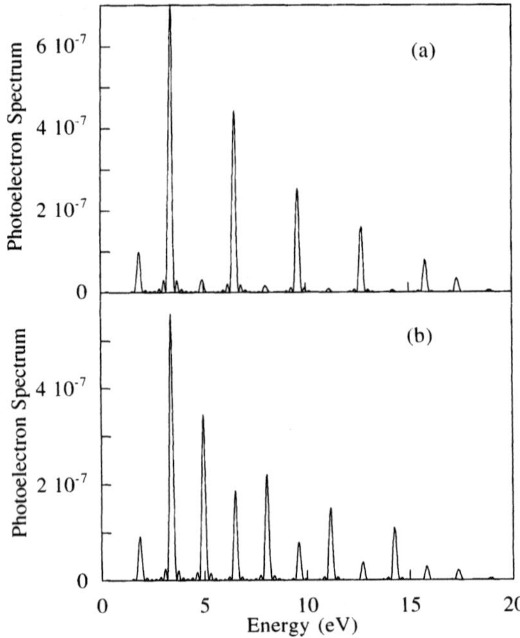

Figure 3. Effect of the harmonic phases on the (5+1)-photoelectron spectrum, with ω_L=1.55 eV, I_L=8.10^{11}W/cm^2. (a) : $(\varphi_{11},\varphi_{13},\varphi_{15},\varphi_{17},\varphi_{19})$ = (0,0,0,0,0); (b) : $(\varphi_{11},\varphi_{13},\varphi_{15},\varphi_{17},\varphi_{19})$ = (0,π,0,π,0).

W/cm^2 and 2. 10^{13} W/cm^2 respectively. Two important features can be noted in these cases. First, one can see an important growth of the low-energy peaks, which can be ascribed to the opening of new channels, associated to ATI, leading directly to the ionization of the target. Here, the laser radiation contributes to the ionization yield as the relative importance of ATI increases with the laser intensity. Then, one can note also the extension of the spectra towards high energies, corresponding to an increase of the number of satellites.

A remarkable result of the present simulation is that the change in the spectra between the regime in which ATI is almost not visible and the one in which ATI is dominant in the low photoelectron energy range takes place in a narrow laser intensity range : here it occurs between I_L=10^{13} W/cm^2 and I_L=2. 10^{13} W/cm^2. This is precisely in this range that quantum interferences can be identified. In this situation, the relative phase φ between the two fields plays an important role and governs the relative heights of the photoelectron peaks when both processes (ATI and LASPI) have comparable probabilities.

3. (N+1)-COLOR PHOTOIONIZATION

We turn now to a presentation of a new class of photoionization experiment, in which one can take advantage of the fact that the harmonic emission exhibit a typical plateau, where successive (odd) harmonics have comparable intensities. Part of the spectrum can be idealized as a Dirac comb. One can use such a finite subset to perform single-photon ionization experiments in atoms, provided the frequencies are larger than the atomic ionization frequency.

296

Starting from a comb with five harmonics and adding a sixth color, namely the IR laser radiation leads to photoelectron spectra such as the ones displayed in the Figures 3. We will exemplify the influence of the harmonic phases in this set of Figures, in which the respective intensities of the laser and of the harmonics are kept fixed as follows : $I_L = 8. \, 10^{11}$ W/cm^2; $I_j = 3. \, 10^8$ W/cm^2, where here $j = 11, ..., 19$.

First is shown in Fig. 3a, the case in which the harmonics have all the same phase, i.e. $\Delta\varphi_{ij} = 0$, where the indices i,j label consecutive (odd-) harmonics. Quite surprisingly, no new satellite peaks show up between the original ones, at least with the linear scale chosen here. In fact their magnitudes differ by almost two orders of magnitudes from the ones originating from the absorption of the harmonics. There is however a notable exception for the two small peaks located on each side of the original distribution. These peaks are the only ones which cannot be reached via interfering paths, at least in the low intensity regime considered here. This indicates that, if the harmonics had the same phase, the presence of the laser would almost not change the photoelectron spectrum. This results from strong destructive interferences which systematically take place for the peaks associated to laser-assisted transitions which can occur via distinct quantum paths. Moreover, the strong cancellations observed are the signature of the fact that the amplitudes associated to the interfering paths have almost equal magnitudes.

We turn now to another extreme case in which the successive harmonics have phases which differ from π, i.e. $\Delta\varphi_{ij} = \pi$, nothing else being changed with respect to the preceding case. The photoelectron spectrum is now dramatically changed, see Fig. (3b). The satellite peaks become dominant, the original ones being notably depleted. This clearly indicates that constructive interferences have now taken place. Comparing the set of Figures 3, helps to evidence the strong influence of the harmonic phases and also hints at the possibility to determine them from the analysis of multicolor photoelectron spectra.

6. CONCLUSION

To conclude, we have shown that interesting new phase-dependent spectroscopy experiments can be envisioned, using higher harmonics. Among these processes we have addressed the possibility of achieving a partial control of the photoelectron yield via multicolor schemes involving the laser and its harmonics. Another outcome has been that the relative phases of successive harmonics, a parameter of importance to discuss the properties of harmonic radiation, can be deduced from the analysis of multicolor spectra. Finally, we would like to mention that one can use the fact that, in most experiments, the duration of the harmonic pulse is much shorter than the laser pulse. In this case, a new class of pump-probe experiments can be developed and can lead to interesting informations on the dynamics of resonant multiphoton ionization. This class of process is under consideration.

Acknowledgments

The Laboratoire de Chimie Physique-Matière et Rayonnement is a Unité de Recherche Associée au CNRS, URA 176.This work was partially supported by the EC contracts ERB CHRXCT 920013 and 940470 . Parts of the computations have been performed at the Centre de Calcul pour la Recherche (CCR, Jussieu, Paris) and at the Institut du Développement et des Ressources en Informatique Scientifique (IDRIS).

REFERENCES

1. A. L'Huillier, L. A. Lompré, G. Mainfray and C. Manus in *Atoms in Strong Fields*, edited by M. Gavrila, Adv. Atom. Molec. Opt. Phys. Suppl. 1, Acad. Press, San Diego, p 139 (1992).
2. R. Haight and D. R. Peale, Phys. Rev. Lett. **70**, 3979 (1993); R. Haight and P. F. Seidler, Appl. Phys. Lett. **65**, 517 (1994).
3. J. Larsson, E. Mevel, R. Zerne, A. L'Huillier, C-G. W ahlström and S. Svanberg, J. Phys. B **28** L51 (1995).
4. J. M. Schins, P. Breger, P. Agostini, R. C. Constantinescu, H. G. Muller, A. Bouhal, G. Grillon, A. Antonetti and A. Mysyrowicz, J. Opt. Soc. Am. B **13**, 197 (1996).
5. T. E. Glover, R. W. Schoenlein, A. H. Chin and C. V. Shank, Phys. Rev. Lett. **76**, 2468 (1996).
6. V. Véniard, R. Taïeb and A. Maquet, Phys. Rev. Lett. **74**, 4161 (1995).
7. R. Taïeb, V. Véniard and A. Maquet, J. Opt. Soc. Am. B **13**, 363 (1996).
8. V. Véniard, R. Taïeb and A. Maquet, Phys. Rev. A **54**, 721 (1996).
9. K. C. Kulander, K. J. Schafer and J. L. Krause in *Atoms in Strong Fields*, edited by M. Gavrila, Adv. Atom. Molec. Opt. Phys. Suppl. 1, Acad. Press, San Diego, p 247 (1992).
10. K. J. Schafer, Comp. Phys. Comm. **63**, 427 (1991).
11. A. Cionga, V. Florescu, A. Maquet and R. Taïeb, Phys. Rev. A **47**, 1830 (1993).